High Energy Phenomena Around Collapsed Stars

NATO ASI Series

Advanced Science Institutes Series

A series presenting the results of activities sponsored by the NATO Science Committee, which aims at the dissemination of advanced scientific and technological knowledge, with a view to strengthening links between scientific communities.

The series is published by an international board of publishers in conjunction with the NATO Scientific Affairs Division

A Life Sciences B Physics	Plenum Publishing Corporation London and New York
C Mathematical and Physical Sciences	D. Reidel Publishing Company Dordrecht, Boston, Lancaster and Tokyo
D Behavioural and Social Sciences E Engineering and Materials Sciences	Martinus Nijhoff Publishers Dordrecht, Boston and Lancaster
F Computer and Systems Sciences G Ecological Sciences	Springer-Verlag Berlin, Heidelberg, New York, London, Paris, and Tokyo

Series C: Mathematical and Physical Sciences Vol. 195

High Energy Phenomena Around Collapsed Stars

edited by

F. Pacini

Arcetri Astrophysical Observatory,
Florence, Italy

D. Reidel Publishing Company

Dordrecht / Boston / Lancaster / Tokyo

Published in cooperation with NATO Scientific Affairs Division

Proceedings of the NATO Advanced Study Institute on
High Energy Phenomena Around Collapsed Stars
Cargèse, Corsica, France,
2-13 September 1985

Library of Congress Cataloging in Publication Data

NATO Advanced Study Institute on High Energy Phenomena around Collapsed Stars (1985 : Cargèse, Corsica)
High energy phenomena around collapsed stars.

(NATO ASI series. Series C, Mathematical and physical sciences; vol. 195)
"Published in cooperation with NATO Scientific Affairs Division."
"Proceedings of the NATO Advanced Study Institute on High Energy Phenomena Around Collapsed Stars, Cargèse, Corsica, France, 2–13 September, 1985"—T.p. verso.
1. Nuclear astrophysics—Congresses. 2. Gravitational collapse—Congresses. 3. Supernovae—Congresses. 4. Neutron stars—Congresses. I. Pacini, F. (Franco) II. North Atlantic Treaty Organization. Scientific Affairs Division. III. Title. IV. Series: NATO ASI series. Series C, Mathematical and physical sciences; vol. 195.
QB463.N39 1985 523.8 87–4381
ISBN 90–277–2453–9

Published by D. Reidel Publishing Company
P.O. Box 17, 3300 AA Dordrecht, Holland

Sold and distributed in the U.S.A. and Canada
by Kluwer Academic Publishers,
101 Philip Drive, Assinippi Park, Norwell, MA 02061, U.S.A.

In all other countries, sold and distributed
by Kluwer Academic Publishers Group,
P.O. Box 322, 3300 AH Dordrecht, Holland

D. Reidel Publishing Company is a member of the Kluwer Academic Publishers Group

---

Printed in The Netherlands

CONTENTS

FOREWORD

This volume contains a series of lectures presented at the NATO Advanced Study Institute on High Energy Phenomena Around Collapsed Stars, held in Cargèse, Corsica from September 2 till September 13, 1985.

The course was planned in collaboration with a Scientific Organizing Committee (C. Cesarsky, France; A. Lyne, U.K.; D. Pines, USA; J. Trümper, W. Germany; E. Van den Heuvel, The Netherlands and L. Woltjer, E.S.O., Münich) and was fully supported by the NATO Scientific Affairs Division. It was organized with the aim of providing students and young researchers with an up-to-date account on the subject of galactic high energy astrophysics and was attended by about 60 researchers from many countries.

The lectures and seminars did represent a complete coverage of our present knowledge and understanding of Supernovae, Supernovae Remnants, Pulsars, sources of high energy photons and particles. Most of them are reproduced in this volume although unfortunately a few speakers chose not to submit their text since they felt that the material was already amply available in the existing literature. I wish to express my gratitude to the Scientific Affairs Division of the North Atlantic Treaty Organization for the generous support given to the Institute and to the lecturers and participants who contributed so much to the success of the course.

Special thanks are also due to Mrs. Fern Torquati Bongianni for her careful work in all phases of the organization of the course and to Miss Marie France Hanseler for arranging the great hospitality of the Institut d'Etudes Scientifiques in the beautiful setting of Cargèse.

Prof. Franco Pacini
Arcetri Astrophysical Observatory
Florence, Italy

STELLAR EVOLUTION AND THE FORMATION OF NEUTRON STARS IN BINARY SYSTEMS

E.P.J. van den Heuvel
Astronomical Institute
University of Amsterdam
The Netherlands

ABSTRACT. The evolution of single stars is reviewed and the possible types of remnants are discussed.
Subsequently, mechanisms for the formation of neutron stars in binary systems are reviewed. In close binary systems, stars in the (zero-age) mass range ~ 10 to > 40 $M_{\odot}$ are expected to leave neutron stars as remnants. In wide binaries and for single stars the lower limit is reduced to about ~ 8 $M_{\odot}$. The precise value of the lower limit depends on various input parameters, especially on the treatment of convection.

Accretion-induced collapse (AIC) of a white dwarf of suitable mass and chemical composition is a second viable mechanism for the formation of neutron stars in binary systems. Evidence is summarized indicating that both in the wide radio pulsar binaries and in the bright galactic bulge X-ray sources that exhibit Quasi Periodic Oscillations (QPO) the neutron stars were formed by AIC.

1. INTRODUCTION

This review consists of two distinct parts. First, in section 2, we provide an overview of the evolution of single stars as a function of stellar mass up till the final nuclear burning stages. The subsequent dynamical evolution of massive stars (> 8 $M_{\odot}$) from the onset of core collapse to the formation of neutron stars is dealt with by Hillebrandt (this volume).

The second part of this review, starting in section 3, is devoted to the evolution of binary systems leading to the formation of compact objects, and further to the final evolution of X-ray binaries and the possible end products of this evolution.

Much of the material to be discussed in sections 3 and 4 has been extensively reviewed in a recent monograph (Lewin and Van den Heuvel, eds. 1983) and in recent review papers (Van den Heuvel and Habets 1985; Van den Heuvel 1986) to which we refer the reader for details and general reading.

F. Pacini (ed.), High Energy Phenomena Around Collapsed Stars, 1–32.

2. THE EVOLUTION OF SINGLE STARS

2.1. Why stars shine and evolve: the virial theorem

A non-rotating globe of ideal monatomic gas without energy sources and in hydrostatic equilibrium obeys the simplest form of the virial theorem

$$2\,E_{th} + E_{pot} = 0 \tag{1}$$

where E_{th} is the thermal energy content of the globe and E_{pot} its gravitational potential energy ("binding energy"). Equation (1) can be derived from the equation of hydrostatic equilibrium, as can be found in any textbook on stellar interiors (e.g.: Chiu 1968; Cox and Giuli 1968; Clayton 1968). E_{th} is given by

$$E_{th} = \frac{3}{2}\,N\,k\,\overline{T} = \frac{3}{2}\,M\,(\mathfrak{R}/\mu)\,\overline{T} \tag{2}$$

where N is the number of particles in the star, k is Boltzmann's constant, $\overline{T}$ is the mean temperature of the gas, M is the mass of the globe, $\mathfrak{R} = k/m_H$ the ideal gas constant, and μ is the mean particle mass, in units of the mass m_H of the hydrogen atom.

E_{pot} is given by

$$E_{pot} = -\,\alpha\,\frac{GM^2}{R} \tag{3}$$

where R is the stellar radius, G is the constant of gravity and α is a constant of proportionality of order unity, which depends on the precise density distribution in the star (see e.g. Cox and Giuli 1968).

Substitution of equations (2) and (3) into eq. (1) yields

$$\overline{T} \simeq \alpha\,\frac{GM}{R}\,\left(\frac{\mu}{3\mathfrak{R}}\right) \tag{4}$$

which shows that for a given stellar mass, the mean interior temperature depends only on the radius, and increases when the star contracts. Equations (1) and (4) imply that a gas globe in hydrostatic equilibrium and without energy sources must contract indefinitely - or until it ceases to obey the ideal gas law - for the following reasons.
Its temperature is above 0 K, which implies that it will radiate, and thus will lose energy. This energy loss goes at the expense of its total energy content which, by using eqs. (1) and (3), is given by:

$$E_{tot} = E_{th} + E_{pot} = \frac{1}{2}\,E_{pot} = -\,\frac{\alpha GM^2}{R} \tag{5}$$

Since E_{tot} is < 0, a decrease in E_{tot} implies according to eq. (5) that R must decrease. The heat losses from the surface therefore force the star to contract. But, according to eq. (4), this contraction implies that the mean interior temperature $\overline{T}$ of the star will increase. Thus, while the star tries to cool itself by radiating away energy from its surface, it gets hotter instead of cooler! Clearly, this cannot be a stable situation: the gas globe, starting out with a very large radius - as part of an interstellar cloud - is forced by its radiation losses

into a vicious circle of contraction and heating. The more it radiates to cool itself, the more it will contract, the hotter it gets and the more it is forced to go on radiating.
It is this "vicious virial circle" which in a nutshell explains how and why an interstellar gas globe, once it has started contracting, must finally end its life as a compact object, either a white dwarf, black dwarf, neutron star or black hole.

In the meantime the globe may spend a considerable amount of time in intermediate stages, which are called "the main-sequence", "the giant branch" etc. These stages are, however, only temporary "stations" in the contraction history of the globe. It is important to realise that stars do not shine "because they are burning nuclear fuel". They shine because they are hot, and they obtained their high interior temperature due to their history of gravitational contraction. Equations (1) and (5) show that half of the gravitational potential energy lost during the contraction of the globe is converted into thermal energy, and the other half - the loss in total energy - is radiated away .

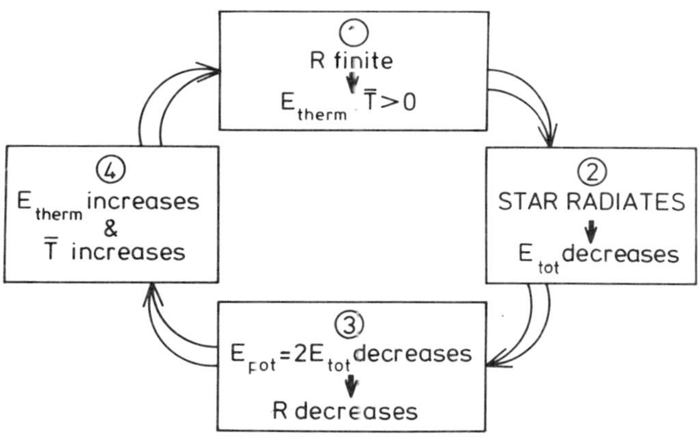

Figure 1. The virial theorem (VT) makes that a gas globe in hydrostatic equilibrium, once it has a finite radius is forced to contract indefinitely. (1) Because of its finite radius it has, according to the VT, a positive temperature $\bar{T}$, and therefore it radiates, causing its total energy to decrease (2) which, according to the VT causes its potential energy and radius to decrease (3) which, according to the VT causes its mean temperature $\bar{T}$ to increase (4) which causes it to radiate even stronger (1), causing it to contract further, etc.

2.2. The onset of nuclear burning: the hydrogen-burning station

The vicious circle of contracting and getting hotter can only be broken if in the stellar interior an energy source appears that produces precisely as much radiation energy as the globe loses from its surface. This occurs when the central temperature of the globe reaches a value of $\sim 10^7$ K and H-fusion begins. In view of the steep temperature dependence of the nuclear energy generation rate:

$$L_{nuc} \propto T^n \tag{6}$$

where n has a value ranging from 4 (p-p reactions) to 12 (CNO cycle), a slight further contraction and temperature increase causes an enormous increase in L_{nuc}, such that the star will be able to reach an equilibrium situation at which

$$L_{nuc,ph} = L_{out} \tag{7}$$

where L_{out} is the heat loss at the outer surface of the star and $L_{nuc,ph}$ is the photon part of the nuclear luminosity. A part $L_{nuc,\nu}$ of the luminosity is lost in the form of neutrinos, which freely escape into space.

When eq. (7) is fulfilled, the contraction stops. The nuclear burning equilibrium that is now reached is stable, since the virial theorem works as a safety valve. If too much heat were to be produced by nuclear burning, the interior temperature would rise, causing an increase in gas pressure which will lead to expansion and cooling of the star.

2.3. Stellar timescales

The three characteristic timescales for the interior and evolution of stars are (see Cox and Giuli 1968; Chiu 1968):

(i) The pulsation timescale ("dynamical timescale") which is the timescale for establishing hydrostatic equilibrium. This is the shortest stellar timescale, given by

$$t_{puls} = 0.04 \sqrt{\bar{\rho}_{\odot}/\bar{\rho}} \text{ (days)} \tag{8}$$

where $\bar{\rho}_{\odot}$ and $\bar{\rho}$ denote the mean density of the sun and of the star, respectively.

(ii) The thermal timescale t_{th}, which is the time required for reaching thermal equilibrium in the star. Roughly this timescale is the time required for the star to emit its thermal energy content at its present luminosity L. Since for main-sequence stars L is given by the mass-luminosity relation

$$L \sim K.M^{3.5} \tag{9}$$

one obtains, by using equations (1) and (3), with $R \propto M^{0.5}$:

$$t_{th} = 3.10^7/M^2 \text{ (yr)} \tag{10}$$

where M is expressed in solar units.

(iii) The nuclear timescale t_{nuc}, which is defined as the time required for the star to exhaust its hydrogen fuel content. Since the amount of fuel is proportional to the mass, and the consumption rate is proportional to the luminosity one has (using eq. (9):

$$t_{nuc} = \text{const.}\ M/L = 10^{10}/M^{2.5}\ \text{(yr)} \qquad (11)$$

where M is in solar units.

2.4. Evolution beyond hydrogen burning

The further evolution of the star after the exhaustion of its hydrogen fuel can be described again by applying the virial theorem, as follows.

When hydrogen is exhausted L_{nuc} drops to zero. However, the star keeps radiating because it is hot. So L_{out} does not decrease. This implies that the star must start contracting again, which will cause its interior temperature to rise.

The contraction will go on until a central temperature is reached at which the next nuclear fuel is ignited. This is helium, which burns at a temperature of $\sim 2.10^8$ K. Information about the sequence of burning stages through which the stellar interior subsequently passes is given in table I (after Arnett 1978; Weaver and Woosley 1980; Trimble 1982). It should be noted that for the radius R in eq. (4) one should in fact read: the radius of the dense burning core. The outer layers of the star, which in most cases comprise some 70 to 80% of the stellar mass, do not take part in the nuclear burning. In the stages beyond hydrogen burning they have a tendency to expand while the core further contracts. The presence of the outer layers hardly affects the value of E_{pot} in eq. (1) as most of the binding energy is concentrated in the core.

TABLE I. Thermonuclear burning stages (after Arnett 1978) and timescales for population I stars with masses of 1 $M_{\odot}$ and 25 $M_{\odot}$, after Iben (1974) and Weaver & Woosley (1980).

Fuel	$T/10^9$ (K)	Ashes	E(erg/g fuel)	Cooling	Time $1M_{\odot}$	(yrs) $25\ M_{\odot}$
^1H	0.02	^4He, ^{14}N	$(5-8).10^{18}$	photons	$1,1.10^{10}$	5.10^6
^4He	0.2	^{12}C, ^{16}O, ^{22}Ne	7.10^{17}	photons	$1,5.10^8$	5.10^5
^{12}C	0.8	^{20}Ne, ^{24}Mg, ^{16}O	$5\ .10^{17}$	neutrinos		60
		^{23}Na, $^{25,26}Mg$		neutrinos		
	0.4	^{20}Ne, ^{23}Na	----	neutrinos	----	
^{20}Ne	1.5	^{16}O, ^{24}Mg, ^{28}Si	$1,1.10^{17}$	neutrinos		1
^{16}O	2	^{28}Si, ^{32}S	5.10^{17}	neutrinos		0.5
^{28}Si	3.5	^{56}Ni, A ~ 56 nuclei	$(0-3).10^{17}$	neutrinos		0.01
^{56}Ni	6-10	n, ^4He, ^1H	-8.10^{18}	neutrinos		
A ~ 56 nuclei	(depends on ρ)	photodisintegration and neutronization				10^{-6}

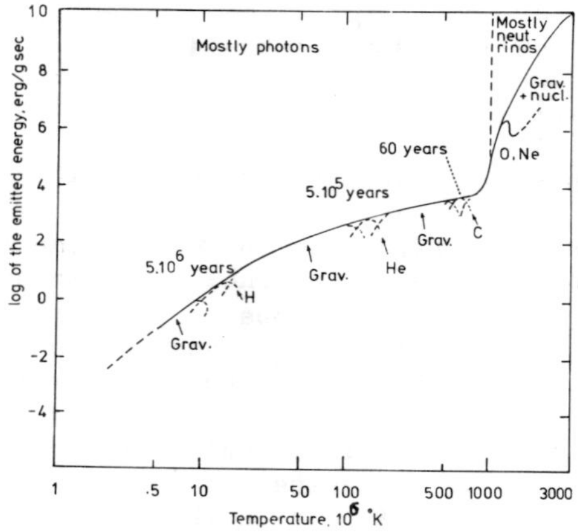

Figure 2. Schematic representation of the evolution of the stellar luminosity as a function of central temperature. The total luminosity on the "photon" branch is equal to the outer luminosity (heat radiation) of the star. It is provided by alternating stages of gravitational contraction and nuclear burning, up till carbon burning. During carbon burning and beyond, most of the energy generated in the interior escapes in the form of neutrinos, causing the duration of these stages to become very short, and the neutrino luminosity to become very large (after Reeves 1968). The timescales are for a star of 25 $M_{\odot}$, cf. Table I.

2.5. Neutrino losses and the duration of the final evolutionary stages

Table I shows that during carbon burning and beyond, most of the nuclear energy generated in the interior is liberated in the form of neutrinos. These represent a pure "leak" of energy from the stellar interior as they freely escape without interaction with the stellar gas. This causes an acceleration of the nuclear burning, because only the fraction $L_{nuc,ph}$ of the liberated energy is of interest for keeping up the hydrostatic equilibrium, and for preventing the star from contracting. Beyond carbon burning the fraction of the energy lost in the form of neutrinos becomes so large that the duration of these stages becomes negligible (of the order of 1 year or less). As an example, the last columns of table I list the duration of the various burning stages for a star of 25 $M_{\odot}$ following from detailed numerical computations by Weaver and Woosley (1980). We will not consider here the stages beyond silicon burning as these will be covered in detail by Hillebrandt (this volume).

2.6. Complications: The occurrence of flashes

Only massive stars ($M > 12\ M_{\odot}$) pass without complications through the sequence of burning stages listed in table I and depicted in figure 2.

For stars with masses below about 7-8 $M_{\odot}$ the occurrence of degeneracy in the core causes complications with the ignition of nuclear fuels. Since for a degenerate gas the pressure only depends on the density and not on the temperature ($P = K_1\rho^{5/3}$ in the non-relativistic case and $P = K_2\rho^{4/3}$ in the relativistic case, where K_1 and K_2 are constants and ρ is the density, cf. Cox & Giuli 1968), degenerate cores have no safety valve. When a nuclear fuel ignites in the degenerate state, the temperature will rise (due to the liberation of energy), but since the pressure is independent of temperature the pressure does not "feel" this temperature rise, such that no stabilizing expansion and cooling will occur. As the nuclear energy generation rate depends very sensitively on the temperature (see eq. (6)), the temperature rise produced by nuclear burning will cause the nuclear luminosity to rise enormously.

As a result ignition of a degenerate fuel causes a runaway nuclear energy generation, and a runaway rise of the interior temperature, producing a so-called "flash". Only when the temperature becomes so large that the degeneracy-boundary is crossed (the double-dashed line in figure 3) the gas will be able to expand and cool.

In stars with $M < 2.8\ M_\odot$ the helium core degenerates during hydrogen shell burning and, when M_{He} reaches $0.45\ M_\odot$, He ignites with a flash; the temperature rises up to $\sim 10^9$ K, upon which the degeneracy is removed. The helium flash is not violent enough to cause disruption of the star (cf. Cox and Giuli 1968).

Stars with masses in the range 2.8 to about 7-8 $M_\odot$ ignite carbon with a flash.

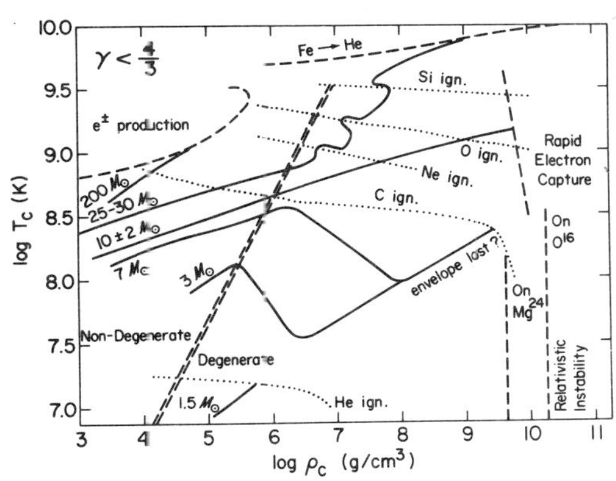

Figure 3. Stellar evolution in the central-temperature - central-density plane after Trimble (1982). Dotted lines show loci of ignition (energy liberated exceeds energy lost by neutrino production, etc.) of helium, carbon, neon, oxygen, and silicon fuels. Double dashed line separates non-degenerate from degenerate conditions. Dashed lines mark off regions of instability due to electron-position pair production and photodisintegration of iron (upper left) and electron capture and relativistic instabilities (far right). Solid lines are the (slightly simplified) evolutionary tracks for stars of 1.5, 3, 7, 8-12, 25-30, and 200 $M_\odot$. Collapse triggered by $e^\pm$ production gives way to collapse triggered by photodisintegration at about 100 $M_\odot$; photodisintegration of iron gives way to electron capture on lighter elements near 12 $M_\odot$.

Figure 3 depicts the evolutionary tracks of the central density and temperature of stars of various masses (cf. Trimble 1982). Only above ~ 8 $M_\odot$ carbon is ignited sufficiently close to the boundary with the non-degenerate region that the star can avoid the carbon flash.
All evidence suggests that in stars of lower mass in the carbon flash sufficient energy is generated to disrupt the entire star in a so-called carbon-deflagration supernova (Mazurek et al. 1974, 1977).
The stellar material is converted into Ni^{56} which decays in 120^d to Fe^{56}. A carbon-deflagration SN may well be identified with a type I SN as both the liberated energy (~ 10^{51} ergs) and expected shape of the lightcurve fit the observations (cf. Chevalier 1981, 1985).

2.7. Competing effects in the 3 - 8 $M_\odot$ region: mass loss; evidence from white dwarfs in clusters

When they are approaching C-ignition, stars in this mass range are on the Asymptotic Giant Branch (AGB) appearing as bright red giants (Iben and Renzini 1983). Such stars lose much mass in the form of stellar winds. White dwarfs have been found in galactic clusters that still contain stars as massive as 5 - 6 $M_\odot$ (Romanishin and Angel 1979). This indicates that stars as massive as at least 6 $M_\odot$ still terminate life as a white dwarf, i.e. are apparently able to shed their envelopes in the AGB phase before carbon violently ignites. This confines the possible mass range for carbon deflagration "supernovae" of single stars to at most 6 - 8 $M_\odot$. Possibly the peculiar Type I supernovae, which are always found in HII regions, are carbon-deflagration supernovae arising from stars in this mass range.

2.8. The fate of stars more massive than about 8 $M_\odot$

In the mass range between 8 and 10-12 $M_\odot$, C-ignition is non-violent, but the O-Ne-Mg core which is formed degenerates. Here, however, after further fuel ignition the treshold for electron-capture is expected to be crossed (see figure 3). This causes the onset of core collapse, as the removal of electrons causes the pressure of the electron-degenerate gas to drop. At time of the collapse the iron-element core is relatively small (~ 0.6 $M_\odot$, cf. Hillebrandt, this volume).
For stars more massive than ~ 10 - 12 $M_\odot$ an iron core of $M > 1.2\ M_\odot$ is formed which collapses as a consequence of the iron-helium transition (see figure 3 and Hillebrandt, this volume).
Thus, for $M > 8\ M_\odot$, stars are expected to leave neutron stars as remnants. At masses $\gtrsim 100\ M_\odot$ stars are expected to undergo an $e^\pm$ pair-production collapse, which might cause complete disruption of the star (see figure 3).

2.9. The evolution of helium stars

As the evolution of the helium core in post main-sequence stars is practically independent of the presence of a hydrogen-rich envelope it is, for the study of the final evolution, practical to study only the evolution of the helium core. Pioneering studies in this field are those of

Arnett (cf. Arnett 1978). The results of recent studies of helium stars by Habets (1985) are shown in figures 4 and 5. Figure 4 shows the interior evolution of a helium star of 3.2 $M_\odot$ which corresponds to a hydrogen star of about 12 $M_\odot$. The figure shows that the heavy-element core formed in the star has a mass of 2.2 $M_\odot$ and will therefore certainly collapse to a neutron star.

Figure 5 shows the evolutionary tracks for helium stars with masses 2.0 - 4.0 $M_\odot$ in the radius vs. core-mass diagram. C denotes C-ignition. The figure shows that in all helium stars with masses > 2.2 $M_\odot$ the heavy-element core becomes larger than the Chandrasekhar mass, such that these stars are expected to leave neutron star remnants.
Table II summarizes the above described expectations for the types of remnants from main-sequence stars of various masses.

TABLE II. Types of final evolutionary products expected as a function of stellar mass.

He core mass	Main-sequence mass	Final product
1.4 - 1.9 $M_\odot$	3 - 8 $M_\odot$	C-flash(disruption?) or CO-white dwarf
1.9 - 2.2 (3.0)$M_\odot$	8 - 10 (12)$M_\odot$	degenerate O-Ne-Mg core Ne-ignition/collapse
>2.2 (3.0) $M_\odot$	> 10 (12) $M_\odot$	collapse of iron core
> 60 $M_\odot$	>100 $M_\odot$	pair-creation instability (No remnant?)

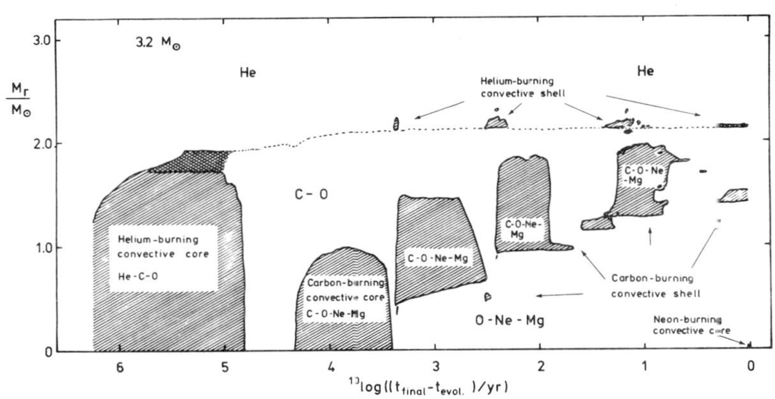

Figure 4. The interior evolution of a helium star of 3.2 $M_\odot$ up to neon ignition, as calculated by Habets (1985). Convective regions are hatched, semi convective regions doubly hatched. The dashed and dotted lines indicate the regions of maximum energy generation in the helium burning shell and in the carbon-burning shell, respectively. These dashed and dotted lines are expected to be the boundaries of the C-O and O-Ne-Mg cores, respectively.

2.10. Discussion

The mass ranges given in table II for the various types of final evolution depend quite strongly on the treatment of convection in the evolutionary computations. For example, Castellani et al. (1985) have pointed out that if convective overshoot is properly taken into account, the helium cores in moderately massive to massive stars become some 20% larger than found in earlier computations.

This would reduce the boundary values of the main-sequence mass ranges given in table II by about the same fraction. For example, the lower mass limit for terminating with a neutron star would move down to about 6.5 $M_{\odot}$.

Also the treatment of semi-convection, important in helium stars, may affect the values of the mass boundaries. Since Habets (1985), however, took the latter effects carefully into account we do not expect this to be a major source of uncertainty. Roughly speaking, an uncertainty of some 20 percent in the mass boundaries in table II can, however, not be excluded.

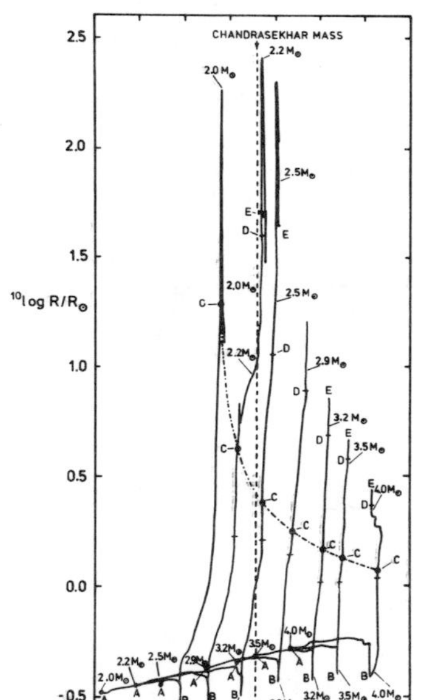

Figure 5. The core-mass vs. radius relation of helium stars in the mass range 2.0 to 4.0 $M_{\odot}$ undergoing C-ignition and neon ignition as calculated by Habets 1985 (the 2.9 $M_{\odot}$ helium star was followed till very close close to neon ignition); the 2.0 $M_{\odot}$ helium star was only followed up to the time at which the convective carbon-burning shell reaches the centre). The open circles mark the onset of either central or off centre convective core-carbon burning (at letter C). The dash-dotted curve has been drawn through these circles. The horizontal marking between B and C indicates the point where carbon burning begins in the radiative core.

3. FORMATION OF NEUTRON STARS IN BINARY SYSTEMS

3.1. Introduction

Before discussing the various ways in which binary systems can evolve to produce neutron stars we briefly review in section 3.2. the observed characteristics of the various types of binary systems that contain neutron stars. In the subsequent section we review the evolution of close binary systems and the types of compact remnants that may be left at the end of the evolution.

3.2. Types of Binaries that contain Compact Stars

Three broad categories of neutron-star binaries are known, as follows.

(i) The Massive X-ray Binaries (MXBs) in which the companion of the neutron star has a mass $M_s > 8 - 10\ M_\odot$. A number of prototype system are listed in table III. For orbital shapes and sizes, see Rappaport and Joss (1983), Joss and Rappaport (1984). In most of these systems the X-ray source shows regular pulsations.

(ii) The Low Mass X-ray Binaries (LMXBs) in which the companion of the neutron star has a mass M_s typically $< 1.2\ M_\odot$ (the only exception is Her X-1 where $M_s \sim 2\ M_\odot$). Orbital periods of a number of characteristic systems are listed in table IV. Only in three of these systems the X-ray source shows regular pulsations.

(iii) The five binary radio pulsars for which the system parameters are listed in table V. In the two systems with highly eccentric orbits, PSR 1913+16 and PSR 2303+46, the companion of the neutron star is most likely also a neutron star (cf. Srinivasan and Van den Heuvel 1982; Van den Heuvel 1984; Stokes et al. 1985). In other systems the companion is most probably a white dwarf (Van den Heuvel and Taam 1984; Van den Heuvel 1984).

A few peculiar binary X-ray sources such as SS 433, Cir X-1 and Cyg X-3 cannot be simply classified in the above scheme, but all evidence suggests that the first two of these belong to the MXBs (see Margon 1984), and Cyg X-3 ($P = 4.8^h$) may be an LMXB. The special feature distinguishing these systems from the others is: an extremely high mass transfer rate.

TABLE III. The standard 'massive' X-ray binaries (top) and some examples of Be-X-ray binaries (bottom).

Source	Optical counterpart	Spectral type	Pulse period(s)	Orbital period(d)	Eccentricity
LMC X-4	Sk-Ph	07 III–V	13.5	1.408 d	$e = 0.00$
Cen X-3	Krz's star	06.5 II–III	4.84	2.087	$e = 0.00$
4U 1700 – 37	HD 153919	06.5 f	. . .	3.412	. . .
SMC X-1	Sk 160	B0Ib	0.714	3.89	$e = 0.00$
4U 1538 – 52	12	B0Iab	529	3.73	$e = 0.00$
Cyg X-1	HD 226868	O9.7Iab	. . .	5.60	. . .
4U 0900 – 40	HD 77581	B0.5Ib	283	8.965	$e = 0.09$
GX 301 – 2	Wra 977	B1.5Ia	696	41.4	$e = 0.47$
4U 0115 + 63	Johns' star	Be	3.61	24.3 d	$e = 0.35$, transient
4U 0352 + 30	X Per	09.5 (III–V)e	835	580 d	Very weak, steady
A 0535 + 26	HD 245770	B0Ve	104	$\geqslant 18$	transient
4U 1145 – 61	Hen 715	B1Vne	292	$\geqslant 35$	Highly variable
4U 1258 – 61	MMV star	B2Vne	272	$\geqslant 13$	Highly variable

TABLE IV. Low-mass X-ray binaries with known orbital periods. The five systems in the lower part of the table have evolved companions.

Name	P_{orb}	Remark
1626 − 67	42 min	7.7s pulsar
2259 − 59	41 min(:)	7s pulsar, eccentr. orbit ($e > 0.1$)
1916 − 05	50 min	burster
1323 − 62	3 h	burster
1636 − 53	3.8 h	
1755 − 31	4.2 h	
Cyg X-3	4.8h	very large radio outbursts; small opposite radio jets
1254 − 69	4.8 h(:)	burster
1822 − 37	5.3 h	
2129 + 47	5.6 h	
1659 − 29	7.1 h	transient
0620 − 00	7.8 h	transient, K-type spectrum
Sco X-1	0.78 d	Radio source, opposite radio jets
Her X-1	1.7 d	1.2s pulsar: A-type spectrum
0921 − 63	9.0 d	F-giant
Cyg X-2	9.8 d	F-giant
GX 1 + 4	⩾ few months	118s pulsar; M6 IIIe spectrum; radio source

TABLE V. Some important properties of the five binary radio pulsars and the single millisecond pulsar, together with estimates of their surface magnetic field strenghts and of the masses of the companions in the binary systems.

Name	P_{orb} (d)	e	Mass function ($M_\odot$)	Most likely companion mass $M_\odot$	P_{pulse} (s)	B_s (G)
PSR 1913+16	0.32	0.617	0.1322	1.40 ± 0.05	0.059	2×10^{10}
PSR 0655+64	1.03	0.000	0.0712	1.00 ± 0.30	0.196	2×10^{10}
PSR 0820+02	1232	0.012	0.00301	0.2 − 0.4	0.865	3.3×10^{11}
PSR 1953+29	~ 117	<0.01	0.00272	0.2 − 0.4	0.0061	$<2.5 \times 10^9$
PSR 2303+46	12.34	0.658	0.2463	1.40	1.066	
PSR 1937+214					0.00155	5×10^8

3.3. Outline of the formation processes of neutron stars in binaries

The neutron stars in X-ray binaries can either have been formed by (i) the direct collapse of the nuclearly exhausted core of an (initially) massive star, or (ii) by the accretion-induced collapse of a white dwarf in an old binary (a similar collapse might also be triggered by the coalescence of two white dwarfs in a very close white dwarf binary, cf. Webbink 1980; 1984; Iben and Tutukov 1984).

The first type of evolution has taken place in the massive X-ray binaries. The reason why these systems were not disrupted by the SN explosion that formed the neutron star, is that at the moment of the explosion the star had already become the less massive component of the system, as a consequence of a preceding stage of large-scale mass transfer (Van den Heuvel and Heise 1972; Tutukov and Yungelson 1973; Van den Heuvel 1974). As was shown by Blaauw (1961), the system is disrupted only if more than half of the system mass is ejected in the explosion, which is not the case here. Figure 6a depicts, as an example, the evolutionary history of a typical B-emission X-ray binary (Habets 1985). The various evolutionary stages are described in the figure caption. This type of close binary evolution, in which the mass transfer starts after the end of core-hydrogen burning, but before helium ignition, is called case B (in the terminology of Kippenhahn and Weigert (KW 1967), see below). Basically the formation of the massive X-ray binaries can be understood in terms of so called "conservative" models such as that depicted in figure 6, in which it is assumed that the total mass and total orbital angular momentum of the system is conserved during phases of mass transfer (i.e. between phases B and C in figure 6a). We refer to reviews available in the literature (Van den Heuvel 1978, 1981 , 1983). The implications of the conservative assumptions for the changes in the orbital parameters are dealt with in section 3.8.

The second neutron-star formation mechanism - by white-dwarf collapse - is expected to have taken place in most of the low-mass X-ray binaries. This is because in the direct collapse of a nuclearly exhausted stellar core always a considerable amount of mass is ejected (see next section). Taking the effects of the impact of the SN shell onto a low-mass companion in a close system into account it turns out that it may be difficult in this case to keep a low-mass companion star bound (cf. van den Heuvel 1978, 1981). On the other hand, in the collapse of a white dwarf not more than ~ 0.1 $M_{\odot}$ (the mass equivalent of the binding energy of the neutron star) needs to be ejected, and disruption of the system is in general not expected. We will now consider these two ways for neutron-star formation in binaries in more detail.

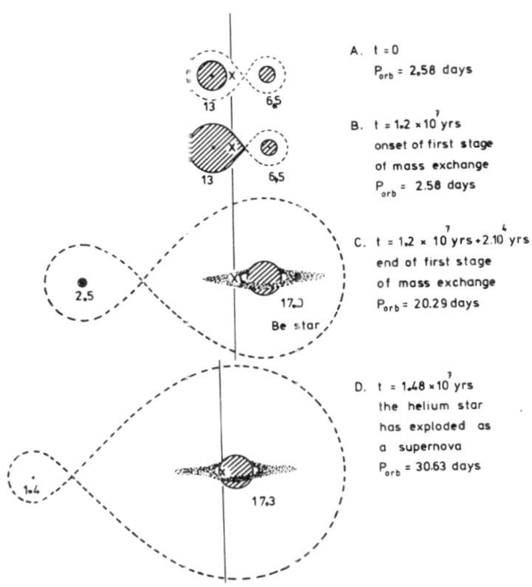

Figure 6a. Conservative evolutionary scenario for the formation of a Be-X-ray binary out of a close pair of early B stars with masses of 13.0 $M_{\odot}$ and 6.5 $M_{\odot}$. The numbers indicate mass ($M_{\odot}$). After the end of the mass transfer the Be star presumably has a circumstellar disc or shell of matter associated with the rapid rotation (induced by the previous accretion of matter with high angular momentum; cf. Habets 1985).

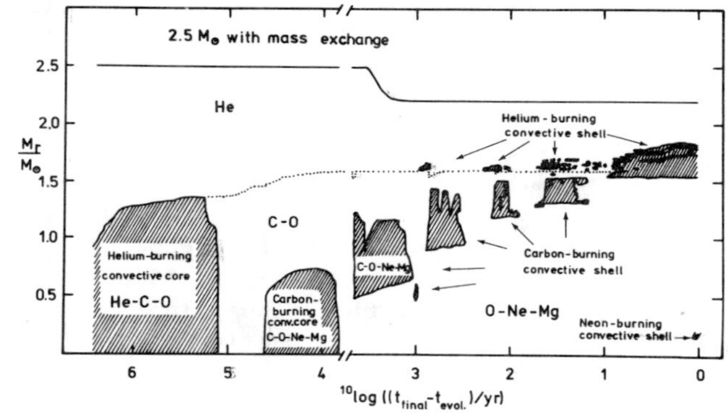

Figure 6b. Interior evolution of a 2.5 $M_\odot$ helium star in the binary of fig. 6a. After carbon-shell ignition, case BB mass transfer starts and 0.3 $M_\odot$ is lost to the companion. The supernova will take place a few years after neon ignition. The meaning of the dashed and dotted lines and of the hatched regions is the same as in Fig. 4.

3.4. Types of close binary evolution

The type of remnant that is produced by close binary evolution, depends on the initial primary mass, mass ratio and orbital period. Figure 7 depicts the three main types A, B and C of close binary evolution as defined in KW 1967, as explained in the figure caption.
Case B and C are statistically most common among the known unevolved close spectroscopic binaries. Case A is relatively rare.
The critical orbital period which separates the cases A and B, and B and C, respectively, depends on the initial mass of the primary star M_1, and (slightly) on the initial mass ratio. Figure 8 depicts the combinations of orbital period and primary mass for which the cases A,B and C occur, for an initial mass ratio unity as derived from Iben's evolutionary tracks (partly after Webbink 1980).

In what follows we will mainly concentrate on the results of case B evolution, as this case is very common, and its results can be described in fairly simple terms. Later on we will extend the discussions also to case C.

In case B, after the first phase of mass transfer - which takes place on a thermal timescale - only the helium core of the primary star is left. Therefore, whether or not the primary star will terminate its life as a white dwarf or as a neutron star, can be discussed simply in terms of the evolution of helium stars, which has been reviewed in the foregoing section.

The correspondence between helium-core mass and initial (hydrogen-rich) primary mass can be simply expressed by analytic forms such as

$$M_{He} = 0.073\ (M_1/M_\odot)^{1.42} \tag{12}$$

(van Beveren 1980; for chemical composition X = 0.70, Z = 0.03).

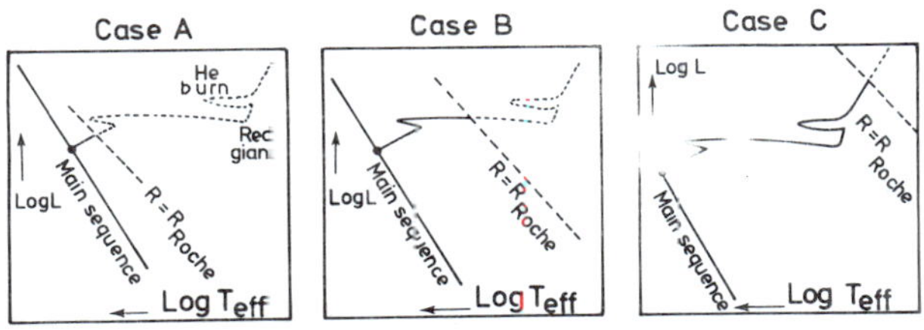

Figure 7. The three basic cases A, B and C of mass exchange in a close-binary system, as defined by Kippenhahn & Weigert (1967), illustrated on the evolutionary track of the primary star in the Hertzsprung-Russell diagram.

3.5. The evolution of helium stars, the occurrence of a second phase of mass transfer, and the lower mass limit for neutron-star formation in a binary

Figures 4 and 6b show that after the end of core carbon burning, several episodes of carbon-shell burning occur around the growing O-Ne-Mg core. The outer radii of helium stars with $M > 3.5\ M_{\odot}$ never become larger than a few solar radii, before the final core collapse. Therefore, in a binary such stars will not undergo a second phase of mass transfer before they explode as a supernova.
However, for lower masses the outer radii of helium stars may become very large during the late evolutionary stages (Paczynski 1971; Arnett 1978). This can be seen in figure 5, based on the calculations by Habets (1985). (The core mass is defined here as the mass inside the helium burning shell). The figure shows that helium stars with $M < 3.5\ M_{\odot}$ may, in a binary, undergo a second phase of mass transfer (so-called case BB mass transfer, cf. Delgado and Thomas 1979).
As this reduces their mass, this second mass-transfer phase may lower the mass limit for neutron-star formation. As an example figure 6b shows the interior evolution of a 2.5 $M_{\odot}$ helium star in the binary of figure 6a, which has an orbital period of $20^{d}.29$ and a companion mass of 17 $M_{\odot}$ (Habets 1985). (Such a system is expected to be the evolutionary product of a binary with initial (zero-age) component masses of 13.0 $M_{\odot}$ plus 6.5 $M_{\odot}$, and an initial orbital period of $2^{d}.58$). The figure shows that during carbon-shell burning the helium star loses about 0.3 $M_{\odot}$ to its companion. Since, however, the core mass grows to $\sim 1.5\ M_{\odot}$, and neon is ignited, this star is still expected to collapse to a neutron star. In a single helium star of 2.5 $M_{\odot}$ the core mass grows to 1.6 $M_{\odot}$, however, which shows that the occurrence of case BB mass transfer slightly lowers the core mass, and therefore increases the lower mass-limit required for neutron-star formation in a binary system.

Habets' calculations show that the cores of helium stars in binaries with $M_{He} > 2.2\ M_{\odot}$ will always collapse to neutron stars (also if still a case BB mass transfer occurs). This implies that, in a case B binary, the initial (hydrogen-rich) stellar mass for reaching collapse to a neutron star is $\sim 10\ M_{\odot}$. For case C and for single stars this limit is lower, since in these stars the helium-core mass can still grow considerably during the later evolutionary stages, as a consequence of hydrogen-shell burning.

3.6. Types of remnants produced, as a function of initial primary mass and orbital period

Figure 8 schematically represents the various types of remnants of primary stars of close binaries that can be produced, as a function of initial primary mass and orbital period, as based on the results of Habets (1985; cf. van den Heuvel 1983, Van den Heuvel and Habets 1985). At the top of the figure also the various types of remnants of single star evolution are indicated: these are practically identical to the remnants of primary stars of case C evolution, with the exception of the O-Ne-Mg white dwarfs, which (most probably) do not occur for single stars. The following points can be noticed in the figure.
1. For initial primary masses > 10-12 $M_{\odot}$ (case B), $\gtrsim$ 8-9 $M_{\odot}$ (Case C) or $\gtrsim$ 8 $M_{\odot}$ (single stars), respectively, a neutron star is produced, as the mass of the helium core is > 2.2 $M_{\odot}$.
2. For a range in initial primary masses below these boundaries, in the cases B and C the remnants are white dwarfs consisting of O, Ne and Mg (products of carbon burning). In case B this occurs for initial primary masses approximately in the range 9 (± 1) to 12 (± 2) solar masses. For Case C: in the range $\sim 8\ M_{\odot}$ to $\sim 9\ M_{\odot}$. These white dwarfs are mainly the products of the later evolution of helium stars in the mass range $\sim$ 1.9 - 2.5 $M_{\odot}$ which lose their envelopes in a second phase - BB or CB - of mass transfer.
(The above mass ranges, as depicted in figure 8, are only approximate, as the precise values of the limits depend not solely on the initial primary mass and orbital period, but also on the initial mass ratio and on the convection theory used; the orbital period and mass ratio determine whether or not cases BB or CB of mass transfer will occur, which may slightly modify the precise values of the limiting masses. A full discussion of the various complicating factors is given in chapter IIIc of Habets 1985).

The interesting point in figure 8 is especially the mass range in which the remnants are O-Ne-Mg white dwarfs. When single stars produce such cores, these are still surrounded by a helium shell and a hydrogen envelope, which leads these cores to grow to the Chandrasekhar mass and to undergo collapse, leaving a neutron star (cf. Sugimoto and Nomoto 1980; Nomoto, 1984 a,b; Hillebrandt, 1984; Hillebrandt et al. 1984, and: this volume). In binaries, the mass transfer (case BB or CB) removes these envelopes, such that the cores may stay behind as O-Ne-Mg white dwarfs. These are interesting objects since in the case of reversed mass transfer in a binary they may be triggered by the accretion to collapse to a neutron star, as was first pointed out by Miyaji et al. (1980) and Sugimoto and Nomoto (1980).

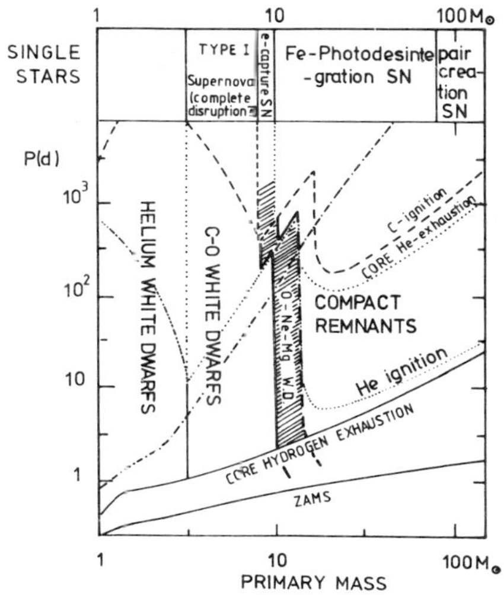

Figure 8. Classification of expected final evolutionary states of primary stars of close binaries as a function of primary mass and initial orbital period (for mass ratio 0.5 and X = 0.70). At the top of the figure the expected final evolution of single stars is indicated. In a subsequent phase of reversed mass-transfer CO white dwarfs may be triggered to explode as a type I supernova (complete disruption, see Nomoto 1982); old CO white dwarfs with a mass close the the Chandrasekhar limit, and O-Ne-Mg white dwarfs may be triggered by accretion to implode to a neutron star.

3.7. Conditions for the occurrence of accretion-induced collapse of a white dwarf in a close binary

The reaction of a white dwarf to accretion of matter depends on a number of aspects, such as (Nomoto 1982):

(i) Composition. C-O white dwarfs will, when their masses increase, evolve to degenerate carbon ignition. This is expected to lead to a nuclear runaway in which part or all of the white dwarf is converted to Ni^{56} (see section 2.6). Under some circumstances (carbon ignition in a shell) a small white dwarf remnant may be left (Taam 1980 a,b; Nomoto 1982; Labay et al. 1983). Many authors believe that this complete or partial nuclear explosion of a C-O white dwarf is to be identified with a type I Supernova (see Chevalier, 1985).

On the other hand, when O-Ne-Mg white dwarfs grow in mass, their core density may grow to the threshold for electron capture, leading to an electron-capture collapse and formation of a neutron star (see section 2.8).

(ii),(iii) Accretion rate and Mass. The accretion rate is probably the

most crucial parameter, as it determines whether or not the white dwarf will grow in mass. A full discussion of all the recent work on this subject and the uncertainties involved is given by Habets (1985). A brief - but rough - summary based mainly on the works by Nomoto (1982, 1984ab) and Mueller and Arnett (1982, 1984) is as follows. At low accretion rates the accumulated hydrogen will when its mass exceeds a critical limit, ignite and burn in a strong nuclear flash. For accretion rates $\leqslant 10^{-9}$ $M_\odot$/yr these flashes are so strong that all (or possibly even more than all) of the accreted matter is expected to be ejected, such that no net growth of the white dwarf occurs.
(If one assumes the accreted matter to be helium, then for $M_{w.d.} > 1.1\ M_\odot$ and low accretion rates, $< 8.10^{-10}$ $M_\odot$/yr, a thermonuclear explosion of the white dwarf can be triggered by pycno-nuclear reactions (Nomoto 1982). If the accreted matter is hydrogen, however, at this rate helium will have difficulty in building up, since (most of) the helium produced is ejected in the strong hydrogen flash; therefore deflagration may only occur here under very special circumstances). Only for hydrogen accretion rates $> 10^{-9}$ to 10^{-8} $M_\odot$/yr (depending on the white dwarf mass) the flashes are so weak that probably much of the accreted matter is retained, and a net growth in mass is possible. For $\dot{M} > 10^{-7}$ $M_\odot$/yr the accreted hydrogen burns steadily in a shell surrounded by a red-giant like envelope. Clearly, the latter is only possible in a wide binary, as in a very close system there is not enough room for a red-giant envelope.A carbon deflagration-explosion requires accretion rates $> 4.10^{-8}$ $M_\odot$/yr, practically irrespective of stellar mass.

3.8. Conservative vs. non-conservative evolution. Reasons for the occurence of common-envelope evolution.

3.8.1. Introduction. So far we have only looked at the final evolutionary states of the primary star after it has lost its hydrogen-rich envelope. We did not look into what happened to the envelope matter, nor what precisely happened with the orbital period. This we will consider here.

Basically two extreme types of assumptions regarding the fate of the exchanged mass and orbital angular momentum can be made: (a) "conservative" evolution, where it is assumed that all mass lost by the one star is captured by the other and also the total orbital angular momentum is conserved, and (b) highly non-conservative evolution, in which all of the envelope mass lost by the star is ejected from the system. We will first consider the implication of the conservative assumptions, and from this derive under which circumstances these assumptions are expected to apply.

3.8.2. Conservative evolution.

(1) Changes in orbit during the mass transfer.

The assumptions of conservation of total mass

$$M_1 + M_2 = M = \text{const.} \qquad (13)$$

and total orbital angular momentum J, given by (for circular orbits):

$$J^2 = \frac{G\, M_1^2\, M_2^2}{M_1 + M_2}\, a \qquad (14)$$

imply that the orbital radius a changes in the conservative case according to the relation

$$\frac{a}{a_o} = \left(\frac{M_1^o \cdot M_2^o}{M_1 \cdot M_2}\right)^2 \qquad (15)$$

where sub- and superscripts zero indicate the initial situation. Eq. (15) in combination with (13) implies that if mass is transferred from the more massive to the less massive component, the orbit shrinks whereas in the opposite case the orbit expands.
The Roche-lobe radius is given by

$$r_L/a = 0.49\, q^{2/3}/[0.6q^{2/3} + \ln\,(1+q^{1/3})] \qquad (16)$$

(Eggleton 1983), where $q = M_1/M_2$.

Figure 9 depicts how the Roche-lobe radius of the initially more massive star changes during conservative mass transfer as a function of the mass M_1 of the mass-losing star. Until a mass ratio $M_1/M_2 = 0.85$ is reached (for decreasing M_1) r_L decreases. For $M_1/M_2 < 0.85$ it increases again.

The change in orbital parameters during the mass transfer can, for example, be seen in figure 6, where the conservative mass exchange of a moderately massive close binary system is depicted (between the phases B and C).

(2) Phases in the evolution of a binary system and their timescales.
The large reduction of the Roche-lobe radius of the more massive component during mass transfer from this star to its companion, as depicted in figure 9, leads to a highly unstable situation, for the following reason.
We first consider stars with radiative envelopes. When mass is taken away from the star both its hydrostatic and thermal equilibrium are disturbed. It will immediately - on a dynamical timescale - restore its hydrostatic equilibrium. But, to restore its thermal equilibrium requires roughly a thermal timescale.
Important is here, that the thermal equilibrium radius - i.e. the radius which it reaches after its thermal equilibrium is restored - does not decrease very much when its mass M_1 decreases. This is depicted by the R_e curve in figure 9.

The figure shows that for decreasing M_1 the Roche-lobe radius r_L can only become larger again than the thermal equilibrium radius if M_1 has become coinsiderably smaller massive than M_2, i.e. when the mass ratio of the system has been reversed.
Since stars with radiative envelopes at first become smaller when mass is taken away from them, but then begin to expand on a thermal

timescale to restore their thermal equilibrium, the timescale of the mass transfer in systems with such primary stars is: **the thermal timescale.**

From the above we conclude that the following phases occur in the evolution of a close binary in which the envelope of the primary is in radiative equilibrium (see figure 9):

a. When neither of the two stars fills its Roche lobe, both stars gradually expand on an **nuclear timescale.**
b. When the primary star begins to overflow its Roche lobe, a phase of rapid mass transfer occurs, on a **thermal timescale,** leading to reversal of the mass ratio.
c. After restoration of the thermal equilibrium, the (now less massive) primary further evolves on a **nuclear timescale.** If it is a helium star, it will not even fill its Roche lobe, unless case BB mass transfer occurs. If it still has a hydrogen-rich envelope and fills its Roche lobe, its radius will gradually expand on a nuclear timescale, leading to stable mass transfer on a **nuclear timescale** since now the Roche lobe expands when the star loses mass.

The system of figure 6, and the massive ($M_1 > 10\ M_\odot$) close binaries in general, are good examples of the conservative evolution in the above-described way. This can be seen from figure 8, where the convective boundary is indicated by the dashed line.

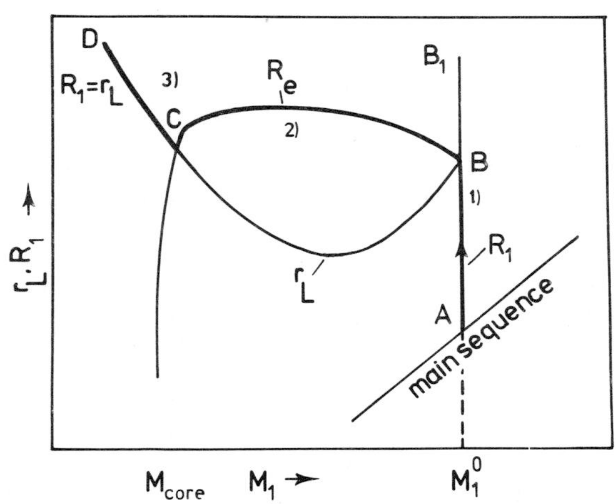

Figure 9. Variation of the thermal-equilibrium radius of the primary component R_e and the radius of its Roche lobe r_L during the mass transfer from the primary to the secondary. Only after the primary has become less massive than the secondary will r_L begin to increase. Consequently, equilibrium is not possible before the primary has lost so much matter that is has become the less massive component (after Paczynski 1970). In stages 1 and 3 the star is expanding on a nuclear timescale, in stage 2 it is transferring mass on a thermal timescale (the envelope of the primary is assumed to be in radiative equilibrium).

3.8.3. Breakdown of the conservative assumptions; the formation of a Common Envelope. The conservatiove assumptions are expected to be a good approximation only if:

a. the mass ratio M_2/M_1 is not too different from unity, i.e. > 0.4 - 0.5.
b. the envelope of the mass-losing star is in radiative equilibrium.

If either of these conditions is not fulfilled one expects that the secondary star will be unable to accept most of the matter transferred to it, and will itself rapidly expand to overflow its Roche lobe. The result is that the transferred matter will form a Common Envelope (CE) around the system in which both stars are embedded.

The reasons for the formation of such a common envelope have been described extensively elsewhere and will not be given here (e.g. see Paczynski 1976; Ostriker 1973; Webbink 1980; Van den Heuvel 1976, 1981; Iben and Tutukov 1984). In the case that the companion has a convective envelope, the reason is the highly unstable character of such an envelope. If mass is taken away from it, a convective envelope has a tendency to expand on semi-dynamical timescale (a timescale intermediate between the thermal and the dynamical timescale, cf. Paczynski and Sienkiewicz 1971).

4. COMMON-ENVELOPE EVOLUTION: THE FINAL EVOLUTION OF MASSIVE X-RAY BINARIES

4.1. Introduction

The MXB's have very small mass ratios and therefore are typical examples of systems that are bound to undergo CE-evolution. CE evolution is a very complex physical process which requires detailed 2-D or 3-D hydrodynamical calculations. Such calculations, in a somewhat simplified form, were carried out for the MXB's by Taam et al. (1978), Delgado (1979) and by Bodenheimer and Taam (1984), and for the origin of Cataclysmic Variable binaries by Meyer and Meyer-Hofmeister (1979).
The secondary star which is moving through the envelope of its companion undergoes very large frictional drag, causing its orbit to rapidly shrink. The orbital energy deposited in the envelope may be sufficient to blow off the envelope, such that a very close binary system remains, consisiting of the secondary star plus the compact core of the primary. The very short orbital periods of the binary radio pulsars PSR 1913+16 and PSR 0655+64 indicate that these systems must be the result of such a type of evolution. The same holds for the Cataclysmic Variable (CV) binaries, which are expected to be the descendants of wide binaries, consisting of a red giant with a degenerate core together with a red dwarf (cf. Meyer and Meyer-Hofmeister 1979).

4.2. Simplified approach to CE evolution

We will follow here a simplified approach based on rather simple energy considerations (see, e.g., Van den Heuvel 1976; Tutukov and Yungelson 1979; Webbink 1980, 1984; Iben and Tutukov 1984).

We assume that the compact star has a mass M_1 and the massive star consists of a compact core of mass M_{2f} and an extended envelope of mass M_{2e}.

To calculate the maximum orbital period with which the system may terminate its evolution, we assume that all of the loss of orbital energy is used to expell the envelope (i.e. that the loss of orbital energy in the form of radiation is negligible). The numerical computations of Bodenheimer and Taam suggest that this is a good approximation.

The above assumptions lead to the following equation for the relation between the original and final orbital radii a_1 and a_2, respectively (cf. Webbink 1984):

$$\frac{G(M_{2f} + M_{2e}) \cdot M_{2e}}{\lambda a_1 r_L} = \frac{G\, M_1 M_{2f}}{2a_2} - \frac{G(M_{2f} + M_{2e})\, M_1}{2\, a_1} \qquad (17)$$

where $r_L a_1$ denotes the Roche lobe radius of the primary star at the onset of the transfer and λ is a weighting factor (< 1) for the gravitational binding energy of core and envelope. Equation (17) yields

$$\frac{a_2}{a_1} = \frac{M_1\, M_{2f}}{M_{2f} + M_{2e}} \,/\, (M_1 + 2\, M_{2e}/\lambda r_L) \qquad (18)$$

With eq. (18) it is simple to calculate the fate of an X-ray binary such as, for example, Cen X-3. Assume $M_1 = 1.4\ M_\odot$, $M_{2e} = 12\ M_\odot$, $M_{2f} = 4\ M_\odot$. In this case: $r_L = 0.6$. We assume $\lambda = 0.5$ (which is a reasonable approximation since stars are rather strongly centrally condensed). With eq. (18) these values result in $a_2/a_1 = 0.0043$. With $a_1 = 17.8\ R_\odot$ (as for Cen X-3) this yields $a_2 = 0.077\ R_\odot$. This value is so small, that it is deep inside the 4 $M_\odot$ helium core of the companion of Cen X-3 (this core has a radius of $\sim 0.4\ R_\odot$).
We therefore are forced to conclude that in Cen X-3 the neutron star will spiral down completely into the center of the companion star. (The same conclusion was reached by Taam et al. (1978) from 2-D hydrodynamical calculations).
In order to avoid such a complete spiral in, $a_2 r_L$ must be $> 0.5\ R_\odot$ which implies that a_1 should have been 13.5 times larger than that of Cen X-3, implying an original orbital period > 104 days.

We thus conclude that the neutron stars in all the "standard" massive X-ray binaries, such as listed in the upper part of table III will end up in the center of their companion stars!
Only for very wide systems, such as some of the B-emission X-ray binaries, such a fate can be avoided, and a binary system can remain, consisting of a helium star plus a neutron star, in a circular orbit. This may for example be the case for 4U1145-61 which has an alleged orbital period of 117^d (Bradt and McClintock 1983), and for X Persei, with P ≃ 587^d days. Figure 10 depicts these two types of results from CE evolution of MXB's.

4.3. Constraints on the initial system parameters of PSR 1913+16 and PSR 2303+46

Helium stars more massive than $\sim 3.5\ M_\odot$ do not expand very much during their further evolution, and will finally explode as supernovae.

If, in such an - assumedly symmetric - explosion more than half of the system mass is ejected, the system will be disrupted, producing two run-away pulsars, as depicted in figure 10b. For a neutron star mass of 1.4 $M_\odot$ this occurs for helium star masses > 4.2 $M_\odot$.

In PSR 1913+16 and PSR 2303+46 the mass of the helium star must thus have been less than 4.2 $M_\odot$, and assuming a spherically symmetric explosion, it is simple, from the presently observed orbital eccentricities of these systems, to calculate the mass of the helium star progenitor of the last-formed neutron star in the system.

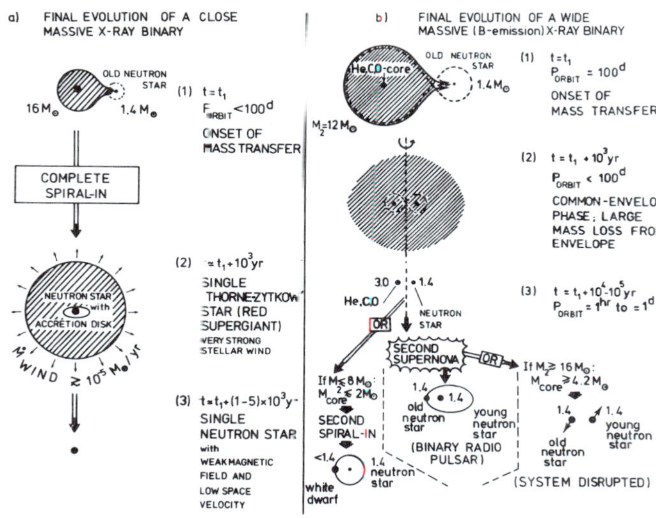

Figure 10. Final evolution of massive X-ray binaries. a. Short-period systems (P_{orb} < 100^d) are expected to evolve towards coalescence during the second phase of mass transfer, leading to a so-called 'Thorne-Zytkow' star, a red supergiant with an accreting neutron star in its center. Due to its very large wind mass loss, this star loses its entire mantle within a few million years, leaving a single old neutron star with a low space velocity (the velocity of the original MXB). b. Wide systems (P > 100^d) leave very close binaries consisting of the old neutron star plus the core of the companion (Cygnus X-3 may be an example). If M_2 was < 8 $M_\odot$, no second SN will occur and a close binary radio pulsar with a white dwarf companion in a circular orbit remains. PSR 0655+64 may have formed in this way. If 8 < $M_2/M_\odot$ < 16, the core explodes as a supernova but has a mass < 4.2 $M_\odot$ such that a bound binary radio pulsar with an eccentric orbit will remain, consisting of two neutron stars. PSR 1913+16 and PSR 2303+46 must have formed in this way. If M_2 > 16 $M_\odot$, the core mass is so large that the system is disrupted in the second supernova and two runaway pulsars are formed. (If their orbital period is < 12 hours, systems such as PSR 0655+64 will evolve towards coalescence leading to the formation of a single rapidly rotating pulsar with a low space velocity. The 1.5 msec pulsar may have formed in this way.).

Celestical mechanics shows (cf. Flannery and Van den Heuvel 1975) that the orbital parameters of the system after the SN explosion are related to those before the explosion by:

$$a_f = \mu_f/(2\mu_f - 1); \; P_f = \mu_f/(2\mu_f - 1)^{3/2}; \; e_f = (1 - \mu_f)/\mu_f \qquad (19)$$

where a_f and P_f are the final semi major axis an orbital period expressed in units of the original one, e_f is the final orbital eccentricity, and

$$\mu_f = (M_1{}^f + M_2{}^f)/(M_1{}^o + M_2{}^o) \qquad (20)$$

where superscripts zero and f denote the initial and final values, respectively.

Assuming a neutron star mass of 1.42 $M_\odot$ in PSR 1913+16, its e-value yields μ_f = 1/1,615 yielding a mass of the exploding helium star of $M_1{}^o$ = 3.16 $M_\odot$, and an orbital period at the time of the explosion of P_o = 1.45^{hr} (a_o = 1.08 $R_\odot$). For PSR 2303+46 these values become $M_1{}^o$ = 3.29 $M_\odot$ and P_o = 1.92 days (a_o = 10.9 $R_\odot$).

If we assume these $M_1{}^o$ values to correspond to helium core masses, the progenitors of both stars must have had masses ~ 12 $M_\odot$.

Adopting this mass and applying eq. (18) with λ = 0.5 we find a_1/a_2 = 161.4 for PSR 1913+16 and a_1/a_2 = 152.8 for PSR 2303+46, yielding original orbital separations and periods for these systems of a_1 = 0.81 AU, P = 72.3 days for the progenitor of PSR 1913+16 and a_1 = 7.71 AU, P = 5.84 yr for the progenitor of PSR 2303+46.

The progenitor of PSR 1913+16 closely resembles one of the wider B-e X-ray binaries, such as 4U1145-61. The progenitor of PSR 2303+46 must have been a system resembling the ζ Aurigae and VV Cephei binaries.

4.4. The progenitor of PSR 0655+64

Like PSR 1913+16 and PSR 1953+29 this pulsar combines a very short pulse period with a relatively weak magnetic field, which strongly suggests that it is "recycled" pulsar, i.e.: that it has been spun up by accretion of matter from its companion (see Van den Heuvel 1984, and Ruderman, this volume, for details). Since the companion is invisible at present but must have overflowed its Roche lobe at an earlier stage, the only possible solution is that it presently has terminated its evolution. The perfect circularity of the orbit indicates that no second supernova explosion took place, which implies that the companion must now be a white dwarf.

The evolution of this systems must thus have been very similar to that of PSR 1913+16 and PSR 2303+46, the only difference being that its companion had a lower mass, i.e. below ~ 8 $M_\odot$.

Figure 10b (left-hand part, after Van den Heuvel and Taam 1984) depicts the evolutionary history of this system.

Assuming a present neutron star mass of 1.4 $M_\odot$, a present companion mass M_{2f} of 0.8 $M_\odot$ and an original companion mass of 5 $M_\odot$, one obtains with eq. (18) an original orbital period and separation of 2.50 yr and 3.43 AU, respectively.

Like the progenitor of PSR 2303+46 this system again shows a close resemblance to the ζ Aurigae binaries.

4.5. The origin of the isolated 1.5 msec binary radio pulsar

If a system such as PSR 0655+64 had been born with a three times shorter orbital period, i.e. $\leqslant 8^h$, then its orbit would rapidly decay by gravitational radiation losses and within $\sim 10^9$ yr the neutron star and the white dwarf would have coalesced leading to the formation of a single rapidly rotating neutron star (Van den Heuvel and Bonsema 1984; Bonsema and Van den Heuvel 1985; Van den Heuvel 1984). The amount of matter accreted in this case is $> 0.12\ M_\odot$, which is sufficient to spin the neutron star up to a rotation period of order 1.5 msec (cf. Alpar et al. 1982). As the magnetic field of the neutron star would, in the meantime, have undergone considerable decay, the resulting object would closely resemble the 1.5 msec pulsar PSR 1937+214. As the massive progenitor system will have had a relatively low runaway velocity, this pulsar is not expected to be a high-velocity object. The same holds for the pulsars that resulted from complete spiral-in, such as depicted in figure 10a.

5. LOW-MASS X-RAY BINARIES

5.1. Mechanisms driving the mass transfer

As the winds from low-mass stars are never very strong (except perhaps when the stars reach the asymptotic red giant branch) in general mass transfer by Roche-lobe overflow is required to power these strong sources.

As the companion is less massive than the neutron star, the mass transfer will be stable as it leads to an increase in orbital separation. (The only exception is Her X-1, where the companion has $M \sim 2\ M_\odot$; here beginning atmospheric Roche-lobe overflow can power the source for several million years before it reaches the Eddington limit, cf. Savonije 1983a).

There are two mechanisms by which Roche-lobe overflow in low-mass systems can be driven (cf. Savonije 1983a):

(i) Angular-momentum losses by gravitational radiation (possibly supplemented by "magnetic braking" (cf. Verbunt and Zwaan 1981)). These losses cause the orbit and the Roche-lobe to gradually shrink, thus inducing the companion to transfer matter. As the characteristic timescale for orbital decay by gravitational radiation (GR) losses is given by (cf. Savonije 1983a):

$$t_{GR} = \frac{(M_1 + M_2)^{1/3}}{2^{1/3}\ M_1 M_2} (P/1.6^h)^{8/3} \cdot 4.10^7 \text{ yr} \qquad (21)$$

This mechanism will only be of importance in systems with $P_{orb} < 12^h$, since only in such systems t_{GR} is smaller than the Hubble time (for $M_1 \simeq M_2 \simeq = 1$, in solar units).

Table IV shows that most of the known low-mass systems belong to this category.

The most extensive analysis of the combined effects of GR losses and magnetic braking for the driving of the mass transfer in low-mass systems was given by Rappaport et al. (1983). As an example figure 11 shows the mass transfer rate and the variations of the orbital parameters with time for a system with an initial companion of 1 $M_\odot$, and magnetic braking according to the Verbunt and Zwaan (1981) formulation (cf. Savonije 1983a):

$$dJ_1/dt = - k^2 M_1 R_1{}^4 (0.5 f^{-2} \cdot 10^{-28}) \omega_1{}^3 \tag{22}$$

where M_1, R_1, ω_1 denote the mass, radius and rotational (= orbital) angular velocity of the companion, and k its radius of gyration. The constant f has to be determined from observations and is about 1.7.

The $\dot{M}$-profile as a function of time in Figure 11 is characteristic for systems in which these angular momentum losses drive the mass transfer. One observes from the figure that although mass loss rates $> 0.5 \times 10^{-8}$ $M_\odot$/yr can be reached for a short interval of time ($< 2.10^7$ yrs), during most of the time ($\sim 10^9$ yrs) the system has $\dot{M}$ in the range 10^{-10} to 10^{-9} $M_\odot$/yr. It should be noticed that although GR losses are not important for $P > 10^h$, magnetic braking may still be important for orbital periods up to about 2-4 days. This does require, however, the low-mass companion to have a radius close to that of its Roche-lobe, implying that it must be an evolved star.

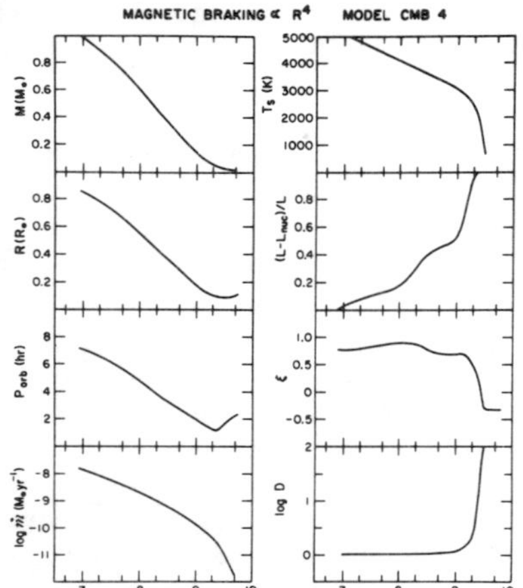

Figure 11. Evolution of close binary with angular momentum losses by gravitational radiation and magnetic braking as calculated by Rappaport et al. (1983). The initial system consisted of a 1 $M_\odot$ main-sequence star and a 1.2 $M_\odot$ compact star with an orbital period of 7.2 hours. Magnetic braking was treated according to the formulation of Verbunt and Zwaan (1981). Plotted are: the evolution with time of the stellar mass, radius, mass-transfer rate, orbital period, as well as some other parameters characterizing the system (C Astrophysical Journal).

(ii) Internal nuclear evolution of the companion star is the only mechanism that can drive the mass transfer in low-mass systems with $P_{orb} > 0^{d}.5$ (since unevolved stars with $M < 1.4\ M_{\odot}$ cannot overflow their Roche-lobes in such systems). As mentioned in section 3.2, in systems such as Cyg X-2 ($P_{orb} = 9^{d}.8$), 2S 0921-63 ($P_{orb} = 9^{d}.0$), and even in Sco X-1 ($P_{orb} = 0^{d}.78$) the companion stars are evolved low-mass stars.

The evolution of binaries in which the mass-losing star is a low-mass (sub-) giant star was studied by Webbink et al. (1983) and Taam (1983). The luminosity in these low-mass ($M \sim M_{\odot}$) giants is generated in a hydrogen-burning shell which surrounds the degenerate helium core of low mass ($M_{He} \simeq 0.2 - 0.4\ M_{\odot}$). During the evolution the core mass of the giant grows, and its radius and luminosity gradually increase. It is this increase in radius which, in a binary system, drives the mass transfer. During this transfer the orbit gradually widens, such that at any time the stellar radius just equals the Roche-lobe radius.

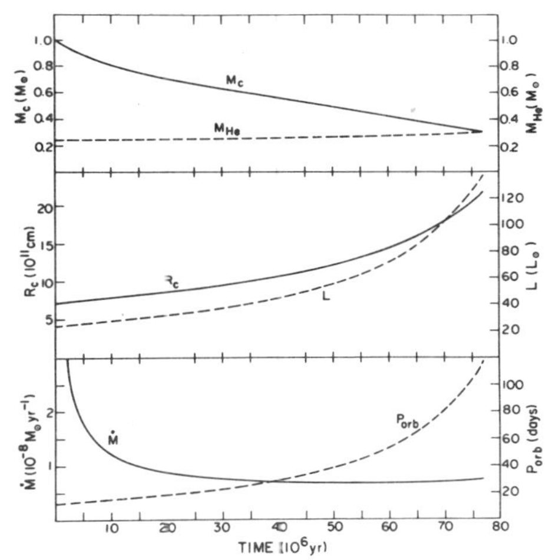

Figure 12. Evolution of a binary with a lower giant-branch secondary component of initial mass 1.0 $M_{\odot}$ and surface composition X = 0.70, Z = 0.02, as calculated by Joss & Rappaport (1983). Plots as functions of time: upper, total mass (M_c) and core mass of the secondary; middle, radius and intrinsic bolometric luminosity of the secondary; lower, mass accretion rate $\dot{M}$ onto the neutron star. Mass of the neutron star was taken to be 1.3 $M_{\odot}$. (C Nature).

Figure 12 represents, as an example, the evolution of a binary with an initial orbital period $P_o = 12.5^{d}$, an initial mass of the giant of 1 $M_{\odot}$ and of the neutron star of 1.3 $M_{\odot}$, calculated by Joss and Rappaport (1983). (Such a system will resemble Cyg X-2 and 2S 0921-63). The figure shows that after a brief initial episode with a mass transfer rate of several times the Eddington limit ($\dot{M}_{Edd} \sim 2.10^{-8}\ M_{\odot}$/yr) the mass transfer rate settles at a value of $\sim 10^{-8}\ M_\odot$/yr for an interval of about 60 million years. During this time the orbital period gradually increases to $\sim 120^d$. In the end only the helium core remains, as a 0.31 $M_\odot$ helium white dwarf (while at the onset of the mass transfer, M_{He} was 0.24 $M_\odot$). It turns out that for systems of this type the average mass transfer rate depends (for a companion mass $M_\odot$, Z = 0.02), in first approximation only on the initial orbital period, roughly as:

$$\langle \dot{M} \rangle = -6 \times 10^{-10}\ (P/\mathrm{day})\ M_\odot/\mathrm{yr}. \qquad (23)$$

(for other companion masses and chemical composition the relation becomes slightly different, cf. Webbink et al. 1983).

5.2. The nature of the strong sources in the galactic bulge: neutron stars with a low-mass giant companion

There are about 8 strong X-ray sources ($L_x \sim 10^{38}$ ergs/s) in the galactic bulge (together with a few dozen weaker sources, $L_x \sim 10^{36} - 10^{37}$ ergs/s). Similarly, in the bulge of M31 there are a few dozen very strong ($L_x \sim 10^{38}$ ergs/s) sources. The mass transfer rates in these systems are typically of order $10^{-8}\ M_\odot$/yr. The only viable model for these old-population sources is that of a binary system in which the companion star is an evolved low-mass star ("giant"), for the following reason (as argued by Webbink et al. 1983). If the systems were binaries in which the mass transfer is driven by gravitational radiation losses and magnetic braking, the systems would, according to figure 11, spend some 50 times longer at accretion rates between 10^{-10} and $10^{-9}\ M_\odot$/yr than at rates between $(0.5 - 2) \times 10^{-8}\ M_\odot$/yr. Consequently, in a steady state one would expect in the bulge of the galaxy some 400 sources with X-ray luminosities between 10^{36} and 10^{37} ergs/s, whereas only a few dozens of such systems are observed. Hence, gravitational radiation (in combination with magnetic braking) predicts an $\dot{M}$-profile as a function of time which disagrees with the observations.

On the other hand, as figure 12 shows, mass transfer from a low-mass giant companion yields a steady high ($10^{-8}\ M_\odot$/yr) mass transfer rate for the entire system lifetime. Therefore, it does not predict a large population of weaker sources. This is a very strong argument in favour of the low-mass giant model for the bulge sources. Thus, one expects the strong bulge sources to be systems typically resembling Cygnus X-2 and 2S0921-63.

5.3. Relation between the wide binary radio pulsars and X-ray binaries with a low-mass giant component; evidence for accretion-induced collapse of old white dwarfs

The orbital characteristics and companion masses of the two wide radio pulsar binaries PSR 0820+02 and PSR 1953+29 (see table V) very closely resemble those of systems that result from the evolution of wide low-mass X-ray binaries considered in the foregoing section, as depicted in figure 12. This was noticed by Joss and Rappaport (1983), Savonije (1983b) and Paczynski (1983).

One paradoxical thing with these systems is, however, that their neutron stars must be younger than $\sim 10^7 - 10^8$ yrs (since surface magnetic fields of neutron stars decay on a timescale of $< (5 - 10) \times 10^7$ yrs (Lyne 1981; Lyne et al. 1985)), whereas the systems themselves must be older than 5.10^9 yrs. The latter follows from the fact that the low-mass companions, which must have started out with $M_s < 1.2\ M_\odot$, have already terminated their evolution and have become helium white dwarfs of 0.2 - 0.4 $M_\odot$ (see table V). For stars with $M < 1.2\ M_\odot$, this requires over 5.10^9 yrs. Thus: we have old binaries that contain young neutron stars. The only way out of this paradox is that the neutron stars in these systems were formed recently - during the mass-transfer process itself - by the accretion-induced collapse of a white dwarf (Helfand et al. 1983; Van den Heuvel and Taam 1984; Van den Heuvel 1984). (We thus imply that at the onset of the mass transfer they still were white dwarfs). As the mass transfer in these systems lasted less than $\sim 7.10^7$ yrs (see figure 12) the neutron stars can indeed have formed quite recently and thus still have retained some magnetic field. Since this seems the only possible way out of the above-mentioned paradox, the two wide radio-pulsar binaries provide very strong observational evidence in support of the possibility that neutron stars can be formed in an old stellar population by the accretion-induced collapse of a white dwarf (as suggested first by Whelan and Iben 1973).

Moreover, just these wide systems provide particularly favourable conditions for such a collapse, since, as mentioned in section 3.7, at their accretion rates of $\sim 10^{-8}\ M_\odot$/yr, the hydrogen accreted by the white dwarfs will burn steadily or in weak flashes, which are unable to cause considerable mass ejection. (For further arguments: see Van den Heuvel and Taam 1984; Van den Heuvel 1984).

Precisely the same arguments as for these systems hold for the pulsating wide low-mass X-ray binary GX 1+4 and for the Quasi Periodic Oscillation sources. In GX 1+4 the companion still is a low-mass red giant, which also must have an age $> 5.10^9$ yrs. This system is an excellent progenitor for systems like the two wide radio-pulsar binaries (Van den Heuvel 1981; Joss and Rappaport 1983). The QPO sources belong to an old population (age (5-10) x 10^9 yrs), but the presence of magnetic fields with a strenghth of order 10^{10} gauss inferred from the QPO's in GX 5-1 and Cygnus X-2 (Van der Klis et al. 1985a; Alpar and Shaham 1985 a,b; Hasinger et al. 1985) and of $\sim 5.10^8$ gauss in Scorpius X-1 and the rapid burster (Middleditch and Priedhorsky 1985; Van der Klis et al. 1985b; Stella et al. 1985) indicates that also these neutron stars are relatively young ($< 10^8$ yrs).

Similar arguments apply also to the two very close - and presumably old - pulsating X-ray binaries 4U 1626-67 (Joss et al. 1978) and 1E 2259+59 (Lipunov and Postnov 1985). (Here, the companion probably was a helium-burning helium star, which drove a white dwarf companion over the Chandrasekhar limit; see Savonije, De Kool and Van den Heuvel 1985, for details).

Thus, we know now at least nine magnetized neutron stars in an old stellar population that provide strong evidence for the formation of neutron stars by accretion-induced collapse of old white dwarfs in binaries (see Taam and Van de Heuvel 1986, for details).

REFERENCES

Alpar, M.A. and Shaham, J. 1985a, IAU Circ. Nr. 4046.

Alpar, M.A. and Shaham, J. 1985b, Nature **316**, 239-241.

Alpar, M.A., Cheng, A.F., Ruderman, M.A., Shaham, J. 1982, Nature **300**, 728.

Arnett, W.D. 1978, in Physics and Astrophysics of Neutron Stars and Black Holes (R. Giacconi and R. Ruffini, eds.), North-Holland Publ. Co., Amsterdam, pp. 356-436.

Blaauw, A. 1961, Bull. Astron. Inst. Netherlands **15**, 265-290.

Bodenheimer, P. and Taam, R.E. 1984, Astroph. J., **280**, 771.

Bonsema, P.F.A. and Van den Heuvel, E.P.J. 1985, Astron. Astrophys., **46**, L3.

Bradt, H.V.D. and McClintock, J.E. 1983, Ann. Rev. Astron. Astrophys. **21**, 13.

Castellani, V., Chieffi, A., Pulone, L. and Tornambe, A. 1985, Astrophys. J. **294**, L31.

Chevalier, R. 1981, Fundamentals of Cosmic Physics, **7**, 1-58.

Chevalier, R. 1985, in: Supernovae, their progenitors and remnants (G. Srinivasan and V. Radhakrishnan, eds.) Indian Acad. Sci., Bangalore, pp. 65-73.

Chiu, H.-J. 1968, Stellar Physics, Blaisdell, London & Toronto.

Clayton, D.D. 1968, Principles of Stellar Evolution and Nucleosynthesis, McGraw Hill, New York.

Cox, J.P. and Giuli, R.T. 1968, Principles of Stellar Structure, I & II, Gordon and Breach.

Delgado, A.J. and Thomas, H.-C. 1981, Astron. Ap. **96**, 142-145.

Eggleton, P.P. 1983, Astroph. J. **268**, 368.

Flannery, B.P. and Van den Heuvel, E.P.J. 1975, Astron. Astrophys. **39**, 61.

Habets, G.M.H.J. 1985, Ph.D. Thesis, University of Amsterdam.

Hasinger, G., Langmeier, A., Sztajno, M., White, N.E. 1985, IAU Circular Nr. 4070, and this volume.

Helfand, D.J., Ruderman, M.A. and Shaham, J. 1983, Nature **304**, 423.

Henrichs, H.F. 1983, in "Accretion-driven Stellar X-ray Sources (W.H.G. Lewin and E.P.J. van den Heuvel;, eds.), Cambridge University Press, p. 393.

Iben, I. 1974, Annual Rev. Astron. Astrophys. **12**, 215-256.

Iben, I. and Renzini, A. 1984, Ann. Rev. Astron. Astrophys. **21**, 271-342.

Iben, I. and Tutukov, A.V. 1984, Astrophys. J. Suppl. Ser. **54**, 335-372.

Joss, P.C., Avni, Y., Rappaport, S. 1978, Astrophys. J. **221**, 645.

Joss, P.C. and Rappaport, S.A. 1983, Nature **304**, 419-421.

Joss, P.C. and Rappaport, S.A.: 1984, Ann. Rev. Astron. Astrophys. **22**, 537-592.

Kippenhahn, R. and Weigert, A. 1967, Zeitschr. f. Astrophys. **65**, 251.

Labay, J., Canal, R. and Isern, J. 1983, Astron. Ap. **117**, L1-4.

Lewin, W.H.G. and Van den Heuvel, E.P.J. (eds.), 1983, Accretion-Driven Stellar X-ray Sources, Cambridge University Press.

Lipunov, V.M. and Postnov, K.A. 1985, Astron. Astrophys. **144**, L13.

Lyne, A.G. 1981, in IAU Symp. nr. 95 "Pulsars", eds. W. Sieber and R. Wielebinski, D. Reidel, Dordrecht, p. 423.

Lyne, A.G., Manchester, R.N., Taylor, J.H. 1985, Mon. Not. Roy. Astron. Soc. **213**, 613.
Margon, B. 1984, Ann. Rev. Astron. Astrophys. **22**, 507-536.
Mazurek, T.J., Meier, D.L. and Cameron, A.G.W. 1974. Astrophys. Sp. Sc. **27**, 261.
Mazurek, T.J., Meier, D.L. and Wheeler, J.C. 1977, Astrophys. J. **213**, 518.
Meijer, F. and Meijer-Hofmeister, E. 1979, Astron. Astrophys. **78**, 167.
Middleditch, J. and Priedhorsky, W. 1985, IAU Circular Nr. 4060.
Miyayi, S., Nomoto, K., Yokoi, K. and Sugimoto, D. 1980, Publ. Astron. Soc. Japan **32**, 303-329.
Mueller, E. and Arnett, W.D. 1982, Astrophys. J. **261**, L109-L115.
Mueller, E. and Arnett, W.D., 1984, preprint Max-Planck-Insitut **108**.
Nomoto, K. 1982, Astrophys. J. **253**, 798.
Nomoto, K. 1984a, Astrophys. J. **277**, 791.
Nomoto, K. 1984b, in: "Collapse and Numerical Relativity" IAU Toulouse; Reidel, Dordrecht, in press.
Nomoto, K. 1984c, in: Stellar Nucleosynthesis (C. Chiosi and A. Renzini, eds.) Reidel, Dordrecht, pp. 239-250.
Ostriker, J.P. 1973, (private conversation).
Paczynski, B. 1970, in: Mass Loss and Evolution of Close Binaries (K. Gyldenkerne and R. West, eds.), Kopenhagen Univ. Publ. Funds, p. 142.
Paczynski, B. 1971, Acta Astron. **21**, 1-14.
Paczynski, B. 1976, in: Structure and Evolution of Close Binaries (P.P. Eggleton, S. Mitton and J. Whelan, eds.). Reidel, Dordrecht, p. 423.
Paczynski, B. 1983, Nature **304**, 421.
Paczynski, B. and Sienkiewicz, R. 1971, Acta Astr. **21**, 73.
Rappaport, S.A. and Joss, P.C. 1983, in: Accretion Driven Stellar X-ray Sources (W. Lewin and E.P.J. van den Heuvel, eds.), Cambridge University Press, pp. 1-39.
Rappaport, S.A. and Van den Heuvel, E.P.J. 1982 in: "Be Stars" (M. Jaschek and H.G. Groth, eds.), Reidel, Dordrecht, p. 327.
Rappaport, S.A., Verbunt, F. and Joss, P.C. 1983, Astrophys. J. **275**, 713.
Reeves, H. 1968, Stellar Evolution and Nucleosynthesis, Gordon and Breach, New York.
Romanishin, W. and Angel, J.R.P. 1980, **235**, 992.
Savonije, G.J. 1983a, in: "Accretion-Driven Stellar X-ray Sources", (W.H.G. Lewin and E.P.J. van den Heuvel, eds.), Cambridge University Press, p. 343.
Savonije, G.J. 1983b, Nature **304**, 422-423.
Savonije, G.J., De Kool, M. and Van den Heuvel, E.P.J., 1985, Astron. Astrophys.
Smarr, L.L. and Blandford, R.D. 1976, Astrophys. J. **207**, 574.
Stella, L., Parmar, A.N., White, N.E., Lewin, W.H.G. and Van Paradijs, J. 1985, IAU Circ. Nr. 4110.
Stokes, G.H., Taylor, J.H. and Dewey, R.J. 1985, Astrophys. J. **294**, L21.
Sugimoto, D. and Nomoto, K. 1980, Space Science Reviews **25**, 155.
Taam, R.E. 1980a, Astrophys. J. **237**, 142.
Taam, R.E. 1980b, Astrophys. J. **242**, 749.

Taam, R.E. 1983, Astroph. J. **270**, 694.
Taam, R.E., Bodenheimer, P. and Ostriker, J.P. 1978, Astrophys. J. **222**, 269.
Taam, R.E. and Van den Heuvel, E.P.J. 1986, Astroph. J. (in press, June 1).
Trimble, V. 1982, Reviews of Mod. Phys., **54**, 1183-1224.
Tutukov, A.V. and Yungelson, L.R. 1973, Nautsnie Inform. **27**, 58.
Tutukov, A.V. and Yungelson, L.R. 1979, in: Mass Loss and Evolution of O-type Stars (P.S. Conti and C. De Loore, eds.). Reidel, Dordrecht, 401-407.
Van Beveren, D. 1980, Ph.D. Thesis, Free University of Brussels.
Van den Heuvel, E.P.J. 1974, Proc. 16th Solvay Conf. on Physics, Univ. of Brussels Press, p. 119.
Van den Heuvel, E.P.J., in: Structure and Evolution of Close Binary Systems, Reidel, Dordrecht, pp. 35-61(1976).
Van den Heuvel, E.P.J. 1977, Annals New York Acad. Sci. **302**, 14.
Van den Heuvel, E.P.J. 1978, in: Physics and Astrophysics of Neutron Stars and Black Holes (eds. R. Giacconi and R. Ruffini), North Holl. Publ. Co. Amsterdam, p. 828-871.
Van den Heuvel, E.P.J. 1981, in: "Pulsars", W. Sieber and R. Wielebinski (eds.), Reidel, Dordrecht, p. 379.
Van den Heuvel. E.P.J. 1983, in "Accretion-Driven Stellar X-ray Sources" (W.H.G. Lewin and E.P.J. van den Heuvel, eds.), Cambridge University Press, p. 303.
Van den Heuvel, E.P.J. 1984, J. Astroph. Astron. **5**, 209.
Van den Heuvel, E.P.J. and Habets, G.J. 1985, in: Supernovae, their Progenitors and Remnants (G. Srinivasan and V. Radhakrishnan, eds.), Indian Acad. Sci., Bangalore, 129-154.
Van den Heuvel, E.P.J. and Heise, J. 1972, Nature Phys. Sci. **239**, 67.
Van den Heuvel, E.P.J. and Bonsema, P.F.A. 1984, Astron. Astrophys. **139**, L16.
Van den Heuvel, E.P.J. and Taam, R.E. 1984, Nature **309**, 235-237.
Van den Heuvel, E.P.J., 1986, in: Evolution of Galactic X-ray Binaries (J. Truemper, W.H.G. Lewin, and W. Brinkmann, eds.), Reidel, Dordrecht, p. 107.
Van der Klis, M., Jansen, F., Van Paradijs, J., Lewin, W.H.G., Van den Heuvel, E.P.J., Truemper, J.E. and Sztanjo, M. 1985a, Nature **316**, 225-230.
Van der Klis, M., Jansen,. F., White, N., Stella, L. and Peacock, A. 1985b, IAU Circular, Nr. 4068.
Verbunt, F. and Zwaan, C. 1981, Astron. Ap. **100**, L7-9.
Weaver, T.A. and Woosley, S.E. 1980, Annals N.Y. Acad.-Sci. **336**, 335. (9th Texas Symp.).
Webbink, R.F., 1980, IAU Coll. **53**, Univ. of Rochester, p. 426.
Webbink, R.F. 1984, Astroph. J. **277**, 355.
Webbink, R.F., Rappaport, S. and Savonije, G.J. 1983, Astroph. J. **270**, 678.
Whelan, J. and Iben, I.Jr. 1973, Astroph. J. **186**, 1007-1014.

OBSERVATIONS OF SUPERNOVAE

Nino Panagia*
Space Telescope Science Institute
3700 San Martin Drive
Baltimore, MD 21218
U.S.A.

ABSTRACT. The observations of supernovae are reviewed and their main implications are shortly discussed.

1. INTRODUCTION

The beginning of the era of modern observations of supernovae (SN) can be set on 1885 when the first spectroscopic measurements of S And, or SN 1885a, in the Andromeda galaxy were made. Since then several hundreds of SNe have been discovered in external galaxies and a number of them have been studied in detail both photometrically and spectroscopically. Until a few years ago observations were limited almost exclusively to the optical domain because the existing observing facilities at other wavelengths (either ground-based such as IR observatories and radio telescopes, or airborne, i.e., satellites, rocket and balloon mounted telescopes) were not sensitive enough for a good detection of the weak fluxes from distant SNe. On the basis of optical information, SNe were subdivided into the two main categories of Type I and Type II supernovae whose properties as on 1978 (e.g. Tammann, 1978) are summarized in Table 1.

The main distinction, and the only reliable one, between the two classes of SNe was based on the properties of their optical spectra: while Type II SNe display lines of hydrogen (Balmer lines: Hα, Hβ, etc.) and possibly neutral helium (e.g. HeI 5875 Å) the optical spectrum of Type I SNe does not show the presence of any H nor He. We see that in all properties Type I SNe appeared to form a very homogeneous class of objects which were found in all types of galaxies. These characteristics have made Type I SNe particularly appealing as standard candles for distance determinations. On the other hand, Type II SNe, which have only been found in spiral galaxies, have properties which vary widely from object to object.

The launch of the International Ultraviolet Explorer (IUE) satellite in early 1978 marked the beginning of a new era for SN studies because of its capability of measuring the ultraviolet emission of objects as faint as $m_B = 15$. Moreover, just around that time other powerful astronomical instruments have become available, such as the Einstein Observatory for X-ray measurements, the VLA for observa-

* Affiliated with the Astrophysics Division, Space Science Department of ESA; on leave from the Istituto di Radioastronomia CNR, Bologna, Italy.

F. Pacini (ed.), High Energy Phenomena Around Collapsed Stars, 33–49.

TABLE 1
"OLD" FACTS ABOUT SUPERNOVAE
(BASED ON OPTICAL OBSERVATIONS)

	Type I	*Type II*
Where	Elliptical, Spiral and Irregular Galaxies	Only in Spiral Galaxies, associated to the Spiral Arms
How Bright	$M_B(max) \sim -20$ ($L \sim 10^{10}\ L_\odot$)	$M_B(max) < -16$ ($L > 3 \times 10^8\ L_\odot$)
Light Curve	Decay Rate $\begin{cases} \sim 0.1 \text{ mag/day}; t < 30^d \\ \sim 0.01 \text{ mag/day}; t > 30^d \end{cases}$	All kinds of behaviour
Optical Spectrum	Broad, unidentified features, more prominent with time; Absence of H and He lines; $v \cong (10-20) \times 10^3$ km s^{-1}	Balmer Lines HeI, NaI(?), etc. $v \cong (5-10) \times 10^3$ km s^{-1}
Behaviour	Very Homogeneous	Heterogeneous
Progenitors	Low Mass Stars	High Mass Stars

TABLE 2
IUE OBSERVATIONS OF SUPERNOVAE

Supernova	*Galaxy*	*Type*	*Period of Observation*	*Number of Exposures*
1978g	IC 5201	II	30 Nov–11 Dec 1978	2
1979c	NGC 4321/M100	II-L	21 Apr–4 Aug 1979	29
1980k	NGC 6946	II-L	30 Oct 80–5 Jan 81	33
1980n	NGC 1316	Ia	11 Dec 80–16 Jan 81	8
1981b	NGC 4536	Ia	9–11 March 1981	7
1982b	NGC 2268	Ia	18 February 1982	3
1983g	NGC 4753	Ia	8–25 April 1983	6
1983n	NGC 5236/M83	Ib	4 Jul–18 Aug 1983	26
1984	NGC 1559	II	13 Aug 1984	1
1985f	NGC 4618	Ib	17 May 1985	1
1985l	NGC 5003	II	28 June–17 July 1985	3

tions at radio wavelengths and a number of new telescopes either dedicated to infrared observations (e.g. UKIRT and IRTF at Mauna Kea) or equipped with new and highly efficient IR instrumentation (e.g. AAT and ESO observatories). As a result, a wealth of new information has become available which, thanks to the coordinated effort of astronomers operating at widely different wavelengths, has provided us with fresh insights as for the properties and the nature of supernovae of both types.

In particular, it has become evident that the class of Type I SNe must be split into two subclasses, "classical" Type I SNe, whose properties are very homogeneous and correspond to those listed in Table 1, and "subluminous" Type SNe which probably form an equally homogeneous group but with distinctly different properties, such as being considerably fainter than classical SNI and being only found associated with young population in spiral galaxies. As for Type II SNe the new observations have made it possible to derive chemical abundances of the pre-supernova atmosphere as well as to directly determine several other properties of the stellar progenitor. These new results will be summarized and discussed in the following sections.

Since most of the new information about SN has been either provided or stimulated by ultraviolet observations it is fair to start by presenting a summary of the IUE observations (Table 2). Until now, eleven SNe have been observed with IUE, of which 5 were Type II and 6 Type I. However, only four SNe, namely 1979c, 1980k, 1981b and 1983n, were bright enough to obtain high quality ultraviolet spectra and/or to follow their time evolution. Therefore, in the following our discussion on individual objects will mostly be concerned with these four SNe while referring to the others just briefly.

2. TYPE I SUPERNOVAE

2.1 Two Classes of Type I Supernovae

As mentioned above observations of Type I SNe have lead to the identification of two distinct subclasses, hereafter called Type Ia and Type Ib, respectively. The class of Type Ia SNe comprises all "classical" SNI including all of those discovered within elliptical and S0 galaxies and a substantial fraction of those discovered in spiral galaxies. On the other hand, "bona fide" Type Ib SNe (see sections 3.3 and 3.4) have only been found in spiral galaxies and apparently associated with a spiral arm. Another basic difference is that in the spectrum of Type Ib SNe the 6150 Å feature is missing while it is the strongest and deepest absorption observed in the spectrum of Type Ia SNe. Moreover, Type Ib appear to be underluminous by about 2 magnitudes relative to SNIa. These spectral peculiarities were already recognized more that 20 years ago by Bertola and collaborators (Bertola, 1964; Bertola *et al.* 1965) for the cases of SN 1962l and 1964l but they had gone unnoticed at that time. Thanks to the new studies in recent years, it is now clear that already on the basis of optical data one must distinguish two classes of Type I SNe. In fact, there are a number of properties at other wavelength ranges (UV, IR, radio) which are dramatically different in the two subtypes: they will be discussed in the following sections.

Here, we first present the observational results for "classical" Type I, or Type Ia, SNe whereas the next two sections will be devoted to the "subluminous" Type I, or Type Ib, supernovae.

2.2 "Classical" Type I, or Type Ia, Supernovae

So far four Type Ia SNe have been observed with IUE (see Table 2). However, none of them was followed for a long time mostly because of pointing constraints of the satellite. Therefore, for all of them we have observations concentrated around the epoch of their maximum light but we know little about their time evolution.

The UV spectrum is clearly not a smooth continuum but rather displays a number of "bands" which are observed for all of the SNe (cf. Fig. 1). The most prominent feature is the emission which peaks at $\sim$ 2950 Å with a half-power width of $\sim$ 100 Å, i.e. $\Delta v \simeq 10^4$ km s^{-1}. Alternatively, this band may be the result of strong absorptions occuring on both sides of the apparent emission, i.e. centered at $\sim$ 2840 Å and $\sim$ 3060 Å and having half-power widths of the order of 100 Å. A similarly prominent emission band is seen at $\lambda \sim$ 1890 Å in the only one spectrum obtained at short wavelengths for SN 1981b. Several other absorptions features can be recognized, which are present at all epochs of observation. Although some of them might be identified with multiplets of Fe I, Fe II and Mg II, no satisfactory identification has been found yet for the majority of the absorptions. Nevertheless, the very fact that the spectrum is virtually the same for the three SNe and at all epochs when observations were made is already an important result. This confirms the homogeneity of properties of Type I SNe which once more show to be all exact replicas of one and the same phenomenon.

After the pionieristic observations of SN 1972e (Kirshner *et al.* 1973) a large sample of Type I SNe has been studied in recent years at IR wavelengths (Elias *et al.* 1981; Elias *et al.* 1985). Two of them, 1980n and 1981b, are in the IUE sample. Again a very close similarity is found in the spectral behaviour and the light variation among the observed SNe. This extends their characteristics of "standard candles" also to the infrared. Another interesting fact is that the light curves in the J, H and K bands display a first minimum 15–20 days after optical maximum, reach a secondary maximum $\sim$ 30 days after maximum and then decay steadily at a rate of $\sim$ 0.04 magnitudes/day. This decay rate is faster than what is observed at optical wavelengths on a long time scale.

Combining different observations, one can reconstruct the complete spectrum of a Type Ia SN from UV to the IR. As an illustration, the spectrum of SN 1981b at an epoch around the optical maximum is shown in Fig. 2. We see that the ultraviolet spectrum declines very steeply with the wavelength. Moreover, the ultraviolet emission is much lower than a black body extrapolation of the optical spectrum. This is just the opposite of what is found for Type II SNe (cf. Section 3). In particular, the UV flux is approximately 10 times lower than the black body curve at the colour temperature appropriate to match the visual spectrum (T = 15800 K, upper curve). Also, in the infrared the observed spectrum is much weaker than a black-body extrapolation from the optical. On the other hand, the IR and UV spectra can smoothly be connected to each other by a lower temperature black-body ($T \sim 9400$ K, lower curve). These results indicate that the opacity in both UV and IR must be much higher than at optical wavelengths, so that the radiation temperature is close to the effective temperature in the visual but reduces to the minimum temperature $T_{min} \sim 0.6\ T_{eff}$ both at shorter and at longer wavelengths.

Within this sample of Type Ia SNe only SN 1981b has been observed extensively in the radio. It has been monitored at 5 GHz with the VLA at regular time intervals within the period March 1981–February 1983 but it has never been detected (Weiler *et al.* 1986), the upper limit to the radio flux being as low as 50

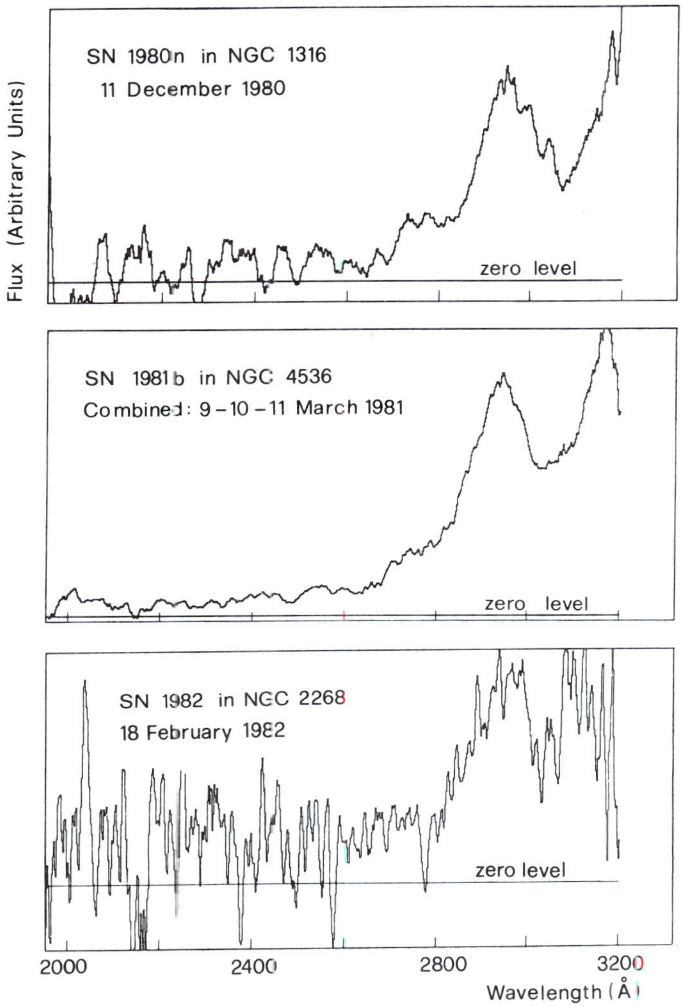

Figure 1. The IUE long wavelength spectra of Type Ia Supernovae. Note the prominent emission feature centered at $\sim$ 2950 Å.

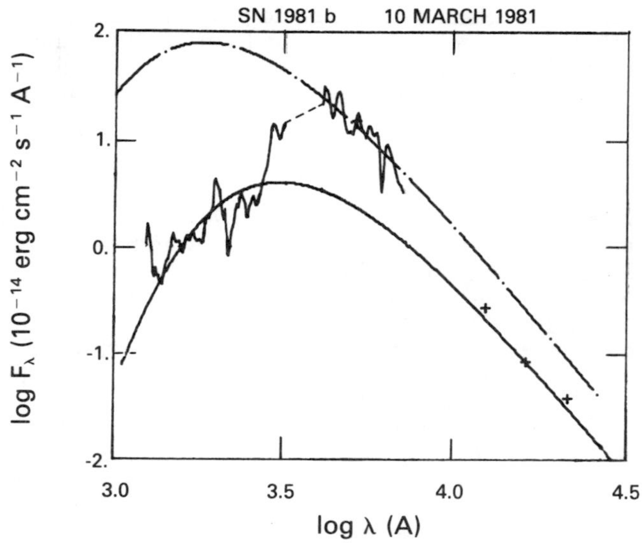

Figure 2. The spectrum on the Type Ia SN 1981b on March 10, 1981, dereddened with E(B–V) = 0.24. The upper curve is a black body at 15800 K which fits the optical spectrum. The lower curve is a black body at 9400 K which connects the IR and UV spectra.

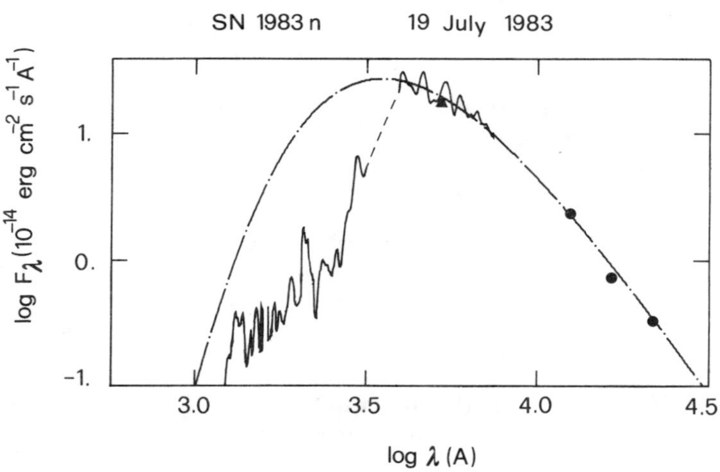

Figure 3. The spectrum of the Type Ib SN 1983n on July 19, 1983, dereddened with E(B–V) = 0.16. Both the UV and the optical spectra have been smoothed with a 100 Å bandwith. The triangle represents the FES photometric point, the dots represent the J, H and K photometric data. The dashed-dotted curve is a black body curve at T = 8300 K.

μJy (1 σ). From a comparison with the radio properties of Type II SNe, Panagia (1984a) argues that the lack of radio emission from SN 1981b implies that its progenitor was a star which in the late phase of its evolution underwent a low mass loss ($\dot{M} < 10^{-7}\ M_{\odot}\ yr^{-1}$) and, therefore, was originally a relatively low mass star ($< 3\ M_{\odot}$).

2.3 SN 1983n: The Prototype of "Subluminous" Type I, or Type Ib, Supernovae

The existence of a separate subclass of Type I SNe has become evident about two years ago when SN 1983n was discovered in the southern spiral galaxy NGC 5236 = M 83 and its properties and evolution were studied in great detail. Discovered on July 3rd, 1983, the IUE observations were started on July 4th: the long wavelength spectrum was remarkably similar to that displayed by SN 1981b on 9–11 March, 1981, i.e. at the epoch of its maximum in the B band (Barbon, Ciatti and Rosino, 1982). In particular, the spectrum showed the prominent "emission" feature centered at $\sim$ 2950 Å as well as a sharp decline of the flux with increasing frequency, which were found to characterize the spectra of all of the Type I SNe observed before (Panagia 1982, 1984b). For this reason and also on the basis of the general appearance of the optical spectrum it was immediately classified as a Type I.

However, it was soon apparent that this was not a "normal" SNI. In fact, when the UV emission reached a maximum between the 8th and the 12th of July the UV spectrum was markedly different fram what it was at the beginning and, therefore, different from the spectrum of classical Type I SNe at maximum. Similarly, in the optical spectrum not only was the 6150 Å feature missing at all epochs but also the other spectral features were quite dissimilar from those of Type Ia SNe. On the other hand, a reasonable resemblance with SNIa spectra can be found if one compares the observations of SN 1983n with those of SN 1981b (which is a prototype of SNIa) at much later epochs (Panagia *et al.* 1986a). In other words, SN 1983n looked like it was "born old" in both UV and optical spectral characteristics. The second basic difference became apparent from a comparison of optical and infrared data: while the other Type I SNe had an IR emission significantly lower than a black body extrapolation of the optical spectrum (Panagia 1984a), SN 1983n displayed an emission which at all epochs can almost perfectly be fitted to a black body curve for all wavelengths longer than, say, 4000 Å (see Fig. 3). In fact, judging from the observed ratios F(2950 Å)/F(FES) and F(J)/F(FES), this supernova is found to be both the reddest and the bluest Type I SN among those ever observed with IUE. This result, together with other evidences discussed in Panagia *et al.* (1986) indicates that the extinction toward SN 1983n is rather modest, i.e. $A_V \lesssim 1^m$. This leads to a third important result: the FES light curve reached a maximum on $\sim$ 17th of July at m(FES) $\simeq$ 11.5 which, corrected for reddening and adopting the largest distance among those proposed for M83 ($\sim$ 8 Mpc; Sandage and Tammann, 1974), corresponds to M(FES) = −18.5. This makes it about 1.5 magnitudes fainter than classical Type I SNe.

The detailed light curves from UV to IR are displayed in Fig. 4 in the form of λF_λ as a function of time. The IR data were collected by Allen *et al.* at several collaborating observatories. Note, in particular, that the IR light curves do not exhibit the secondary maximum which is found in all Type Ia SNe. As already apparent in Figure 3, most of the energy at all epochs is emitted at optical wavelengths. In particular, only 13% of the total luminosity was emitted shortward of 3400 Å at the epoch when the UV emission attained a maximum. Moreover,

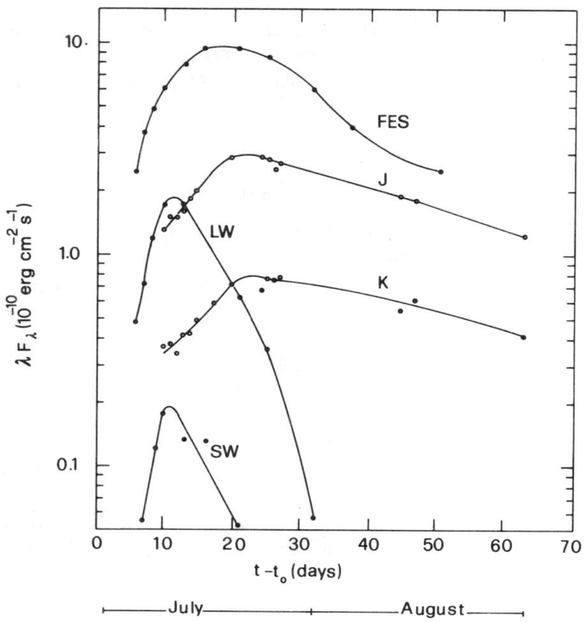

Figure 4. The light curves of SN1983n from the UV to the IR, in the form of log λF_λ as a function of the time elapsed since the explosion. The curves denoted as SW and LW represent the behaviour of the average flux measured with short and the long wavelength IUE cameras, respectively.

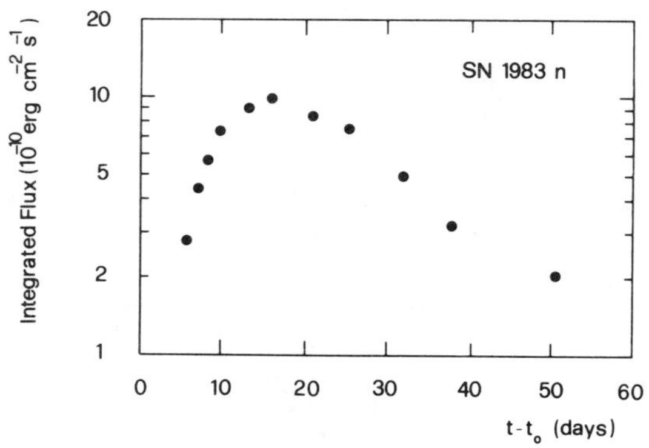

Figure 5. The bolometric light curve of SN1983n obtained by direct integration of the spectra dereddened with $E(B-V) = 0.16$.

there is no indication of any stronger emission in the UV at very early epochs: this implies that the initial radius of the SN, i.e. the radius the stellar progenitor had when the shock front reached the photosphere, was definitely much less than 10^{13} cm and probably lower than 10^{12} cm, and rules out a red supergiant as a possible progenitor of this supernova.

The bolometric light curve presented in Figure 5 is computed by direct integration of the spectrum over all wavelengths: this is the first time that this can be done reliably for a Type I SN. A comparison of the bolometric light curve with the results of model calculations involving the radioactive decay of ^{56}Ni into ^{56}Co as the main source for the luminosity (e.g. Arnett, 1982) leads to estimates of the ^{56}Ni mass in the range 0.1–0.5 $M_\odot$ for an adopted distance of 8 Mpc (and much less if the distance is smaller). This implies that models which involve the complete incineration of a white dwarf to transform it entirely into ^{56}Ni can be ruled out for this supernova.

SN 1983n was also observed at radio wavelengths with the VLA and detected as early as July 6th (Sramek, Panagia and Weiler 1984). The maximum flux density was detected on July 27th (S = 28 mJy) followed by a steady decay at a rate of the type $S \propto t^{-1.6}$. Observations made at 20 cm suggest a power law spectrum with a strongly non-thermal spectral index $\alpha \sim 1$. In fact, its radio emission is at a comparable level as it is observed in Type II SNe (i.e. 1979c and 1980k) with the difference that in SN 1983n the time decay is faster, $t^{-1.6}$ instead of $t^{-0.7}$, and the spectrum is steeper (spectral index 1 instead of ~ 0.7; Weiler *et al.* 1986). In order to both produce relativistic electrons at the shock front and to account for the initial strong absorption, the presence of a circumstellar envelope is needed, whose extent implies a steady mass loss rate of $\sim 3 \times 10^{-6}$ $M_\odot$ yr^{-1} for an assumed wind velocity of 10 km s^{-1} (Chevalier, 1984, Sramek *et al.* 1984). On the other hand, the SN progenitor could not be responsible for this wind because the bolometric light curve excludes that it was a red supergiant. The most plausible picture is that of a SN produced in a binary system: both stars had to be originally rather massive (> 5 $M_\odot$) with the primary, which eventually goes supernova, about 1–2 $M_\odot$ more massive than the companion. The primary evolves first and ends up as a massive white dwarf, a little below the Chandrasekhar limit ($M_{wd} \sim 1.35$ $M_\odot$). It slowly accretes mass from the companion which, meanwhile, has reached the red supergiant phase. Thus, when the WD exceeds the critical mass and explodes as a SN, the explosion occurs within the envelope of the companion: the observed radio emission originates from the interaction of the ejecta with that envelope. Within this frame, it is clear that a strong radio emission from Type I SNe is possible only when the involved binary system is relatively massive.

2.4 The Subclass of Type Ib Supernovae

As mentioned before, SNe with characteristics similar to those of SN 1983n were already noticed more than 20 years ago by Asiago astronomers. However, only the wealth of information on SN 1983n has made the recognition of this new subclass a reality. Since its discovery, much attention has been devoted to an appropriate identification and classification of Type I SNe with the lack of the 6150 Å feature from the optical spectrum being the main criterion to identify a Type Ib SN. The renewed interest has led to the recognition (for old events) and to the discovery (for new events) of quite a number of Type Ib SNe. Table 3 lists the SNe which at the time of writing (November 1985) have been recognized to belong to this class. Such a list is bound to be incomplete, however, because

TABLE 3
RECOGNIZED SUBLUMINOUS TYPE I SUPERNOVAE (SNIb)

Supernova	Galaxy	$M_{max}^{(1)}$
1962l	NGC 1073	−18.1
1964l	NGC 3938	−17.7
1982r	NGC 1187	< −17.5
1983i	NGC 4051	< −17.4
1983n	NGC 5236	−18.0
1984l	NGC 991	−18.3
1985f	NGC 4618	< −14

(1) No allowance is made for extinction.

careful analyses of old material may, and probably will, lead to the identification of more members. The absolute magnitudes given in Table 3 have been derived from the apparent magnitudes at maximum as reported in the Barbon *et al.* (1984) Revised Supernova Catalogue and using the distances either taken from Sandage ant Tammann (1981) or deduced assuming $H_o = 50$ km s^{-1} Mpc^{-1}. No allowance has been made for extinction; however, since these SNe are likely to be intrinsically red, the interstellar reddening correction should be rather modest in most cases.

Of special interest is the case of SN 1984l in NGC 991, which has been observed quite extensively at optical, infrared and radio wavelengths (e.g. Wheeler and Levreault, 1985; Elias *et al.* 1985; Panagia *et al.* 1986b). A detailed comparison with SN 1983n reveals that SN 1984l is virtually identical to SN 1983n in all aspects. In particular, its absolute luminosity at all observed wavelengths has been found to be the same as for SN 1983n to within 30% (Panagia *et al.* 1986b). These results suggest that Type Ib SNe form a class at least as homogeneous as that as Type Ia SNe. In fact, the IR light curves of three well studied Type Ib SNe (1983n, 1983i and 1984l; Elias *et al.* 1985) are very similar to each other and quite different from that of Type Ia SNe in both having a slower time decay and not displaying a secondary peak. These characteristics make it easy to recognize Type Ib SNe on the basis of photometric measurements alone with no need for detailed spectral information. Note that this is not possible with optical photometry because in this frequency range the time variations of the two subtypes are rather similar to each other.

Another interesting case is SN 1985f which was discovered early this year in the bar of the spiral galaxy NGC 4618 (Filippenko and Sargent, 1985). The tentative classification of this SN as Type Ib is based on the close resemblance of its spectrum with those of SN 1984l obtained at epochs about 13 months after maximum (Kirshner, 1985): they both display broad emissions of neutral elements (OI, NaI, possibly MgI) which were never observed in the spectra of Type Ia Sne either at similar epochs or at all as it is case for the OI line at 6300 A.

The properties of subluminous Type I SNe (SNIb) have summarized by Panagia *et al.* (1986b) as follows:

General

1. They are found only in spiral galaxies.
2. They appear to be associated with HII regions or spiral arms.

Optical

1. They are subluminous by 1.5 – 2 magnitudes.
2. They are missing the characteristic absorption feature at 6150 Å.
3. They show an unidentified "doublet" feature at 6300 and 6500 Å.
4. They appear to be standard optical candles.

Infrared

1. Their H-band light curves reaches a maximum soon after optical maximum and do not display the expected secondary maximum at $(t - t_o) \sim 30$ days.
2. They exhibit a much slower decay of their H-band light (dH/dt (Type Ib) ~ 0.025 mag/day whereas dH/dt (Type Ia) ~ 0.047 mag/day).
3. They have "bluer" J–K, and to a lesser extent H–K, colors and their colors evolve differently than for SNIa.
4. They may be standard infrared candles.

Radio

1. They are significant and distinct (from Type II radio SNe—see Weiler *et al.* 1986) emitters of radio radiation.
2. They may be standard radio candles.

In order to gain further insight as for the nature of Type Ib SNe it is worth considering the frequency of Type I SNe as a function of the absolute magnitude at maximum (Fig. 6). The values of $M(\max)$ have been estimated by using the data collected in the Revised Supernova Catalogue of Barbon et al (1984) which contain all SNe discovered until the end of 1983. Only those SNe whose magnitude at maximum was reliably determined were selected: this amounts to 14 events in elliptical and S0 galaxies and 27 in spirals. The distances were either taken from Sandage and Tammann (1981) or estimated from the radial velocity assuming $H_o = 50$ km s^{-1} Mpc^{-1}. No reddenig correction was applied in any case. We see that, while in elliptical and S0 galaxies the distribution is sharply peaked around $M_B \sim -20^m$, for spiral galaxies the distribution has a main peak around $M_B \sim -20^m$ but extends considerably toward low luminosities. This extension is centered around $M_B \sim -18^m$ and contains about 1/3 of the total in a sample of 27 well studied Type I SNe in spiral galaxies.

Explicitely marked are the known Type Ib SNe for which a good measurement of m(MAX) was available. To get a better feeling about the involved statistics, Type Ib SNe either that were discovered after 1983 or for which only lower limits to the luminosity at maximum are available were also included in the figure (dashed boxes). We see that in spiral galaxies the Type Ib SNe display magnitudes ~ 1.5m fainter than Type Ia SNe. These, in turn, appear to follow exactly the same distribution as Type Ia in early type galaxies. From this it is tempting to conclude that most, if not all, Type I SNe that have exploded in spiral galaxies and had maximum magnitudes fainter than −19 were Type Ib events. Moreover, it is apparent that Type Ib SNe occur only in spirals and not in early type galaxies. This result agrees perfectly well with the idea that SNIb represent the end products of the evolution of rather massive binary systems in which neither of the two components was massive enough to explode as a Type II SN. Despite all possible uncertainties, from Fig. 6 we can conclude that at least 10%, and possibly as much as 30%, of the Type I SNe discovered in spiral galaxies are Type Ib events. Since they are fainter than SNIa and since the SN searches are essentially magnitude limited, the true percentage of Type Ib events in spiral galaxies may be considerably higher. In particular, if we adopt $\Delta M = 1.5$ as the magnitude difference

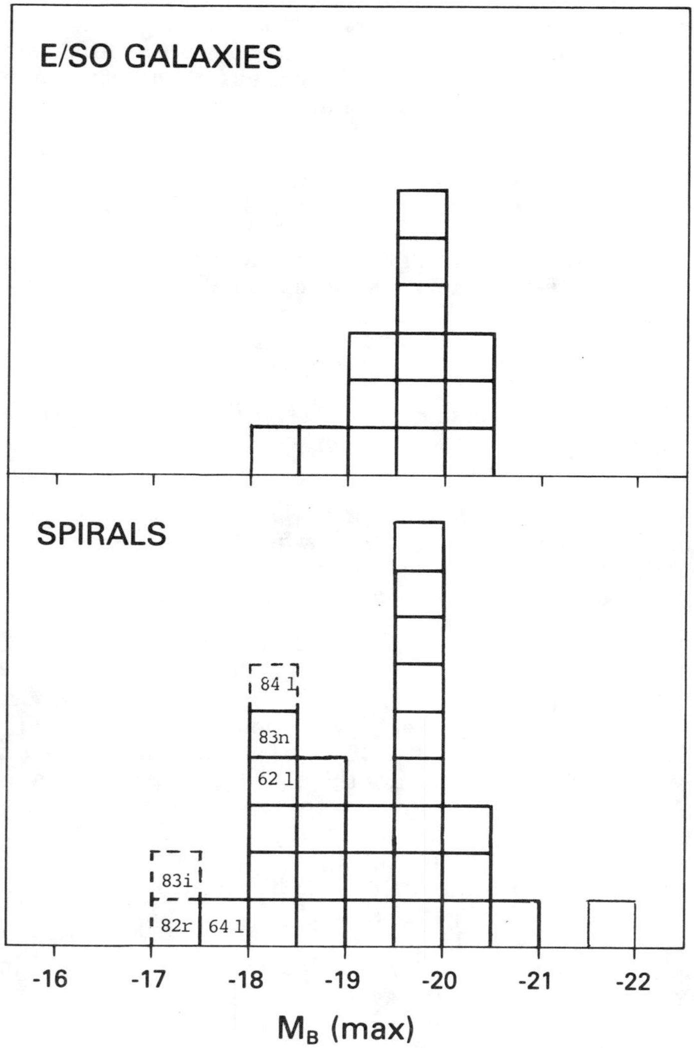

Figure 6. The distribution of Type I SNe discovered and classified by 1983 as a function of the absolute B magnitude for E/S0 galaxies (upper panel) and for spiral galaxies (lower panel). Type Ib supernovae (see Table 3) are explicitely marked. Dashed boxes denote SNe that either have been discovered after 1983 or have only measurements after maximum.

between the two subtypes, the observed fraction 10%< f(SNIb) < 30% transforms into true percentages of SNIb events between 30% and 60% of the total Type I SN explosions. Moreover, since in spiral galaxies one observes approximately as many Type II events as Type I (Tammann, 1978), it is clear that Type Ib SNe not only are a substantial fraction of the Type I explosions but they must account also for quite a fraction of all SN events in spiral galaxies, say, between 20% and 40%.

3. TYPE II SUPERNOVAE

Among the five Type II SNe which have been observed with IUE, for both the first one and the last two the information is meagre because they were rather weak. For this reason, the present discussion is limited to the other two SNe (1979c and 1980k) for which the data are quite comprehensive (Benvenuti *et al.* 1982).

Figure 7 displays the UV spectrum and the fluxes in the U, B, V bands of SN 1980k in NGC 6946 obtained on 9 November 1980, i.e. about 10 days after the optical maximum. The UV flux is higher than the extrapolation of the optical spectrum with a black body curve: a clear excess can be seen for $\lambda < 2000$ Å. Such an excess is found at all epochs although it is less pronounced at early times. The characteristics of a high continuous UV emission with an excess at shortest wavelengths was also found for 1979c in NGC 4321 (Panagia *et al.* 1980). Benvenuti, Dopita and D'Odorico (1984) show that the excess may be explained in terms of two-photon emission arising from the upper layers of the ejecta. Alternatively, Fransson (1984) suggests, that the UV excess may be photospheric radiation which has been Compton-scattered by energetic, thermal, electrons ($T \sim 10^9$ K) at the shock front where the ejecta interact with pre-existing circumstellar material. This model is able to explain both the extra radiation observed at short wavelengths and the high ionization implied by some emission lines (e.g. NIV] 1486 Å, CIV 1550 Å, etc.) observed in the spectrum of SN 1979c (Panagia *et al.* 1980; see Fig. 8): therefore, this explanation seems more plausible.

From a comparison of the line profiles observed in the UV and in the visual with theoretical calculations (Fransson *et al.* 1984) it is concluded that the UV emission lines of highly ionized species are produced in the upper atmosphere. as well as Hα or Mg II λ 2800 Å, although just in the outermost layers where density is lower and the ionizing radiation flux is higher. The line profiles imply an expansion velocity of 8400 km s^{-1} (Fransson *et al.* 1984), which is only marginally lower than that measured in the optical (i.e. 9200 km s^{-1}; Panagia *et al.* 1980). From an analysis of the NV 1240 Å, NIV] 1486 Å, CIV 1550 Å, NIII] 1750 Å and C III] 1909 Å line intensities the abundance ratio of nitrogen to carbon has been estimated to be $N/C \sim 8$ (Panagia, 1980; Fransson *et al.* 1984) i.e. $\sim$ 30 times higher than the cosmic value. This strong enhancement of nitrogen relative to carbon suggests that the pre-supernova star was a massive supergiant which had undergone a long period of mass loss, thereby exposing CNO processed material.

As for observations at other frequencies, both SNe have been detected in the radio and are currently being monitored with the VLA at wavelengths from 2 to 20 cm (Weiler *et al.* 1981, 1982, 1983, 1986; see also Weiler, this volume). The time behaviour is qualitatively the same in the two SNe: the "light" curve is characterized by a steep rise, with time scale of a few months and which occurs at earlier times for higher frequencies, followed by a slow decay. The high brightness temperature and the rather steep spectrum at late times ($S_\nu \propto \nu^{-0.7}$) indicate a non-thermal origin for the radio emission. Also, the early and steep rise of the

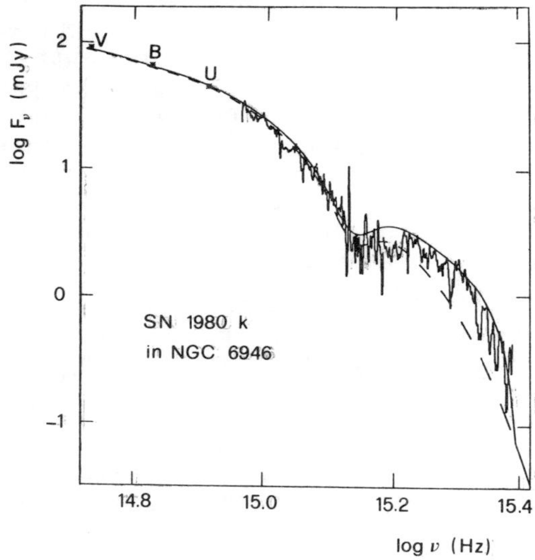

Figure 7. The spectrum of SN 1980k on November 9, 1980. The dashed curve represents the best fit to the optical and near ultraviolet data in terms of a black body at 9900 K and a reddening of E(B–V) = 0.32. The solid line is the spectrum obtained by including also radiation produced by two photon emission (Benvenuti *et al.* 1984).

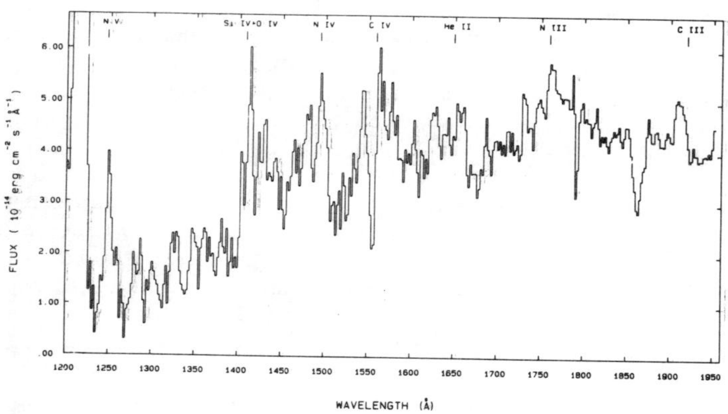

Figure 8. The short wavelength IUE spectrum of SN 1979c on April 24, 1979. The positions of the identified emission lines are shown redshifted by z(NGC 4321) = 0.0054.

radio emission implies that it originates at the outer layers of the SN envelope because, otherwise, the high f-f- opacity of the ejecta would have prevented any signal from becoming detectable for several years. These data are best explained in terms of a model in which the intrinsic emission is due to the synchrotron process and decreases steadily with time ($\propto t^{-0.7}$). The acceleration of the relativistic electrons may either be produced by turbulence at the shock front of the SN, where the ejecta interact with pre-existing wind material shed by the red Supergiant progenitor (Chevalier, 1981), or be provided by a newly-born plerion (Pacini and Salvati, 1981; Shklovskii, 1981; Salvati, 1983). From a detailed analysis of the radio light curves the former explanation seems to be more likely appropriate (Weiler *et al.* 1986). The presence of wind material which is responsible for the f-f absorption accounts for both the steep rising branch of the radio light-curve and the delay of the rise at lower frequencies. From the amount of f-f absorption the mass loss rate of the stellar progenitor is estimated to be about 6×10^{-5} and 3×10^{-6} $M_\odot$ yr^{-1} for SN 1979c and 1980k, respectively, if a velocity of v(RSG) = 10 km s^{-1} is assumed for the stellar wind from the parent red supergiant.

At IR wavelengths SN 1980k is the one Type II SN which has extensively been observed (Dwek *et al.* 1983). During the first couple of months after maximum an IR excess has been revealed for this SN. Such an excess radiation is likely to be the result of extended atmosphere effects in the expanding ejecta. After several months the near IR emission becomes more prominent and is consistent with thermal emission from dust at ~ 800 K. A similar result has been found for SN 1979c at about 260 days past the optical maximum (Merrill, 1980), which also can be interpreted in terms of heated dust (Bode and Evans, 1980; Dwek 1983).

In the X-ray domain, observations with the Copernicus satellite of SN 1979c and with the Einstein Observatory of both SN 1979c and 1980k have been made (Panagia *et al.* 1980; Palumbo *et al.* 1981; Canizares *et al.* 1982). Only the latter has been detected about 40 days after the optical maximum and was already below the detection limit one month later (Canizares *et al.* 1982). The two mechanisms which may explain the observed X-ray emission are: a) Inverse Compton effect on optical photons of the SN, and b) Thermal f-f- emission of the hot shocked gas. Both mechanisms imply a fast decay of the X-ray emission which is consistent with observations. Actually, it is likely that both mechanisms operate to produce X-rays and have comparable importance. The missed detection of SN 1979c at X-ray frequencies (Panagia *et al.* 1980; Palumbo *et al.* 1981) can be explained as due to the quick decline of the X-ray flux which may have gone below the detection limit already at the time of the first Einstein observation (about two months past optical maximum).

Finally, we must mention that SN 1980k was observed with the COS-B satellite from November 4 until December 29, 1980, but no detection was achieved yielding to an upper limit to the flux $F(> 100MeV) < 10^{-6}$ photons cm^{-2} s^{-1} (Cavallo *et al.* 1981). This result indicates that the acceleration of very energetic particles is not efficient at early phases and suggests that a supernova remnant may emit gamma-rays mostly at the end of its life rather than at the beginning (Cavallo *et al.* 1981).

It is clear that, although qualitatively very similar to each other, these two SNe were considerably different in the absolute luminosity. In fact, after allowance for extinction SN 1979c at maximum appeared to be only about one magnitude fainter than SN 1980k even if the distance to NGC 4321 is almost three times greater than to NGC 6946. Therefore, SN 1979c became more than one magnitude intrinsically brighter than SN 1980k. In general, the spread of properties among different Type

II SNe is even larger, not only in absolute luminosity but also in time evolution, i.e. light curve and colours.

REFERENCES

Arnett, W. D., 1982, *Ap. J.*, **253**, 785.

Barbon, R., Ciatti, F., Rosino, L., 1982, *Astr. Ap.*, **116**, 35.

Barbon, R., Cappallaro, E., Ciatti, F., Turatto, M., Kowal, C. T., 1984, *Astr. Ap. Suppl.*, **58**, 735.

Benvenuti, P., Sanz Fernandez de Cordoba, L., Wamsteker, W., Macchetto, F., Palumbo, G. C., Panagia, N., 1982, ESA SP-1046.

Benvenuti, P., Dopita, M., D'Odorico, S., 1984, private communication.

Bertola, F. 1964, *Ann. Astrophys.*, **27**, 319.

Bertola, F., Mammano, A., Perinotto, M., 1965, Contr. Asiago Obs. No. 174.

Bode, M. F., Evans, A. 1980, *Monthly Notices Roy. Astron. Soc.*, **193**, 21 p.

Canizares, C. R., Kriss, G. A., Feigelson, E. D., 1982, *Ap. J. (Letters)*, **253**, L17.

Cavallo, G., Caraveo, P. A., Bignami, G. F., 1981, 17th International Cosmic Ray Conference, Paris, V. 9, XG 5.1-6, p. 80.

Chevalier, R. A., 1981, *Ap. J.*, **251**, 259.

Chevalier, R. A., 1984, *Ap. J. (Letters)*, **285**, L63.

Dwek, E., A'Hearn, M. F., Becklin, E. E., Brown, R. H., Capps, R. W., Dinerstein, H. L., Gatley, I., Morrison, D., Telesco, C. M., Togunaka, A. T., Werner, M. W., Wynn-Williams, C. G., 1983, *Ap. J.*, **274**, 168.

Dwek, E., 1983, *Ap. J.*, **274**, 175.

Elias, J. H., Frogel, J. A. Hackwell, J. A., Persson, S. E. 1981, *Ap. J. (Letters)*, **251**, L13.

Elias, J. H., Matthews, K., Neugebauer, G., Persson, S. E., 1985, *Ap. J.*, **296**, 379.

Filippenko, A. V., Sargent, W. L. W., 1985, *Nature*, **316**, 407.

Fransson, C., 1984, *Physica Scripta*, **T7**, 50.

Fransson, C., Benvenuti, P., Gordon, C., Hempe, K., Palumbo, G. G. C., Panagia, N., Reimers, D., Wamsteker, W., 1984, *Astr. Ap.*, **132**, 1.

Kirshner, R. P., Willner, S. P., Becklin, E. E., Neugebauer, G., Oke, J. B., 1973, *Ap. J. (Letters)*, **187**, L97.

Kirshner, R. P., 1985, private communication.

Merrill, K. M. 1980, I.A.U. Circ. No. 3444.

Pacini, F., Salvati, M., 1981, *Ap. J. (Letters)*, **245**, L107.

Palumbo, G. G. C., Maccacaro, T., Panagia, N., Vettolani, G., Zamorani, G., 1981, *Ap. J.*, **247**, 484.

Panagia, N., 1980, Proc. 2nd European IUE Conference, ESA SP-157, p. XXVII–XXXI.

Panagia, N., 1982, Proc. 3rd European IUE Conference, ESA SP-176, p. 31.

Panagia, N., 1984a, *Physica Scripta*, **T7**, 15.

Panagia, N., 1984b, Proc. 4th European IUE Conference, ESA SP-218, p. 15-18.

Panagia, N., *et al.* 1980, *Monthly Notices Roy. Astron. Soc.*, **192**, 861.

Panagia, N. *et al.* 1986a, in preparation.

Panagia, N., Sramek, R. A., Weiler, K. W., 1986b, *Ap. J. (Letters)*, in press.

Salvati, M., 1983, I.A.U. Symp. No. 101 "Supernova Remnants and their X-ray Emission," eds. J. Danziger and P. Gorenstein, D. Reidel (Dordrecht-Holland), p. 177–182.

Sandage, A., Tammann, G. A. 1974, *Ap. J.*, **194**, 559.
Sandage, A., Tammann, G. A.. 1981, in "A Revised Shapley-Ames Catalog of Bright Galaxies," (Washington, D.C., Carnegie Inst.), Publ. 635.
Shklovskii, I. S., 1981, *Sov. Astron. Lett.*, **7**, 263.
Sramek, R. A., Panagia, N., Weiler, K. W., 1984, *Ap. J. (Letters)*, **285**, L59.
Tammann, G. A., 1978, *Mem. S.A.It.*, **49**, 315.
Weiler, K. W., van der Hulst, J. M., Sramek, R. A., Panagia, N., 1981, *Ap. J. (Letters)*, **243**, L151.
Weiler, K. W., Sramek, R. A., van der Hulst, J. M., Panagia, N. 1982, in "Supernovae," eds. Rees, M. J. and Stoneham, R. J., D. Reidel (Dordrecht-Holland), p. 281.
Weiler, K. W., Sramek, R. A., van der Hulst, J. M., Panagia, N. 1983, I.A.U. Symp. No. 101. "Supernova Remnants and their X-ray Emission," eds. J. Danziger and P. Gorenstein, D. Reidel (Dordrecht-Holland), p. 171-176.
Weiler, K. W., Sramek, R. A., Panagia, N., van der Hulst, J. M., Salvati, M., 1986, *Ap. J.*, in press.

THE DYNAMICS OF SUPERNOVA ENVELOPES

Claes Fransson
Stockholm Observatory
S-133 00 Saltsjöbaden, Sweden

ABSTRACT. The dynamics and radiative emission of supernovae are reviewed from the supernova explosion up to the early remnant stage.

1. INTRODUCTION

Supernovae are important from several points of view. They provide the universe with most of the heavy elements, inject a major part of the energy to the interstellar medium, can be used as cosmological distance indicators and represent the final stage of the evolution of massive stars. To understand and exploit these various aspects it is, however, necessary to have a good understanding of the physics of the different supernova stages. In this review I give a summary of the dynamics and radiative processes from the time of the collapse until the early remnant phase, as well as their use as cosmological distance indicators. For similar reviews I refer to Chevalier (1981), Trimble (1982, 1983) and the complementary review by Fransson (1985), which includes a more detailed discussion of the interaction with the circumstellar medium and the remnant stage.

2. PROGENITORS AND THE SUPERNOVA ENVIRONMENT

Observationally two types of supernovae can be distinguished. Here I just summarize the main characteristics of these, for a more detailed description see e.g. the review by Trimble (1982). Type I supernovae are characterized by the absence of hydrogen lines, exponential light curves and the occurrence in both elliptical and spiral galaxies. Type II:s on the other hand, show very strong hydrogen lines in their spectra, which become increasingly stronger with time, have light curves of less regular shape than Type I:s, often with a plateau lasting about two months and occur in spiral galaxies only. The presence of Type I:s in ellipticals indicate that at least these Type I supernovae must originate from low mass stars ($< 2\ M_{o}$), while the prevalence of Type

F. Pacini (ed.), High Energy Phenomena Around Collapsed Stars, 51–71.

II:s to the inner spiral arms (Maza and van den Bergh, 1976) indicate a mass larger than ~6 M_o.

As is shown in the contribution by Hillebrandt in this volume, detailed calculations of the structure and dynamics of supernova explosions have indicated that the dividing line between Type I and II supernovae probably occurs at 8-10 M_o. As we will see later, most of the observational properties of the supernova explosion depends on the structure of the envelope of the supernova progenitor and the density of the surrounding medium. Its place in the HR-diagram is therefore of crucial importance. Evolutionary calculations with no mass loss indicate that massive stars end their lives as red supergiants with low surface temperatures and very extended envelopes (10^{14} cm or more). However, supergiants in our own Galaxy in general, have strong winds with mass loss rates, in the range 10^{-6}-10^{-4} M_O/yr, and velocities of the order of 10 km/s. That this can have strong influence on the properties of the envelope of the supernova progenitor has been shown by calculations by Chiosi and by Maeder (see Chiosi and Maeder (1986) for a review). They find that for massive, luminous stars (more than 20-30 M_o) with high mass loss rates, the star can loose most of its outer, hydrogen rich envelope before the carbon burning stage, so instead of ending its life as a red supergiant it may collapse as a compact hot Wolf-Rayet star with no hydrogen envelope (radius less than 10^{12} cm). Later, it will be argued that this may explain some of the properties of Cas A.

Direct evidence for a large amount of circumstellar matter around supernovae comes from radio observations with VLA of a number of recent supernovae. Since the first observed radio supernova, the Type II supernova SN 1979c (Weiler et al. 1981), several others have been discovered (Weiler et al. 1985). The most interesting property of these is that the peak of the radio emission occurred at a considerable later time than the optical maximum. For SN 1979c the delay was about one year at 6 cm and for SN 1980k about 45 days. The radio emission was first seen at short wavelengths and later at longer, indicating an optical depth effect.

Pacini and Salvati (1981) proposed the exciting idea that the emission comes from the plerionic nebula powered by the pulsar, assumed to be the result of a Type II explosion. However, the very large optical depth to free-free absorption of the supernova ejecta at these wavelengths makes this interpretation less likely. There has been some attempts to explain this by strong filamentation of the ejecta (Bandiera et al. 1983), but this has to be very efficient, with a density outside the filaments less than 10^{-6} of the mean density. Also, the energy needed to break up the ejecta at this early stage must be of the same order as the total energy in the explosion. It is, however, possible that this scenario may be relevant at a later stage, e.g. for the radio supernovae discovered in M83 by Cowan and Branch (1985).

A more plausible model was proposed by Chevalier (1982b), who found that the radio observations could be well explained if the radio emission is produced close to the supernova shock wave, as it propagates out through the circumstellar gas around the supernova. The turn-on of the radio emission then occurs as a result of the decreasing free-free optical depth of the circumstellar gas as the supernova shock wave expands. The optical depth of the wind is given by

$$\tau_{ff} \propto (\dot{M}/u)^2 \, T_e^{-1.5} \, \lambda^2 \, R_s^{-3}$$

where $\dot{M}$ is the mass loss rate, u the wind velocity, T_e the temperature of the circumstellar gas, R_s the shock radius (estimated from the expansion velocity) and λ the wavelength. This form explains both the rapid turn-on of the emission, as well as its wavelength dependence. If the temperature of the wind is known, this model allows a determination of the quantity $\dot{M}/u$. However, the temperature of the wind can not be taken as that of an ordinary HII region, since the radiation from the supernova shock wave ionizes and heats the wind (Sect. 5). Detailed calculations (Lundqvist and Fransson, 1985) have shown that the UV and X-ray radiation from the shock may heat the wind to 2×10^4–2×10^5 K, depending on the epoch and the density of the wind. This heating increases the derived mass loss rate $\propto T_e^{3/4}$, which is a factor of ~2. For SN 1979c a mass loss rate $\dot{M}=1.5\times10^{-4}$ M_o/yr is found for u=10 km/s, and for SN 1980k $\dot{M}=2\times10^{-5}$ M_o/yr. These two supernovae and SN 1970g are the only ones which have been well-observed in radio, so judging from this small sample, a strong mass loss of the supernova progenitor seems to be a general feature of Type II supernovae.

Of the three observed Type I:s, two have been detected in the radio. Although observed for two years, the apparently normal Type I SN 1981b has not been detected. Scaling from the other detected cases, the mass loss rate must have been less than about 10^{-7} M_o/yr (Panagia, 1985). For the other two, a mass loss rate of 2×10^{-6} M_o/yr (all for a wind velocity of 10 km/s) was inferred for SN 1983n (Chevalier, 1984) and the flux of SN 1984l is also consistent with this value (Panagia et al. 1985). Panagia (1985) has suggested that the radio bright Type I supernovae belong to a common class, with the designation Type I_{SL} (SL for subluminous), and could comprise as much as 25–50 % of all Type I:s. The characteristic feature of these is that they occur only in spiral galaxies, are subluminous by 1.5–2 magnitudes and lack the Si I feature at 6150 Å (Uomoto and Kirshner, 1985) If the supernova is due to an exploding white dwarf, the most likely origin of the circumstellar gas is not the supernova progenitor itself, but rather its companion star. Cameron and Iben (1985) explains the circumstellar gas as due to mass lost from the binary system in the form of a common envelope surrounding the system. Another possibility, which means a completely different origin of the progenitor, is based on a collapsing, massive Wolf-Rayet star. As explained in the next section, this type of explosion could give a light curve very similar to a subluminous Type I. The radio emission could in that case come from the interaction with the rapid wind. To get the same density, a mass loss rate of ~2×10^{-4} M_o/yr would be required, for a wind velocity of ~2000 km/s.

For the Type II:s, the mass loss rates are at the high end of those observed for red supergiants in the Galaxy (Zuckerman, 1980). There is also a number of stars in our Galaxy and in the Magellanic Clouds which have very high mass loss rates, 10^{-4}-10^{-1} M_o/yr, so called super-winds (see de Jager, 1980). Most of the stars with these abnormal mass loss rates are, however, very massive and therefore hardly typical for supernovae. Dopita et al. (1984) have suggested that the narrow component of the Hα line seen in the supernova in NGC 3169 could have its origin in a wind of this type, and estimate a mass loss rate of more than 10^{-4} M_o/yr.

Although an ordinary stellar wind in the red supergiant stage is expected and fits most of the observations, it is also possible that the circumstellar gas could be due to a more short-lived, sporadic event just before the supernova explosion. Woosley et al. (1980) found that in models with a mass of 10 M_o, the core underwent a series of flashes in the Ne shell, which developed into shock waves ejecting the hydrogen envelope at a velocity of 100-500 km/s five years before the supernova explosion. The radio observations, however, give a minimum duration of the high mass loss stage of $tV_s/u \sim 10^2$ $t(u/100\ km/s)^{-1}$ yrs (where V_s is the shock velocity and t the time since the explosion in years). Since SN 1979c is still observed after six years, this implies that this particular scenario is ruled out. The total mass of the circumstellar gas around SN 1979c can be estimated from the radio observations, independent of the wind velocity, as at least 0.6 M_o. Similar instabilities during the carbon burning stage can, however, not be excluded, and it is possible that mass ejections resembling those suggested to give rise to planetary nebulae (Tuchman et al. 1979) could be responsible for the observed gas. There is an indication from the most recent observations of a decrease in the radio flux of SN 1979c, as compared to that expected for a uniform stellar wind, possibly due to a decrease in the circumstellar gas density. Also the fact that the radio flux from SN 1970g was not detected down to a very low level after 12 years (Weiler et al. 1985), indicates that the interaction with the high density region may be a relatively short-lived period. SN 1950b and SN 1957d in M83, on the other hand, are still seen after twenty years (Cowan and Branch, 1985). It is interesting that a duration of ~10 years corresponds to a wind with a velocity of ~10 km/s, lasting during the carbon burning phase of a 10-15 M_o red supergiant ($\sim 10^{4-5}$ yrs).

Summarizing this discussion, circumstellar matter can thus be expected around both Type I supernovae and Type II supernovae. For Type I:s mainly due to mass loss from the binary system either in the form of a wind from the companion star or due to ejection of a common envelope, and for Type II:s due to a strong stellar wind or sporadic mass ejections in the red supergiant phase.

3. OUTBREAK OF THE SHOCK AND SUPERNOVA LIGHT CURVES

3.1. Type II supernovae

After the hydrodynamic collapse and subsequent bounce (if this occurs), a strong shock wave is formed in the central region of the star. The exact details are yet controversial (see Hillebrandt, this volume), but the main result for the subsequent evolution is that about 10^{51} ergs of thermal and kinetic energy is deposited outside the iron core. After the shock wave has passed this region, neutrino losses and nuclear dissociation losses will be negligible. Since the streaming velocity of the photons is very small, the expansion of the hot bubble in the center will be essentially adiabatic, and as the pressure is dominated by photons the adiabatic index will be close to 4/3. The structure of the shock wave at this time has been studied in detail by Weaver (1976), who finds that the shock is mediated by scattering of photons by the unshocked gas in front of the shock. The optical thickness, τ_s, of the shock transition is determined by the condition that the diffusion time scale, t_{diff}, over the shock thickness, Δ_s, is equal to the time it takes the shock to transverse this region, t_{flow}. Thus we have

$$t_{diff} = (\Delta_s/\lambda_{mfp})^2\lambda_{mfp}/c = 3\Delta_s\tau_s/c = \Delta_s/V_s$$

or

$$\tau_s = c/3V_s \sim 10-20$$

Here V_s is the shock velocity and $\lambda_{mfp}=1/3\kappa\rho$ is the mean free path for the photons. The density jump behind the shock is $(\gamma+1)/(\gamma-1)=7$.

Numerical simulations of the hydrodynamics of the explosion has been done by several groups. For illustrative purposes I will here briefly describe the calculation of the explosion of a 15 M_o star by Weaver and Woosley (1980). After the bounce the shock propagates through the core and mantle region in about 500 secs. As the shock sweeps up matter, it is decelerated from more than 10^4 km/s to about 10^3 km/s. It then enters the very steep density gradient between the mantle and envelope, where the density falls by six orders of magnitude. In a sharp density gradient the shock velocity increases roughly as $V_s \propto \rho^{-0.2}$ (Chevalier, 1976a and references therein), and consequently the shock velocity increases again to 1.2×10^4 km/s. The subsequent evolution depends on the ratio of the envelope to core/mantle mass. If most of the mass resides in the envelope, as is indicated by mass conserving evolutionary calculations, the shock will be decelerated (Chevalier, 1976a). In order to halt the rapidly expanding inner mantle, a reverse shock must then propagate inwards through this material. This has the important consequence that a condensed region of processed core/mantel material will be present in the center of the remnant. On the other hand, if the mass of the envelope is small, e.g. due to mass loss, the blast wave will not be substantially decelerated and the remnant will evolve to a shell structure, similar to that of the Sedov solution.

Approximately one day after the explosion, the shock enters the final density gradient at the surface of the star. The pressure will then decrease dramatically, and this message will be propagated back through the whole remnant as a rarefaction wave, giving a final acceleration because of the pressure gradient in the wave. This will set up the final configuration with a velocity proportional to the radius, which is a general feature of these calculations. The basic reason for this is the same as for a non-linear wave to steepen into a saw-tooth shape (see Landau-Lifshitz, 1959). The density structure now consists of a central region of fairly high density, containing the processed material, a large envelope of fairly constant density and an outer sharp density gradient, where $\rho \propto r^{-n}$, with n=7-12. Still further out we find the circumstellar gas with $\rho \propto r^{-2}$.

Since the pressure inside the supernova is dominated by radiation, the temperature behind the shock is given by $\sigma T^4 = 9/14\ \rho_o V_s^2$ or $T = 0.96\ 10^8\ \rho_o^{1/4} V_9^{1/2}$ K, where ρ_o is the density in front of the shock and V_{s9} the shock velocity in units of 10^9 cm/s. The density in the envelope decreases from $\sim 10^{-8}$ g/cm^3 to $\sim 10^{-13}$ g/cm^3 at the photosphere. The temperature behind the shock is therefore $3 \times 10^4 - 10^6$ K, much less than that of a gas pressure dominated shock due to the large number of photons available compared to the number of ions.

When the shock comes within a distance equal to τ_s from the photosphere (as defined by the optical depth to electron scattering), the first photons will leak out. The temperature of these will be close to that of the shock wave as it approaches the surface, $2 \times 10^4 - 10^5$ K, depending on the density and extent of the outer envelope. In the last section it was argued that the processes responsible for the mass loss could modify the structure of the outer envelope considerably. In particular, the location of the $\tau = 1$ point could be quite different from the hydrostatic case, and one has to include this in the calculation of the lightcurve. For a mass loss rate of $10^{-4}\ \dot{M}_{-4}\ M_o$/yr and a wind velocity of $10\ u_6$ km/s, the optical depth through the wind at a radius $10^{14}\ R_{14}$ cm is $2\ \dot{M}_{-4}\ u_6^{-1}\ R_{14}^{-1}$, implying that the wind itself may effect the light curve for high mass loss rates and moderate photospheric radii.

Although the shock wave has reached the surface of the star, the energy loss of the hot gas in the interior of the supernova will take place on the diffusion time scale for the whole envelope, given by

$$t_{diff} = 3\kappa\rho R^2/c = 1.9 \times 10^7\ M/R_{15}\ \text{sec}$$

where M is in solar masses (compare the expression derived earlier). Since the expansion time scale is $10^6\ R_{15}/V_9$ secs, the diffusion velocity of the photons is therefore small compared to the expansion, for radii less than a few times 10^{15} cm. Therefore, the evolution is nearly adiabatic, and if the gas is radiation dominated, the internal energy (i.e. the radiation) will decrease as

$$E_{int} = pV/(\gamma-1) \propto \rho^{\gamma-1} \propto R^{3(1-\gamma)} = R^{-1}$$

(V is here the volume and p the pressure). Therefore, if we start with a small initial radius most of the internal energy is converted into kinetic energy before the photons escape and a very dim supernova results (Colgate and White, 1966). Chevalier (1976a) and Falk and Arnett (1977) found that in order to reproduce the observed light curves, a radius of $\sim 10^{14}$ cm was needed, consistent with that of red supergiants.

An example of a dim supernova would be an exploding Wolf-Rayet star, which has lost most of its hydrogen envelope, having a radius less than 10^{12} cm. Chevalier (1976b) has suggested that this may be the explanation for the low brightness of Cas A, which must have been more than four magnitudes less luminous than a 'standard' Type II supernova. Furthermore, Lamb (1978) finds that the large N/C ratio (Chevalier and Kirshner, 1978) in the slow moving flocculi can be explained as a result of CNO processed material, which has been lost in a stellar wind, as first suggested by van den Bergh (1971). A requirement for a faint explosion is that there is no additional energy source than the thermal and kinetic energy from the shock. Even for a Type II explosion a small amount (0.1-0.4 $M_\odot$) of radioactive ^{56}Ni is likely to be produced (Weaver and Woosley, 1980). This could, then, result in a light curve similar to the subluminous class of Type I:s, and it may actually be quite difficult to separate this type of (basically Type II supernova) explosions from ordinary Type I:s. As has already been discussed, the presence of the radio emission can then rather naturally be explained by this model.

The detailed calculation of the light curve has to be done numerically (Grassberg et al. 1971, Chevalier, 1976a, Falk and Arnett, 1977, Weaver and Woosley, 1980), but the essential features can be understood from the hydrodynamic structure of the supernova envelope. The light curve can be divided in three stages, the initial peak, the plateau phase and the late decline. The initial peak, lasting 10-20 days, is a result of the the arrival of the shock at the surface. During this time the gas cools from the shock temperature to ~6000 K. At this temperature hydrogen starts to recombine and consequently the opacity drops rapidly. This marks the beginning of the plateau phase. The effective temperature will now be roughly constant, ~6000 K, as well as the density of the recombining gas at the photosphere ($\sim 10^{11}$ cm^{-3}). Since the ejecta is freely expanding with $\rho \propto t^{-3}$, this means that the photosphere has to move inward to deeper layers with higher density, to compensate for the expansion. Because of the expansion and the density structure of the ejecta, the absolute radius of the photosphere is, however, roughly constant, which explains the fairly constant luminosity during this epoch. This occurs only if $\rho \propto r^{-3}$ near the photosphere, and since there is reason to expect that the density structure can vary between individual supernovae, it should not be surprising that the observations show a range of behaviour in this phase, from the linear subclass to the plateau type.

When the photosphere has receded through the whole envelope (after ~100 days), the final decline phase starts. The luminosity in this phase depends sensitively on whether a high density core region exists or not, and how much radioactive material has been synthesized. If a dense central region is present, it can, in the same way as for Type I:s, trap the γ-rays from the radioactive ^{56}Ni produced in the explosion, and an exponential light curve will result. That the late part of the light curve really is powered by radioactivity is quite clear in at least the well-studied cases (Barbon et al. 1984). However, in the case of strong mass loss of the progenitor there may not be sufficient mass to decelerate the core-mantle material, and a more shell-like structure results. In this case, it is likely that the late light curve may drop considerably faster.

3.2. Type I supernovae

In Type I:s the explosion is due to a nuclear runaway in a degenerate core, the most popular progenitor being an accreting white dwarf in a binary system, which exceeds the Chandrasekhar mass (see Nomoto, 1984). The structure of the nuclear burning front, which propagates through the star, can either be that of a supersonic detonation front or a subsonic deflagration wave (see Mazurek and Wheeler, 1980). In the case of a detonation, the whole star is transformed into ^{56}Ni, which then decays first to ^{56}Co ($t_{½}$=6.1 days) and then to ^{56}Fe ($t_{½}$=77 days). In the deflagration model, the subsonic burning front dies out before it reaches the surface, and only part of the star (<0.5 $M_{\odot}$) is converted into nickel. Observations of early time spectra (Branch et al. 1983) indicate that considerable quantities of Ca and Si are present, strongly favouring the deflagration model.

Since the radius of the progenitor star is very small, adiabatic losses would make this type of explosions very faint, if it were not for the presence of an additional energy source in the form of radioactive decay. In the initial stages the gamma rays produced by the nickel decay are thermalized by inelastic Compton scattering. For each decay 1.78 MeV in γ-rays is emitted. At epochs later than about 15 days the decay of ^{56}Co with the emission of 3.51 MeV in γ-rays and 0.14 MeV in positrons dominates. In the early phases, the envelope is opaque to the γ-rays, but later it will gradually become transparent. The interaction of the γ-rays and the gas will produce energetic electrons, which are decelerated by repeated ionizations, excitations and elastic scattering by thermal electrons (Meyerott, 1980, Axelrod, 1980). The positrons emitted by the Co decay undergo a similar process. The degree of positron trapping is, however, uncertain, since even a weak magnetic field will decrease the mean free path considerably. As long as the fraction of free electrons is larger than ~10^{-3}, most of the energy goes into heating of the thermal electrons and into ionizations. The cooling of the gas is then due to collisional excitations by the thermal electrons. For a mainly neutral gas, a large fraction of the energy may go directly into excitations by the non-thermal, secondary electrons. This has not been included in the calculations mentioned above. Axelrod found that for temperatures above ~4×10^3 K excitation of optical, permitted and forbidden lines dominate the cooling (Sect. 4), but for lower temperatures fine structure transitions in the infra-red of low-lying levels take over. This results in a fairly flat cooling curve down to temperatures of a few hundred degrees. Therefore, if the temperature falls below ~4×10^3 K, the gas will probably undergo a thermal instability, emitting most of its residual energy in the infra-red. In this connection it is very intersting that a strong fine structure transition of [Fe II] at 1.644 μm has been observed for SN 1983n by Graham et al. (1985). Even if the temperature falls to a very low level, direct excitations of higher levels by the non-thermal secondary electrons may occur. Therefore, in this case a considerable fraction of the energy may be emitted in the optical region of the spectrum, closely resembling the excitation conditions of aurorae.

4. ATMOSPHERES

A central problem for progress in several areas of supernova research is the need for good model atmospheres. One example is the comparison of hydrodynamic models with observations, which today is mainly based on broad band photometry, and little use has been made of the large quantity of information contained in the spectral data. Whereas photometry can give us some rough information about the temperature of the photosphere, the line spectrum is central to our understanding of the structure of the transparent envelope, and therefore to the interior of the supernova. Supernova atmospheres are here rather unique compared to other stellar atmospheres, since the spectral lines of the latter either come from a small interval, where the optical depth is close to unity (if it is a collision dominated line), or is loosely coupled to the local conditions (if it is a scattering line). In contrast, because of the very large velocity gradients, in supernova atmospheres each frequency of the line comes from a given layer of the supernova, and one can thus probe the whole envelope. The hydrodynamic structure is therefore directly reflected in the line profile. By observing lines of different elements it is, thus, possible to study the nucleosynthesis, which has taken place during the various evolutionary phases of the pre-supernova and in the explosion. Finally, for the application of supernovae as distance indicators, in particular the Baade-Wesselink method, good model atmospheres are crucial (see Sect. 6.2.).

The large velocities in the supernova atmospheres poses both problems and simplifications to the calculation. The simplifications arise, because when a photon is emitted in a narrow spectral line, the Doppler effect rapidly shifts the photon out of the line width, as seen by gas moving at different velocities. Therefore, if the velocity gradient is large, the emission and absorption in the line can be considered as a local process, and one does not have to solve a complicated equation of transfer. This approximation is known as the Sobolev approximation (Sobolev, 1957). The optical depth is given by $\tau = \sigma n_i \Delta \ell$, where σ is the cross section, n_i the number density of absorbing atoms, and $\Delta\ell$ the width of the region, which has a velocity within half a thermal line width (v_{th}) from the line center. It is then easy to see that $\Delta\ell=v_{th}/(dv(r)/dz)$, where $dv(r)/dz$ is the velocity gradient in the line of sight to the observer in the z direction. Therefore,

$$\tau = \sigma n_i v_{th}(dv(r)/dz)^{-1}.$$

The points with a given frequency, ν, as seen by the observer, represents a surface given by $(\nu-\nu_o)/\nu_o=(z/r)v(r)/c$, where ν_o is the rest frequency. For a homologous expansion $v(r)\propto r$, and the constant frequency surfaces are perpendicular to the line of sight. From this it is obvious how the different parts of the line profile represents the conditions at various radii of the envelope, and can, therefore, directly give information about the interior structure of the remnant with regard to nuclear processing etc. It is interesting, and for qualitative purposes very useful, that one can in these cases obtain analytic solutions for the line profiles (Fransson, 1984a). For resonance lines, where the two-level atom is a good approximation, this solution agrees very well with numerical calculations, and has been used for SN 1979c to model the region where the UV lines arise.

For the optical lines and continuum, the situation is far more complicated since the populations of the various levels have to be calculated self-consistently for the atmosphere. In the simplest case one assumes LTE and it is then simple to calculate the populations. This approach has been followed by Branch and his co-workers (see Branch 1980 for a review). Because of the low densities of the supernova envelope ($\sim10^{12}$ cm^{-3} at the time of the explosion and $\sim10^{9}$ cm^{-3} after a year), the collisional coupling between the various levels is, however, very weak and non-LTE effects will be important, unless the background continuum is very close to Planckian. The populations of the different levels then have to be calculated from the statistical equations, supplemented by the equation of radiative equilibrium. The continuum causes an additional problem, since for lines arising close to the photosphere, there can be a strong interaction between the lines and continuum. The simple Sobolev approximation then breaks down and a much more sophisticated treatment is necessary. This is also the case when distinct features, like absorption edges of the continuum are calculated, and when there exist large gradients in density and temperature. To treat these cases other methods are needed, e.g. the ones formulated in the co-moving frame of the fluid (see Mihalas, 1978). A special problem is line blending, which is best treated with the co-moving frame formalism. This blending is due to both the large width of the lines (only $\sim c/v\sim30$ lines are needed to fill the whole spectrum!), and the large number of possible lines due to eg Fe I and II. That this can be quite severe is illustrated by the LTE-calculations by Karp et al. (1977), where several thousands of lines contributed to the opacity.

Calculations of supernova spectra taking a varying number of these effects into account have been done by several authors for both Type I and II supernovae. Branch and co-workers (1980) have used the Sobolev approximation together with the LTE assumption to calculate synthetic spectra of both Type I:s and II:s. The electron temperature is obtained from the colour temperature of the continuum, and the number density of a given ion is treated as a parameter obtained from the observation. The other lines from the same ion are then calculated from the LTE assumption. For Type I supernovae, in particular, the agreement with the observed spectra has been remarkably good, taking the many simplifications into account, and a good example of this analysis is provided by SN 1981b (Branch et al. 1983). One should, however, remember that only if several lines from the same ion exist can a real test of the LTE-assumption be made. The main use of the method has been in identifying the different lines in the very complex Type I spectra. The blending of the lines due to their large widths causes an additional problem. A considerable improvement of this approach is due to Harkness (1985), who solved the co-moving frame equations together with the equation of radiative equilibrium, allowing for relativistic effects and a radially varying structure. The different level populations were, however, still assumed to be in LTE. Another problem was the specification of the inner boundary conditions, since it is not clear that even at maximum there is a clear photosphere. Also the deposition and thermalization of the gamma rays due to the ^{56}Ni and ^{56}Co decay is complex, in particular outside the photosphere. This model has been used to compare the Type I deflagration model of Nomoto et al. (1984) with the observations of SN 1981b. An interesting result is that the agreement with the observations become considerably better for models where the metals are mixed radially, rather than for the stratified, onion type structure. This is also in agreement with Branch et al. (1985), and may indicate that large scale convective motions are important in connection with the deflagration wave.

An opposite approach has been taken by Axelrod (1980), whose main interest was the late time spectra of Type I:s (after several hundred days). His models use the nebular approximation, i.e. that most of the electrons are in their ground states so that continuous absorption from excited levels can be neglected. For the radiative transfer in the lines, the Sobolev approximation is used and the statistical equations are solved for the many level populations together with the energy equation. The energy input at this late time comes mainly from the ^{56}Co positrons, which are assumed to be trapped by the weak magnetic field. The positron energy is assumed to be thermalized by the electrons. Direct excitations (except ionization) are neglected, which is likely to be valid in the initial epochs, but not at late times (several hundred days). The ions included were Fe I-VI and Co I-III. A major problem in these models are the large uncertainties in the atomic data, in particular the collision cross sections. Another possible source of uncertainty is the neglect of line fluorescence. The most interesting feature of his model spectra is the prediction that certain cobolt and iron lines should vary in time due to the $^{56}\mathrm{Co} \rightarrow {}^{56}\mathrm{Fe}$ decay. In particular, Axelrod identifies a line complex near 6000 Å in SN 1972e as due to Co III, behaving in the predicted way. The quality of these spectra are, however, rather low and there are severe problems with line blending, and the identification therefore somewhat uncertain. The high quality spectra of SN 1981b (Branch et al. 1983) did not show any clear evidence for Co at late time, although the presence of Co is compatible with the spectral fit. Also in the early time spectra by Branch et al. there is a line attributed to Co II at 3500 Å, but unfortunately, this is right in the region where neither IUE nor ground based detectors are very sensitive.

Branch has also applied his spectral synthesis method to Type II supernovae (Branch et al. 1981). Even if most of the lines could be identified, the fit was rather unsuccessful. This was especially evident for the Balmer lines, which are dominated by emission, and can then obviously not be modelled by pure scattering. This illustrates well the need in Type II supernovae to take non-LTE effects into account, especially in the late stages (more than a month after the explosion). The Balmer lines, and in particular Hα, are interesting for several reasons. During the first month after the outburst the spectrum is mainly a featureless continuum, and only after this time does the Hα line appear. It is then gradually growing stronger with time, and after half a year most of the energy may come out in this single line (see eg Branch et al. 1981). Another remarkable thing is that the maximum width of the line is fairly constant (Fransson, 1984b), indicating that at least the high velocity part arises close to the shock. The Balmer decrement is also quite steep with Hα/Hβ=5-10. A simple non-LTE model for the Balmer lines was proposed already by Kirshner and Kwan (1975), who suggested that the emission was due to recombinations from the continuum, following photoionization from the n=2 level of hydrogen by the Balmer continuum. A problem with this interpretation, may, however, be that there does not seem to be enough photons in the continuum at late times to provide the ionizations. The constant width of the line may also be difficult to explain, unless the envelope is concentrated to a shell close to the shock. One possibility to give the necessary excitation is by radioactive excitation, either directly as in Type I:s, or indirectly due to photoionization by the thermalized radiation from the core. For this method to work the γ-ray optical depth must be high enough, which may be a problem, unless the density distribution is fairly concentrated to the center (Sect. 3.1.). A third contribution to the Hα flux may be the X-ray flux from the reverse shock (Sect. 5), which is absorbed by the cooling gas behind the shock (Chevalier and Fransson, 1985).

To study the spectral formation in both Type I and Type II supernova envelopes one thus needs to make a complete non-LTE model, taking the velocity field and the effects of the extended atmosphere into account. This is, however, a very difficult problem and only a few steps in this direction have been taken yet, mainly for the determination of the continuum spectrum for a given hydrodynamic structure and effective temperature at the very early stages (Sect. 6.2.).

5. INTERACTION WITH THE CIRCUMSTELLAR MEDIUM

With the arrival of the supernova shock wave at the photosphere the remnant phase starts. The evolution of the shock wave is now determined mainly by the interaction of the expanding supernova ejecta and the circumstellar and interstellar material. Before 1979 this was known as the free expansion phase and was thought to be observationally quite uninteresting, with the decaying continuum emission of the central regions of the supernova as the main characteristic. Both radio emission and X-ray emission were thought to be too weak to be observed. The observations of SN 1979c in NGC 4321 (M 100) followed by SN 1980k in NGC 6946 a year later, changed this radically. Not only in radio as was discussed in Sect. 2, but also in the X-ray, UV, optical and infrared, observations could be combined into a fairly coherent picture. Since I have recently reviewed this phase in detail (Fransson, 1985), I will here only summarize the main features.

The main ingredient in this scenario is the interaction of the expanding supernova ejecta with the circumstellar gas. This interaction results in a hot region of shocked gas, separating these media. Because of symmetry, there has to be one shock propagating outwards into the circumstellar gas, and one reverse shock back into the ejecta. Chevalier (1982a) has discussed the dynamics of this region in detail, and found a very useful similarity solution under the assumption that the density of the supernova ejecta, into which the reverse shock propagates, can be described by a power law, $\rho_e(r)\propto r^{-n}$. If the circumstellar medium is constant, its density will be given by $\rho_{cs}(r)=\dot{M}/(4\pi r^2 u)$. Since there are no characteristic length or time scales, a similarity solution can be found. The fact that there should be a simple scaling of the whole structure implies, in particular, that the ratio of the ejecta density immediately in front of the reverse shock, to that immediately in front of the blast wave should be independent of time. Since $\rho_e(R_s)\propto t^{-3}(R_s/V_o t)^{-n}$, we get $\rho_e(R_s)/\rho_{cs}(R_s)\propto R_s^{(2-n)}t^{(n-3)}=$ constant, where R_s is the radius of the shock and V_o is the velocity of the reference point. Thus we find that

$$R_s \propto t^{(n-3)/(n-2)}$$

and the velocity of the blast wave $V_s=dR_s/dt=(n-3)/(n-2)\ R_s/t$. The velocity of the unshocked ejecta is $V_e=R_s/t$, since it is freely expanding. Therefore the velocity of the reverse shock is $V_r=V_e-V_s=R_s/(n-2)t=V_s/(n-3)$. Consequently, the temperature behind the reverse shock is smaller than that of the blast wave by a factor $(n-3)^2$. The temperature and density structure of the shocked gas are, of course, very sensitive to cooling, which is important, especially for the reverse shock, but also in the initial phases for the blast wave. The basic time variation of the variables are, however, not affected since it is set by the density dependence of the circumstellar gas and supernova ejecta and the total energy in the explosion.

The temperature of the gas behind the blast wave is $1.36 \times 10^9 V_{s9}^2$, where V_{s9} is the shock velocity in 10^9 cm/s. Hydrodynamic calculations (Chevalier and Jones, priv. comm.) and observations of the width of the Hα line indicates that n=9-12. The temperature behind the reverse shock is therefore $(1-2) \times 10^7$ K and the density ~40 times that behind the blast wave. Most of the cooling of the blast wave is done by Compton scattering of soft photons from the photosphere by the nearly relativistic electrons. This process boosts the energy of the photons at the expense of the electrons, and can be shown to give rise to a power-law extension of the photospheric radiation into the UV and EUV, with a spectral index of 1.5-2.5 (Fransson (1982)). This probably explains the UV excess seen in SN 1979c and SN 1980k (Panagia, 1982). Because of the low temperature and high density, the gas behind the reverse shock will cool efficiently by line and bound-free emission. The cooling is fairly sensitive to the density gradient, but it is likely that the reverse shock will be radiative for more than a year after the explosion.

The EUV and X-ray emission from the shocks is important as an ionizing source for both the ejecta and the circumstellar gas. The fraction absorbed by the ejecta is there degraded into UV line emission (Fransson, 1984), explaining the presence of the strong lines seen with IUE (Panagia et al. 1980, Fransson et al. 1984). The outgoing flux will ionize and heat the surrounding gas to a temperature of $4 \times 10^4 - 10^5$ K (Lundqvist and Fransson, 1985), which is important to take into account when deriving mass loss rates from the radio observations (Sect. 2). Another interesting consequence of this ionization is that strong P-Cygni shaped absorption lines are expected for the UV resonance lines. These should be easily observable with the Space Telescope (and may already have been observed with IUE). In the future these may become a powerful diagnostic of the structure of the circumstellar gas, complementary to the radio observations. In particular, any pre-acceleration of the wind by the strong radiation field from the supernova should be seen in these lines.

The subsequent dynamic evolution of the remnant depends on the duration and mass loss rate in the various evolutionary phases of the progenitor. In addition, the dynamics of the reverse shock is sensitive to the density structure of the supernova envelope. Berman and Kahn (1983) have discussed the case of a reverse shock propagating into a uniform envelope, and finds that cooling is probably still important after several years. An interesting possibility is that as the shock propagates deeper and deeper into the ejecta more metal rich matter is encountered, which increases the cooling, so that a radiative shock may persist many years after the explosion. In the same way as was discussed in the last section, this may be favourable for the production of dust in this metal rich environment.

Since the red supergiant stage is preceded by a blue supergiant phase, with a mass loss of approximately the same rate but with a wind velocity of 2000-3000 km/s, the density is much smaller and a wind blown cavity in the interstellar medium may be formed outside the slow wind

(Weaver et al. 1977). The interaction of the supernova with its surroundings is therefore likely to be quite complex.

6. SUPERNOVAE AS COSMOLOGICAL DISTANCE INDICATORS

Due to their large luminosities, supernovae have long been recognized as potential distance indicators. The different methods existing can be divided into two different classes, depending on whether the supernova is used as a standard candle, or if the available observations are used to calculate the absolute luminosity or diameter in the context of a physical model. For a detailed comparison of these different methods see the different contributions in Bartel (1985).

6.1. Standard candles

From photometric studies it is well known that both the shape and the maximum absolute magnitude differs considerably between different Type II supernovae, making them unsuitable as standard candles (Barbon et al. 1979). Also the lower luminosity ($M_B \sim -19$) and the fact that they occur only in spiral galaxies, often in dust rich regions, make them less valuable in this respect.

Type I supernovae, on the other hand, show a remarkable similarity in light curves (Barbon et al. 1973), and in spectroscopic properties, both in UV, optical and infra-red (Panagia 1985, Branch et al. 1983, Elias et al. 1981). Therefore, they seem to be well suited for the purpose. The maximum magnitude of about -20 also means that they can be observed at very large distances. There are, however, some complications. When absolute magnitudes are compared (for a given value of H_o), it is found that Type I:s in elliptical and S0 galaxies show little scatter (dispersion $0^m.43$), whereas those found in spirals have a broad distribution, with a dispersion of more than one magnitude (Tamman, 1982). The simplest explanation for this difference is that it is due to internal absorption within the spirals, and that only elliptical galaxies with a small dust content should be used. For a reduced sample of 12 Type I:s in elliptical and S0:s, Cadonau et al. (1985) finds that the maximum magnitude $M_B^o(max)=-19.69\pm0.13+5\log(H_o/50)$, and $\sigma=0.45$. It is, however, disturbing that in an investigation of a sample of elliptical galaxies, Sparks et al. (1985) finds that a large fraction of the galaxies have dust lanes with sizes of several kpc, which could introduce systematic errors. Also the growing suspicion that Type I supernovae probably consist of two spectroscopically and photometrically somewhat different classes (Sect. 2), is of some concern. The subluminous 'peculiars', however, seem to be confined to spiral galaxies only. To obtain the value of H_o one also need to calibrate the absolute luminosity of the Type I:s. This has been done by Sandage and Tamman (1982), assuming that the brightest stars in two nearby galaxies have the same absolute magnitude as in M 31.The calibration, however, suffers from problems of internal absorption in these galaxies. With the most likely value for the reddening, the result quoted is $M_B=-20.0\pm0.4$, corresponding to $H_o=43^{+10}_{-7}$ km/s Mpc^{-1}.

6.2. The Baade-Wesselink method

The principle of this method is that the absolute, bolometric luminosity of the supernova can be calculated from $L_B=4\pi R_p^2 \sigma T_{eff}^4$, where R_p is the photospheric radius, T_{eff} the effective temperature and σ the Stefan-Boltzmann constant. Since all the quantities in principle can be derived from the observations, the absolute luminosity can be directly calculated, giving the distance. The photospheric radius is obtained from $R_p=R_o+V(t-t_o)$, where R_o is the initial radius of the SN-precurser, V the expansion velocity of the photosphere measured from the optical spectra, and t_o the time of the explosion. Most models of the pre-supernova indicates that the radius is less than ~10^{14} cm, and this can then be neglected after a few days, when the radius is ~10^{15} cm. By observing the supernova at several epochs the uncertainty in the explosion date can be eliminated. The main uncertainty, therefore, is in the determination of the effective temperature. In the first approximation this is obtained directly by fitting a black-body spectrum to the continuum for the different days. This works quite well for Type II supernovae, and has been applied to several supernovae (Kirshner and Kwan, 1974, Branch et al. 1981). Typically one obtains values of H_o~60 km/s/Mpc.

The assumption of a black-body spectrum for the continuum is, however, only a rough approximation and several effects could contribute to a systematic deviation from this. Wagoner (1981) has pointed out that the fact that electron scattering is likely to be important could dilute the spectrum, as the photon is scattered many times from the true photosphere. It is then a well-known result from the theory of stellar atmospheres (see e.g. Mihalas, 1978 p. 148) that the flux is diluted by a factor $(\kappa_a/(\kappa_a+\kappa_s))^{\frac{1}{2}}$, where κ_a and κ_s are the absorption and scattering opacities. This result can also be obtained by a random walk argument similar to that given by Felten and Rees (1972). In a scattering dominated atmosphere the flux could therefore be overestimated by a large factor, and consequently also the distance. Branch (1985) has pointed out that, if this dilution is varying with time, this could go in either direction.

The only way of getting around this problem is to make a detailed model atmosphere of the supernova. As we have already discussed, this is a complicated problem, although some progress has been made (Hershkowitz et al. 1985). In particular, the observed Balmer jump could give some indication of the importance of scattering. A problem with these models is that they depend on the density structure of the supernova envelope. Up to now this has been inferred from hydrodynamical models, but it would certainly be more satisfying if it could be obtained directly from the observations. A detailed modelling of the line profiles at several epochs would therefore be very important.

For Type I:s the continuum level (if there is one) is very difficult to determine without realistic model atmospheres (Harkness, 1985), and the reliability is doubtful. Even the expansion velocities are in this case uncertain as well as the structure of the ejecta and the excitation conditions.

6.3. ^{56}Ni decay in Type I supernovae.

Current models of Type I supernovae predict that 0.5-1.4 M_o of ^{56}Ni is produced in the explosion (Sect. 3.2.). Since there is a direct correspondence between the amount of ^{56}Ni and the supernova luminosity, the distance to the supernova can be derived (Arnett et al. 1985). In the analytical model by Arnett (1982), the peak luminosity is given by

$$L(max) = 1.36x10^{43}\ e^{-t(max)/114d}\ M_{Ni}\ erg/s,$$

where M_{Ni} is the ^{56}Ni mass in solar masses. This expression assumes that ^{56}Ni dominates the decay of ^{56}Co, which is valid only for $t(max) \lesssim 15$ days. Since this is independent of the opacity, it is probably quite insensitive to the model and thus suitable for a comparison with the observations. The width of the peak depends on the parameter $\kappa M_{tot}/V_{max}$, where κ is the opacity, M_{tot} the total mass and V_{max} the maximum velocity in the envelope. This parameter, however, depends sensitively on both the opacity and the hydrodynamic structure which are uncertain.

The main problem with this method is the amount of ^{56}Ni produced. In detonation models the whole supernova (1.4 M_o) is converted to ^{56}Ni, but in the more popular deflagration models this quantity depends on the detailed physics of the deflagration front and can be anywhere between 0.5-1 M_o. A second problem is the determination of the bolometric luminosity, since a large fraction of the energy output is outside the visible wavelength range. From detailed multifrequency observations from UV to IR, Panagia et al. (1985) have been able to calculate the bolometric flux with good accuracy for SN 1983n. Correcting the flux for extinction, they find that $0.55(H_o/50)^{-2}$ M_o of ^{56}Ni is produced. At the moment it is probably more correct to do it in this direction, but with a better understanding of the deflagration wave this may become a powerful method for distance determinations in the future. The general constraints on H_o by this method have been discussed by Arnett et al. (1985).

6.4. Angular diameters of radio supernovae.

A new way of estimating the distances was introduced by the VLBI observations of SN 1979c (Bartel et al. 1985). The first observation of this supernova took place in Dec 1982, about 3.5 years after the explosion, revealing an extended source with a diameter of 1.05 milli arcsec. Subsequent observations showed an increase in the size, and in May 1984 the diameter was 2.36 milli arcsec. Since the absolute diameter can, in principle be calculated from the expansion velocity and the time since the explosion, the distance can be derived.

Unfortunately, the exercise is not so easy, and there are a number of effects, which must be taken into account. First, the measured diameter is subject to a systematic uncertainty, since the source is not resolved and one therefore has to assume a model distribution for the flux. Marsher (1985) has shown that this introduces a factor of ~1.5 uncertainty in the measured size. The most reasonable assumption is probably a thin shell, close to the shock. In addition, as we have already seen, the supernova is not freely expanding and one has to take the deceleration of the shock into account. Chevalier and Fransson (1985) have discussed how one can use observations at other wavelengths in combination with the similarity solution to allow for this. The most important source of information is the evolution of the velocity of the Hα line at late times, giving information about the density structure of the ejecta. Another question is whether the radio emission comes from a region very close to the shock or from a wide region outside it.

The distance Bartel et al. derive corresponds to a value of $H_o=65^{+35}_{-25}$ km/s. With a longer time base the observations should become better, and it may prove to be a competitive method in the future.

6.5. X-rays from young supernovae.

Also X-ray observations can be used for distance determinations (Chevalier and Fransson 1985). This idea is based on the fact that the reverse shock into the supernova ejecta is radiative a long time after the explosion (see Sect. 5). Therefore, all the shock energy, $\propto \dot{M}V^3/u$, will be radiated in the form of soft X-rays. $\dot{M}/u$ can be obtained from radio observations, V from the optical line profiles, so if the density gradient can be calculated from other observations, like the evolution of the Hα line or from the temperature of the X-rays, the total luminosity can be calculated. The attractive feature is that this is independent of complications like the atomic data, chemical composition and the detailed structure of the shock. These are important only for the form of the spectrum, not for the integrated luminosity. On the negative side is the fact that since the shock is radiative, the dense, cool shell between us and the reverse shock will absorb part of the X-ray flux. From spectral X-ray observations the column density should, however, be possible to derive and the flux corrected for this in the standard way. If the X-ray observations can be made sufficiently late, the cool shell will become optically thin, eliminating this complication. Also the density gradient in the ejecta introduces some uncertainty, although this can in principle be obtained from other observations.

SN 1980k was detected by Einstein (Canizares et al. 1982) with a luminosity of $2\times10^{39}(D/10\ \mathrm{Mpc})^2$ erg/s. Unfortunately, the detection was close to the sensitivity limit and the uncertainty in the flux is large. Only some very crude spectral information is available. It is, however, encouraging that the distance obtained to NGC 6946 is consistent with other methods. A more serious test of the method has to wait for the launch of ROSAT or AXAF. In general, X-ray observations of young supernova remnants are very important both for understanding the interaction with the circumstellar medium and the ejecta. With these new instruments the observations should be fairly easy, and the main problem will be the unpredictable nature of supernovae.

Acknowledgement: I am very grateful to Kerstin Palm-Johansson and Eva Jurlander for help with the manuscript, to Prof. Franco Pacini for interesting discussions on the beach, and to Claes-Ingvar Björnsson for helpful comments on the manuscript.

REFERENCES

Arnett, W.D. 1982, Ap. J. 253, 785.
Arnett, W.D., Branch, D., and Wheeler, J.C. 1985, Nature 314, 337.
Axelrod, T.S. 1980, in Type I Supernovae, ed. J.C. Wheeler, Univ. of Texas, p. 80.
Bandiera, R., Pacini, F., and Salvati, M. 1983, Astr. Ap. 126, 7.
Barbon, R., Capellaro, E., and Turatto, M. 1984, Astr. Ap. 135, 27.
Barbon, R., Ciatti, F., and Rosino, L. 1973, Astr. Ap. 25, 241.
Barbon, R., Ciatti, F., and Rosino, L. 1979, Astr. Ap. 72, 287.
Bartel, N. (ed.) 1985, Supernovae as Distance Indicators, Springer.
Bartel, N., Rogers, A.E.E., Shapiro, I.I., Gorenstein, M.V., Gwinn, C.R., Marcaide, J.M., and Weiler, K.W. 1985, Nature 318, 25.
van den Bergh, S. 1971, Ap. J. 165, 259.
Berman, N.M., and Kahn, F.D. 1983, M.N.R.A.S. 205, 303.
Branch, D. 1980, in Supernova Spectra, ed. R. Meyerott and G.H. Gillespie, N.Y. American Institute of Physics, p. 39.
Branch, D. 1985, in Supernovae as Distance Indicators, ed. N. Bartel, Springer, p. 138.
Branch, D., Doggett, J.B., Nomoto, K., and Thielemann, F.-K. 1985, Ap. J. 294, 619.
Branch, D., Falk, S.W., McCall, M.L., Rybski, P., Uomoto, A.K., and Wills, B.J. 1981, Ap. J. 244, 780.
Branch, D., Lacy, C.H., McCall, M.L., Sutherland, P.G., Wheeler, J.C., and Wills, B.J. 1983, Ap. J. 270, 123.
Cadonau, R., Sandage, A., and Tamman, G.A. 1985, in Supernovae as Distance Indicators, ed. N. Bartel, Springer, p. 151.
Cameron, A.G.W., and Iben, I. Jr. 1985, preprint.
Canizares, C.R., Kriss, G.A., and Feigelson, E.E. 1982, Ap. J. (Letters) 253, L17.
Chevalier, R.A. 1976a, Ap. J. 207, 872.
Chevalier, R.A. 1976b, Ap. J. 208, 826.
Chevalier, R.A. 1981, Fundamentals of Cosmic Physics 7, 1.
Chevalier, R.A. 1982a, Ap. J. 258, 790.
Chevalier, R.A. 1982b, Ap. J. 259, 302.
Chevalier, R.A. 1984, Ap. J. (Letters) 265, L63.
Chevalier, R.A., and Fransson, C. 1985, in Supernovae as Distance Indicators, ed. N. Bartel, Springer, p. 123.
Chevalier, R.A., and Kirshner, R.P. 1978, Ap. J. 219, 931.

Chiosi, C., and Maeder, A. 1986, Ann. Rev. Astron. Astrophys. 24, in press.
Colgate, S.A., and McKee, C. 1969, Ap. J. 157, 623.
Colgate, S.A., and White, R.H. 1966, Ap. J. 143, 626.
Cowan, J.J., and Branch, D. 1985, in Supernovae as Distance Indicators, ed. N. Bartel, Springer, p. 75.
Dopita, M.A., Evans, R., Cohen, M., and Schwartz, R.D. 1984, Ap. J. (Letters) 287, L69.
Elias, J.H., Frogel, J.A., Hackwell, J.A., and Persson, S.E. 1981, Ap. J. (Letters) 251, L13.
Falk, S.W., and Arnett, W.D. 1977, Ap. J. Suppl. 33, 515.
Felten, J.E., and Rees, M.J. 1972, Astr. Ap. 17, 226.
Fransson, C. 1982, Astr. Ap. 111, 150.
Fransson, C. 1984a, Astr. Ap. 132, 155.
Fransson, C. 1984b, Astr. Ap. 133, 264.
Fransson, C. 1985, in Radiation Hydrodynamics In Stars and Compact Objects, ed. D. Mihalas and K.H. Winkler, Springer, in press.
Fransson, C., Benvenuti, P., Gordon, C., Hempe, K., Palumbo, G.G.C., Panagia, N., Reimers, D., and Wamstecker, W. 1984, Astr. Ap. 132, 1.
Graham, J.R., Meikle, P.S., Allen, D.A., Longmore, A.J., and Williams, P.M. 1985, submitted to M.N.R.A.S.
Grassberg, E.K., Imshennik, V.S., and Nadyozhin, D.K. 1971, Astrophys. Space Sci. 10, 28.
Harkness, R. 1985, in Supernovae as Distance Indicators, ed. N. Bartel, Springer, p. 183.
Hershkowitz, S., Linder, E., and Wagoner, R.A. 1985, preprint.
de Jager, C. 1980, The Brightest Stars, Reidel.
Karp, A.H., Lasher, G., Chan, K.L., and Salpeter, E.E. 1977, Ap. J. 214, 161.
Kirshner, R.P., and Kwan, J. 1974, Ap. J. 193, 27.
Kirshner, R.P., and Kwan, J. 1975, Ap. J. 197, 415.
Kronberg, P.P., Bierman, P., and Schwab, F.R. 1985, Ap. J. 291, 693.
Lamb, S.A. 1978, Ap. J. 220, 186.
Landau, L., and Lifshitz, E. 1959, Fluid Dynamics, Addison-Wesley.
Lundqvist, P., and Fransson, C. 1985, in preparation.
Marsher, A.P. 1985, in Supernovae as Distance Indicators, ed. N. Bartel, Springer, p. 130.
Maza, J., and van den Bergh, S. 1976, Ap. J. 204, 519.
Mazurek, T.J. and Wheeler, J.C. 1980, Fundamentals of Cosmic Physics 5, 193.
Meyerott, R.E. 1980, Ap. J. 239, 257.
Mihalas, D. 1978, Stellar Atmospheres, 2nd ed., San Francisco, Freeman.
Nomoto, K. 1984, in Stellar Nucleosynhtesis, ed. C. Chiosi and A. Renzini, Reidel, p. 205.
Nomoto, K., Thielemann, F.-K., and Yokoi, K. 1984, Ap. J. 268, 644.
Pacini, F., and Salvati, M. 1981, Ap. J. (Letters) 245, L107.
Panagia, N. 1982, Proc. of the Third European IUE Conference, ESA SP-176, p. 31.
Panagia, N. 1985, in Supernovae as Distance Indicators, ed. N. Bartel, Springer, p. 14.

Panagia, N. et al. 1980, M.N.R.A.S. 192, 861.
Panagia, N., Sramek, R.A., and Weiler, K.W. 1985, submitted to Ap. J. (Letters).
Panagia, N. et al. 1985, in preparation.
Sandage, A.R., and Tamman, G.A. 1982, Ap. J. 256, 339.
Sobolev, V. 1957, Sov. Astron. 1, 678.
Sparks, W.B., Wall, J.V., Thorne, D.J., Jorden, P.R., van Breda, I.G., Rudd, P.J., and Jorgensen, M.E. 1985, M.N.R.A.S. 217, 87.
Sutherland, P.G., and Wheeler, J.C. 1984, Ap. J. 280, 282.
Tamman, G. 1982, in Supernovae: A Survey of Current Research, ed. M.J. Rees and R.J. Stoneham, Reidel, p. 371.
Trimble, V. 1982, Rev. Mod. Phys. 54, 1183.
Trimble, V. 1983, Rev. Mod. Phys. 55, 511.
Tuchman, Y., Sack, N., and Barkat, S. 1979, Ap. J. 234, 217.
Uomoto, A., and Kirshner, R.P. 1985, Astr. Ap. 149, L7.
Wagoner, R.V. 1981, Ap. J. (Letters) 250, L65.
Weaver, T.A. 1976, Ap. J. Suppl. 32, 233.
Weaver, T.A., and Woosley, S.E. 1980, in Supernova Spectra, ed. R. Meyerott and G.H. Gillespie, American Institute of Physics, p. 15.
Weaver, R., McCray, R., Castor, J., Shapiro, P., and Moore, R. 1977, Ap. J. 218, 377.
Weiler, K.W., van der Hulst, J.M., Sramek, R.A., and Panagia, N. 1981, Ap. J. (Letters) 243, L151.
Weiler, K.W., Sramek, R.A., Panagia, N., van der Hulst, J.M., and Salvati, M. 1985, submitted to Ap. J.
Woosley, S.E., Weaver, T.A., and Taam, R.E. 1980, in Type I Supernovae, ed. J.C. Wheeler, Univ. of Texas, p. 96.
Zuckerman, B. 1980, Ann. Rev. Aston. Astrophys. 18, 263.

STELLAR COLLAPSE AND SUPERNOVA EXPLOSIONS

Wolfgang Hillebrandt
Max-Planck-Institut für Physik und Astrophysik
Institut für Astrophysik
Karl-Schwarzschild-Str. 1, D-8046 Garching
Federal Republic of Germany

ABSTRACT. The present status of core-collapse models of type II supernovae and the formation of neutron stars is discussed. A survey is given of the physics of stellar collapse, the numerical methods are described that have been applied to this problem, and finally the results of recent numerical simulations are summarized. It is demonstrated that although the general ideas are quite simple and convincing, the problem is still far from being solved. This is at least partially due to poorly known input physics, but it may also be that the right mechanism has not yet been discovered.

1. INTRODUCTION AND STATEMENT OF THE PROBLEM

If we compute the dynamics of a collapsing stellar core we hope to be able to answer among others the following basic questions: How are neutron stars or black holes formed? What is the physical mechanism that causes stars to explode as (type II?) supernovae? Can we explain the abundances of heavy elements and their isotopes?

Since stellar matter is well described by an ideal non-viscous fluid the equations that have to be solved then read

$$\frac{1}{r^2}\frac{\partial}{\partial r}\left[r^2\frac{\partial\Phi}{\partial r}\right] = 4\pi G\rho \text{ (Poisson equation)}, \tag{1.1}$$

$$\frac{\partial \ln\rho}{\partial t} + \frac{u-u_g}{r}\frac{\partial \ln\rho}{\partial \ln r} + \frac{1}{r^2}\frac{\partial(r^2 u)}{\partial r} = 0 \text{ (continuity)}, \tag{1.2}$$

$$\frac{\partial u}{\partial t} + (u-u_g)\frac{\partial u}{\partial r} + \frac{p}{\rho}\frac{\partial \ln p}{\partial r} + \frac{\partial\Phi}{\partial r} = 0 \left[\begin{matrix}\text{equation}\\ \text{of motion}\end{matrix}\right], \tag{1.3}$$

$$\frac{\partial \ln s}{\partial t} + \frac{u-u_g}{r}\frac{\partial \ln s}{\partial \ln r} = \frac{\dot\epsilon}{k_B T_s} \text{(entropy conservation)}, \tag{1.4}$$

where we have assumed that the collapsing star is spherically symmetric and that the relativistic effects can be neglected. Here u is the matter

F. Pacini (ed.), High Energy Phenomena Around Collapsed Stars, 73–104.

velocity in a coordinate system moving with arbitrary velocity u_g in r-direction, $\partial/\partial t$ are time derivatives with respect to the moving coordinate system, Φ is the gravitational potential, ρ the mass density, p the pressure, s the specific entropy and $\dot{\epsilon}$ denotes energy sources and sinks such as non-equilibrium nuclear interactions and neutrinos. If, in particular, the choice $u_g \equiv u$ is made the equations are given in a co-moving (Lagrange) system of coordinates and the radial coordinate r can be replaced by a mass-coordinate

$$m(r) = 4\pi \int_0^r \rho r^2 dr. \tag{1.5}$$

If $u_g \equiv 0$, on the other hand, the equations are given in a fixed (Eulerian) system of coordinates. Note that equations (1.1) through (1.4) are only valid if dissipative processes such as shocks are absent and they have to be modified if shocks are present. Equations (1.1) through (1.4) can in principle be solved if initial values $\rho(r,to)$, $u(r,to)$ and $s(r,to)$ are known, an equation of state is known meaning that the pressure and specific entropy are given as a function of density, temperature and composition, and finally equations that describe the changes in the composition as a function of time are known. The latter also determine $\dot{\epsilon}$. Additional complications arise from the facts that neutrinos are not always advected with the matter, so transport equations have to be solved, that general relativity may become important, and that non-spherical effects such as rotation and large scale convection may have to be included. We will come back to those questions later.

Despite the fact that the problem is very well defined, solutions cannot be obtained in a straightforward way, the main reason being that the solutions are very sensitive to a lot of details among which the initial conditions and the input physics seem to be extremely important. Moreover, even the numerical method applied in order to compute stellar collapse seems to matter. This is not surprising because the gravitational binding energy of a neutron star is of the order of a few times 10^{53}erg whereas the energy given to the ejecta of a type II supernova is only of the order of 10^{51}erg. We therefore try to model a very inefficient process with numerical methods and the results should be taken with care.

2. THE PHYSICS OF STELLAR COLLAPSE

2.1. Initial models

Stars can only collapse if their core mass exceeds the Chandrasekhar mass, given by

$$M_{CH} \simeq 5.27\ (y_e)^2\ M_\odot,$$

where y_e is the electron concentration and $M_\odot$ is the mass of the sun, and if no significant amounts of nuclear fuel (e.g. He, C) are present in their cores. Candidates for core collapse are therefore either massive stars ($M \gtrsim 8 M_\odot$ on the main sequence) at the end of hydrostatic evolution or O-Ne-Mg white dwarfs accreting matter from a companion star.

Stellar evolution models predict that stars in the mass range from 8 $M_\odot$ to about 10 or 12 $M_\odot$ first form either O-Ne-Mg or Ne-Si cores with a mass close to M_{CH}, which then contract due to electron captures, and nuclear burning incinerates a certain fraction of the core mass into iron-group nuclei (see e.g., Nomoto (1984), Woosley et al. (1980), Wilson et al. (1985)). More massive stars, on the other hand, burn all their nuclear fuel quietly and their iron-core masses range from 1.5 $M_\odot$ to approximately 2.2 $M_\odot$ (Arnett, 1977a; Wilson et al., 1985). The upper mass limit for this type of evolution is not well known and may be somewhere between 50 $M_\odot$ and 100 $M_\odot$. Stars beyond that limit are believed to be disrupted by explosive oxygen-burning and are not candidates for core collapse.

The properties of stellar cores at the onset of collapse can be summarized as follows.

Central densities are typically around 10^{10}g cm^{-3}, temperatures are around $(8-10)\times10^9$K, specific entropies range from 0.8 to 1.2 k_B/nucleon, and the core radii are a few 10^8 cm.

Simple estimates show that under such conditions electrons dominate the gas pressure. Since the electron chemical potential is much larger than their rest mass, they are relativistic particles and therefore the adiabatic index $\gamma:=(\partial \ln p/\partial \ln \rho)_{s,Y_e}$ is close to the critical value 4/3. Effectively, γ is even below 4/3 because either e^--captures on nuclei and free protons lower the pressure, the dominant instability for low-entropy stellar cores, or photo-dissociations of heavy nuclei reduce the internal energy, which become important if the entropy per nucleon exceeds 1 k_B. There are several problems and uncertainties which affect this general picture. The core mass and its entropy will turn out to be of crucial importance for the outcome of numerical simulations of stellar collapse. These quantities in turn depend upon the way convective transport and mixing is modelled, on certain thermonuclear reaction rates such as $^{12}C(\alpha,\gamma)^{16}O$ which has been revised recently, on weak interaction rates and neutrino losses, mass loss, etc.. Pre-collapse stellar models, therefore, should still be considered with some care.

2.2. The equation of state (EOS)

2.2.1. <u>Nuclear statistical equilibrium (NSE)</u>. Computations of the equation of state are greatly simplified by the fact that strong and electromagnetic interactions are in equilibrium at temperatures above roughly 5×10^9K. This can be demonstrated by a simple example. Let us consider the reaction $^{52}Cr+\alpha \rightleftharpoons {}^{56}Fe+\gamma$. At a temperature of 8×10^9K and a density of 10^{10}g cm^{-3} the reaction time for α-particles with ^{52}Cr is approximately 10^{-6}s and the reaction time for photodissociation of ^{56}Fe is around 5×10^{-9}s. These numbers have to be compared with the collapse time-scale which is on the order of 10^{-1}s. It is therefore an excellent approximation to assume an equilibrium composition of nuclei and free nucleons for a given total number of neutrons and protons (NSE). The method of computation is then (in principle) straightforward. One first writes down an expression for the free energy F, which is a function of temperature, volume and composition. Next F is minimized with respect to the abundances at fixed temperature, volume and neutron and proton number. Finally all variables are obtained by usual thermodynamic relations such as $p=-\partial F/\partial V$. Different approaches to obtain the equation of state differ only in models that are applied in order to compute F.

2.2.2. <u>The "low"-density EOS.</u> We will first discuss the EOS at densities below a few times 10^{12} g cm^{-3}. Then nuclei can be considered as being Boltzmann-particles and their free energy density is given by

$$f_i = n_i k_B T \left\{ 1 + \ln\left[\frac{1}{n_i} \left[\frac{m_i k_B T}{2\pi\hbar^2}\right]^{3/2} \omega_i(T) \right] \right\} +$$

$$+ m_i c^2 n_i + (\text{interaction terms}), \qquad (2.2.2.1)$$

where n_i denotes the number density of a species i, m_i its mass and ω_i its partition function, all other quantities having their usual meaning. Interaction terms should describe Coulomb (lattice) interactions. They are normally small and may be neglected. If we define an "effective" chemical potential by

$$\exp\{-\mu_i'/k_B T\} := \frac{1}{n_i} \left[\frac{m_i k_B T}{2\pi\hbar^2}\right]^{3/2} \qquad (2.2.2.2)$$

this is related to the true chemical potential by

$$\mu_i := (\partial f_i/\partial n_i)_{T,ye} = \mu_i' + m_i c^2 + \begin{bmatrix}\text{interaction}\\ \text{terms}\end{bmatrix} \qquad (2.2.2.3)$$

and the equilibrium conditions read

$$\beta\mu_{ZA}'^{(J)} = (A-Z)\eta_n + Z\eta_p + \beta B_{ZA}^{(J)} + \ldots \qquad (2.2.2.4)$$

where $\beta := 1/k_B T$, $\eta_{n,p} := \beta\mu_{n,p}$. J denotes an excited state of a nucleus with Z protons and A-Z=N neutrons. The binding energy of the nucleus in its excited state J is given by

$$B_{ZA}^{(J)} = (A-Z) m_n c^2 + Z m_p c^2 - m_{ZA}^{(J)} . \qquad (2.2.2.5)$$

and n and p denote neutrons and protons, respectively. The number density of a nucleus (A, Z) is then given by

$$n_{ZA} = \sum_J n_{ZA}^{(J)}$$

$$= \omega(\beta) \left[\frac{2\pi m_{ZA}^{(0)}}{h^2\beta}\right]^{3/2} e^{\{\beta B_{ZA}^{(0)} + N\mu_n + Z\mu_p\}} \qquad (2.2.2.6)$$

with the partition function $\omega(\beta)$ defined by

$$\omega(\beta) := \sum_J (2I_J + 1)\, e^{-\beta\{B_{ZA}^{(0)} - B_{ZA}^{(J)}\}}$$

$$= \sum_J (2I_J + 1)\ e^{-\beta E_J} \qquad (2.2.2.7)$$

where now (0) indicates ground state properties. E_J is the excitation energy and the first factor takes care of spin degeneracies. Together with the Fermi-gas expressions for the number densities of free neutrons and protons and by imposing the conditions of baryon number conservation and charge neutrality equations (2.2.2.6) determine the composition of the nucleonic component of the matter once ground state binding energies and nuclear partition functions are known from nuclear physics. Problems arise, however, because these properties have to be known for neutron-rich nuclei which are very unstable in the laboratory. Therefore extrapolations from known nuclei are commonly used in computations of the EOS of supernova matter (see, e.g., El Eid and Hillebrandt (1980), Mazurek et al. (1979)). It also has turned out that in order to determine the neutron and proton chemical potentials, and thus their number densities, correctly, rather large networks with soveral hundred nuclear species have to be considered. This makes such computations quite expensive and difficult to handle within the framework of hydrodynamic studies. Therefore either tabulations or simplified versions based on the assumption that the nuclear composition can be described by one characteristic heavy nucleus are used in numerical models.

Finally, contributions from leptons (electrons and positrons) and photons have to be added to the nuclear part of the equation of state.

2.2.3. The "high"-density EOS. Besides these more practical questions the Boltzmann-gas approximation faces fundamental problems once the density exceeds a few times 10^{12} g cm^{-3}. Firstly nucleon-nucleon and nucleon-nucleus interactions become important and cannot be treated in a consistent way. Secondly the assumption that nuclei can be described as being point-like particles breaks down because their average separation becomes comparable to their size. Consequently the Boltzmann-gas EOS should be replaced by a self-consistent microscopic model at high densities.

A feasible way is to apply the temperature-dependent Hartree-Fock method (Bonche and Vautherin (1981), Hillebrandt and Wolff (1985), Wolff (1985)) or simplifications of it like the Thomas-Fermi method (Barranco and Buchler (1980), Ogasawara and Sato (1982)) or the compressible liquid drop model (Lamb et al. (1978, 1983)). In these approaches the general idea is to put a fixed number of neutrons and protons into a spherical box of radius R_c ("Wigner-Seitz spheres"). For a given temperature the thermodynamic properties are then computed from a model Hamiltonian.

The Hartree-Fock method may serve as an example. In second quantization the model Hamiltonian reads

$$H = \sum_{i,j} \langle i|t|j\rangle\, a_i^+ a_j + \tfrac{1}{2} \sum_{i,j,k,\ell} \langle ij|V|k\ell\rangle\, a_i^+ a_j^+ a_k a_\ell , \qquad (2.2.3.1)$$

where t is the kinetic energy operator, V an effective nucleon-nucleon interaction, and a^+ and a are nucleon creation and annihilation operators, respectively. In mean field approximation the one-body density operator is given by

$$\hat{\rho} = \frac{1}{z} \exp\{-\sum_i \alpha_i\, a_i^+ a_i\}, \tag{2.2.3.2}$$

where $z=\sum_{n_1,n_2,\ldots 0,1} \langle n_1 n_2 \ldots | \exp\{-\sum_i \alpha_i a_i^+ a_i\} | n_1 n_2 \ldots\rangle$

is the grand partition function and the sum is over all completely antisymmetrized states. The α_i are variational parameters. The grand partition function can also be written as

$$Z = \prod_i (1 + e^{-\alpha_i}), \tag{2.2.3.3}$$

which is more convenient.

From (2.2.3.2) the occupation probability for a single-particle state i can be obtained

$$n_i = Tr(\hat{\rho}\, a_i^+ a_i) = (1 + e^{\alpha_i})^{-1}, \tag{2.2.3.4}$$

and the energy and the entropy become

$$E = Tr(\hat{\rho}\, H) =$$

$$= \sum_i \langle i|t|i\rangle n_i + \tfrac{1}{2}\sum_{i,j} \langle ij|V|ij-ji\rangle n_i n_j, \tag{2.2.3.5}$$

$$S = -\,Tr\,(\hat{\rho}\, \ln\hat{\rho}) =$$

$$= -\sum_i \{n_i \ln n_i + (1 - n_i)\, \ln\,(1 - n_i)\}. \tag{2.2.3.6}$$

Finally the particle number is

$$N_k = Tr\,(\hat{N}_k\, \hat{\rho}) = \sum_i n_i^k\ , \tag{2.2.3.7}$$

where k denotes either neutrons or protons. Next the grand canonical potential

$$G = E - T\cdot S - \mu_k N_k\ , \tag{2.2.3.8}$$

where μ_k are neutron and proton chemical potentials, has to be minimized with respect to the parameters α_i and the single-particle wave functions for fixed neutron and proton number in the cell. The variation with respect to the wave functions yields the well-known Hartree-Fock equations

$$W|\phi_m\rangle = e_m|\phi_m\rangle \tag{2.2.3.9}$$

where $\langle i|W|j\rangle = \langle i|t|j\rangle + \sum_\ell \langle i\ell|V|j\ell-e\ell\rangle n_\ell$,

and e_m are single-particle energies.

The variation of G with respect to α_i relates occupation probabilities and chemical potentials

$$n_m = [1 + \exp((e_m - \mu_k)/k_BT)]^{-1}. \qquad (2.2.3.10)$$

These equations can be solved in the following way. First one has to choose an effective interaction V. Since, per definition, the Hartree-Fock method is an independent-particle approach nucleon-nucleon correlations have to be modelled by V. Skyrme-type forces have been used in computations of the EOS because they are tailored to fit the properties of finite nuclei and also symmetric nuclear matter and neutron matter in the Hartree-Fock model (see e.g. Hillebrandt and Wolff (1985) and references therein). It is obvious that the results should depend on the particular choice of V. Next the Hartree-Fock equations (2.2.3.9) have to be solved with periodic boundary conditions for the single-particle wave functions, and then G, E and S can be computed. The pressure is obtained from $p=-\partial G/\partial \Omega$, where Ω is the volume of the cell and finally lepton and photon contributions are added.

An example for an EOS obtained in this way is displayed in figures 1 to 3. Note that for low entropies ($S \simeq 1k_B$/nucleon) characterizing the matter of collapsing stellar cores, the adiabatic index is always close to its critical value 4/3 up to roughly nuclear matter density. This means that the collapse will proceed to nuclear matter density unless non-equilibrium processes increase the entropy considerably or the critical value of γ is increased by rotation.

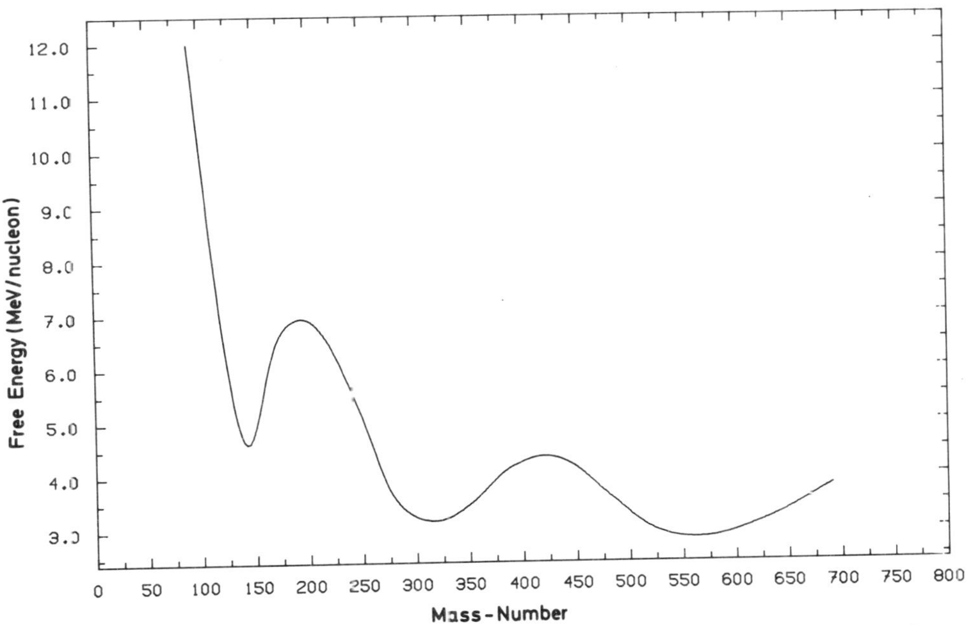

Figure 1: Free energy of a Wigner-Seitz cell versus mass number for a fixed temperature (2.5MeV), density (6×10^{13}g cm^{-3}), and electron concentration (0.3). The configuration with ~560 nucleons has the lowest free energy.

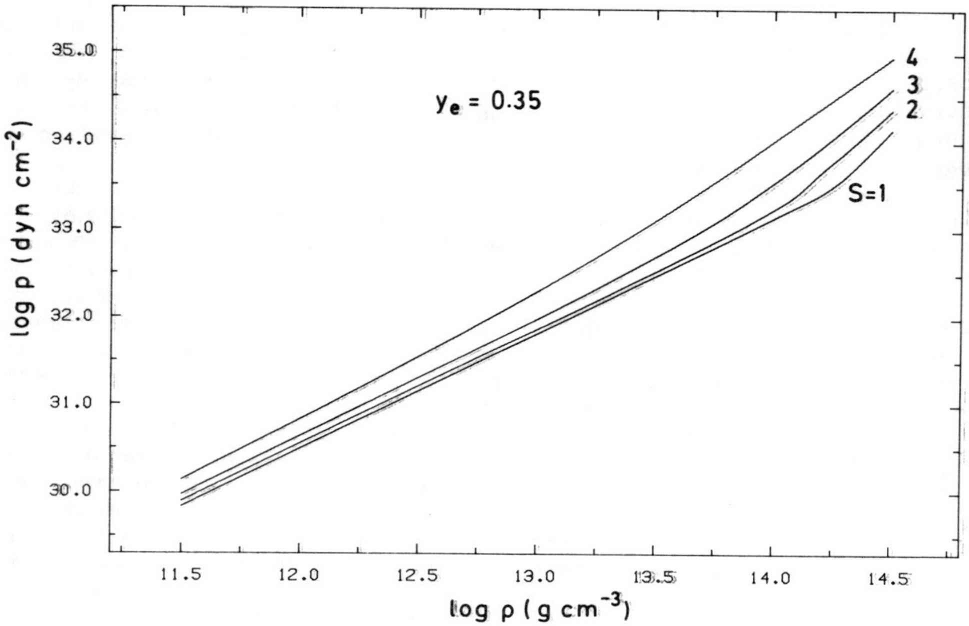

Figure 3 and figure 4: Pressure and temperature, respectively, versus density for constant values of the entropy in units of k_B/nucleon and an electron concentration of 0.35 (from Hillebrandt et al. (1984)).

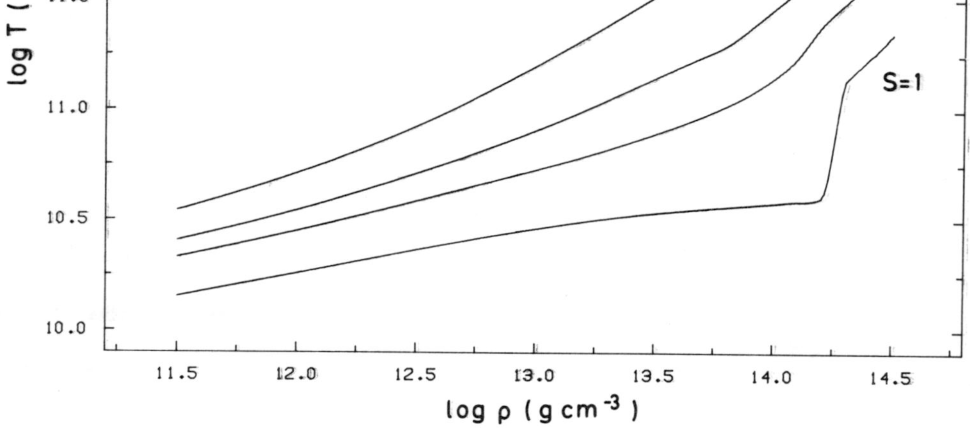

The method outlined here can, of course, also be applied to densities beyond nuclear matter density where nuclei dissolve into a homogeneous fluid of nucleons. But since the parameters of the effective forces have been fitted to known nuclei they gradually lose reliability with increasing density. Fortunately the adiabatic index just beyond nuclear matter density turns out to be of the order of 3 and therefore a collapsing core will not overshoot this density by much.

In conclusion, a few remarks should be made. The methods that have been applied in order to compute the equation of state of a collapsing star may give good results for the energy but one has to keep in mind that the adiabatic index which governs the dynamics is the second derivative of this quantity. So small changes in the energy can lead to significant modifications of γ. Besides the uncertainties coming from the effective nucleon-nucleon interaction additional problems arise from the one-nucleus and the Wigner-Seitz approximation. As can be seen in fig. 1 the minima in the free energy are quite flat and even different minima are only separated by barriers of the order of k_BT. In reality, therefore, the matter of collapsing star will be a mixture of many nuclear species even at high densities. Secondly the Wigner-Seitz approximation becomes questionable near nuclear matter density where the nuclei almost touch each other. A self-consistent model should, therefore, include lattice structures and nuclear deformations. The discussion about the structure of matter based upon simple models, namely whether or no a "bubble"-phase forms, whether or not nucleons form two-dimensional ("lasagne") or one-dimensional ("spagetti") structures seems, therefore, to be premature.

2.3. Weak interaction rates

So far we have considered only physical processes which are not explicitely time-dependent in a collapsing star because their time scales are so short that they can be treated by an equilibrium assumption. The same is not true for weak interaction processes such as

$$e^- + \begin{Bmatrix} p \\ A(Z,N) \end{Bmatrix} \rightarrow \begin{Bmatrix} n \\ (Z-1,N+1) \end{Bmatrix} + \nu_e ,$$

$$\nu_e + n \rightarrow p + e^- ,$$
$$\nu_e + e^- \rightarrow \nu_e + e^- \quad , \text{ etc.},$$

for which reaction times and collapse time are of the same order of magnitude until the density exceeds several times 10^{12}g cm^{-3}. Of particular importance are reactions which change the electron concentration y_e. The rate of change of y_e can be written as

$$\dot{y}_e = dy_e/dt = \lambda_{\nu n} y_n - \lambda_{e^- p}\, y_p - \sum_i \lambda_{e^- i} y_i \ . \qquad (2.3.1)$$

where the y's denote abundances and the λ's reaction rates. Similarly the change in neutrino concentration is given by

$$\dot{y}_\nu = -\dot{y}_e - q_\nu \ . \qquad (2.3.2)$$

where q_ν describes ν-losses or gains from neutrino transport. At high entropies (or temperatures) also positrons, anti-neutrinos and μ- and τ-neutrinos will be present, but they can be treated in the same way. During collapse, on the other hand, reactions involving positrons can be neglected, because electrons are degenerate and therefore the positron concentration is very small. Moreover, since in a neutron-rich NSE composition the most abundant nuclei possess a closed neutron shell, their electron capture rates are much lower than those of free protons (Fuller et al. (1982)) and the latter will dominate eq. (2.3.1). This does not mean that electron captures on heavy nuclei can completely be omitted because the change in entropy due to non-equilibrium weak interactions depends also on those rates.

2.3.1. <u>Some important e^--capture rates</u>. Electron captures on free protons can be computed numerically exactly and in the limit of small neutron and proton momentum ($|\bar{p}_n| \ll m_n c$, $|\bar{p}_p| \ll m_p c$) the rate is given by the following expression (see e.g. Bruenn (1985))

$$\lambda_{e^-p} = \frac{m_e^5 c^4}{2\pi^3 \hbar^7} \left[|G_F|^2 + 3|G_{GT}|^2 \right] \times$$

$$f_{pn} \int_Q^\infty d\omega F_e(\omega+Q)\ (\omega+Q)^2 \left[1 - \frac{m_e^2 c^4}{(\omega+Q)^2} \right]^{1/2} . \qquad (2.3.1.1)$$

where $Q = (m_n - m_p)c^2 = 1.29$ MeV,

$f_{np} = \dfrac{n_n - n_p}{e^{(\mu_p - \mu_n)/k_B T} - 1}$ is the baryon phase space factor,

$F_e(\omega+Q) = [1 + e^{(\omega+Q-\mu_e)/k_B T}]^{-1}$ is the lepton phase space factor, and

$G_F = gC_V = 1.4\times10^{-49}$ erg cm^3 and $G_{GT} = gC_A = -1.2 G_F$ are Fermi and Gamow-Teller coupling constants, respectively.

Analytic approximations of eq. (2.3.1.1) have also been computed (see e.g. Takahashi et al. (1978)).

Electron captures on nuclei are much more difficult to handle because nuclear transition matrix elements cannot be determined by experiment and thermally excited states have to be considered. Denoting by i initial states, by f final states and setting $h=c=m_c=1$ the general expression for the reaction rate is

$$\lambda = \sum_i (2J_i+1) e^{-E_i/k_B T} \lambda_i \Big/ \sum_i (2J_i+1) e^{-E_i/k_B T} . \qquad (2.3.1.2)$$

where E_i is the excitation energy of state i, J_i its spin and λ_i the transition rate from state i:

$$\lambda_i = \frac{1}{2\pi^3} \sum_{\Omega,\Omega'} \sum_f \left[G^*_\Omega G_{\Omega'} (\int\Omega)_{if} (\int\Omega')_{if} + \text{c.c.} \right] g_{\Omega\Omega'} \cdot$$

$$(2.3.1.3)$$

Here Ω and Ω' denote β-decay nuclear operators, $G_{\Omega,\Omega'}$ are Fermi and

Gamow-Teller coupling constants. $(\int\Omega)_{if}$ are nuclear transition matrix elements and $g_{\Omega\Omega'}$ are lepton phase space factors. For allowed and first-forbidden transitions this expression may be replaced by

$$\lambda_i = \frac{1}{2\pi^3} \sum_{\Omega} G_{\Omega}^2 \int_{E_{min}}^{\infty} |M_{\Omega}(E)|_i^2 \, g_{\Omega} dE, \tag{2.3.1.4}$$

where now $|M_{\Omega}|_i^2$ is the so-called β-strength function defined as the energetical average of the squared nuclear matrix element times the final nuclear level density; and $E=E_f-E_i$.

The β-strength function may now be obtained from a model (e.g. "gross theory"; Takahashi et al. (1978)) or may be calculated using shell model wave functions (Fuller et al. (1982), Mathews et al. (1983)). It should be kept in mind, however, that those approaches are probably accurate only to within a factor of about five.

2.3.2. Neutrino interactions. Important processes include neutrino-absorption by free neutrons, neutrino-electron scattering, neutrino-nucleus coherent scattering and neutrino-neutron scattering. The corresponding reaction rates or scattering cross-sections can again be computed (see e.g. Tubbs and Schramm (1975), Yueh and Buchler (1976), Tubbs (1979), Bruenn (1985)). For example the rate for $\nu_e+n\rightarrow e^-+p$ is given by

$$\lambda_{\nu_e n}(\omega) = \frac{m_e^5 c^4}{2\pi^3\hbar^7} [\,|G_F|^2 + 3|G_{GT}|^2] \; f_{np} \; [1 - F_e(\omega+Q)] \times$$

$$\times \; (\omega+Q)^2 \left[1 - \frac{m_e^2 c^4}{(\omega+Q)^2}\right]^{1/2}, \tag{2.3.2.1}$$

where ω is the neutrino energy and all other quantities have been explained above.

2.4. Neutrino transport and entropy generation

2.4.1. Some general considerations. In a collapsing star neutrinos are produced mainly by e^--captures on free protons. Therefore we will have ν_e's only and their energy will be roughly 2/3 of the electron Fermi energy ($\gtrsim$10-20MeV). Neutrino opacities are dominated by coherent scattering off heavy nuclei:

$$\sigma \simeq \sigma_o N^2 \left[\frac{E_\nu}{\text{MeV}}\right]^2, \quad \sigma_o = 10^{-44} \text{ cm}^2 , \tag{2.4.1.1}$$

where we have assumed a Weinberg-angle $\sin^2\Theta_w=0.25$ and N is the neutron number of the nucleus.

Neutrinos are predominantly produced at densities between 10^{11} and 10^{12}g cm^{-3} where the typical heavy nucleus has about 80 to 100 nucleons and 50 neutrons. The mean free path of neutrinos is therefore

$$\bar{\ell}_\nu \simeq \frac{1}{n_A\sigma} \simeq 10^7 \text{ cm} \left[\frac{\tau}{10^{12}\text{g cm}^{-3}}\right]^{-1} \frac{A}{N^2} \left[\frac{E_\nu}{10\text{MeV}}\right]^{-2}. \tag{2.4.1.2}$$

where n_A is the number density of an average nucleus with mass number A. From eq. (2.4.1.2) we see that matter is opaque once the density exceeds $\sim 5\times10^{11}$g cm^{-3}. Since, on the other hand, the diffusion time scale is given by

$$\tau_{diff} \simeq \frac{d^2}{\frac{1}{3}\, c\ell_\nu}\,, \quad d \simeq 10^7 \text{ cm}, \tag{2.4.1.3}$$

we in addition find $\tau_{diff} \simeq 2s \gg \tau_{collapse} \simeq 10^{-3}s$, which means that neutrinos will be trapped and will move with the matter.

Neutrino-electron scattering, finally, will equilibrate the neutrino distribution function, and at densities beyond 10^{12}g cm^{-3} we expect that also weak interactions approach an equilibrium ("β-equilibrium"). As an important consequence the entropy of stellar matter should be constant above 10^{12}g cm^{-3} and the collapse should be adiabatic unless shocks heat the matter.

2.4.2. Neutrino transport equations. During collapse and after rebound at nuclear matter density we will always find regions in the star where neutrinos are either streaming freely or diffusing outwards. So in principle we would have to solve the Boltzmann transport equation which in Lagrange coordinates and in spherical symmetry up to terms of first order in u/c reads (Castor (1972))

$$\frac{1}{c}\frac{\partial f}{\partial t} + \frac{\mu}{r^2}\frac{\partial}{\partial r}(r^2 f) + \frac{1}{r}\frac{\partial}{\partial \mu}\left[(1-\mu^2)f\right] -$$

$$- \frac{f}{c}\frac{\partial \ell n\rho}{\partial t} - \frac{1}{c}\left[\left[\frac{\partial \ell n\rho}{\partial t} + \frac{3u}{c}\right]\mu^2 - \frac{u}{r}\right] +$$

$$+ \frac{1}{c}\left[\left[\frac{\partial \ell n\rho}{\partial t} + \frac{3u}{c}\right]\mu^2 - \frac{u}{r}\right]\frac{1}{\omega^3}\frac{\partial}{\partial \omega}(\omega^4 f) +$$

$$+ \frac{\partial}{\partial \mu}\left\{\left[\frac{1}{c}\left[\frac{\partial \ell n\rho}{\partial t} + \frac{3u}{r}\right]\mu(1-\mu^2)f\right\}\right.$$

$$= \text{Source terms (emission, absorption, scattering)}. \tag{2.4.2.1}$$

Here $f=f(t,r,\mu,\omega)$ is the neutrino occupation probability: $dn = f\,\frac{\omega^2\, d\omega dV d\Omega}{(2\pi\hbar c)^3}$; ω is the neutrino energy, μ the cosine of the angle between the propagation direction of the neutrino and the radius vector $\vec{r}$, and $d\Omega$ the solid angle at μ.

This transport equation, however, has never been used in core-collapse computations but only approximations to it, some of which we will discuss in section 3, where also arguments in favour of simplifications are given.

2.4.3. Entropy generation. As we have mentioned earlier only weak interaction processes are not in equilibrium during the initial phase of the collapse and can therefore change the entropy.

In a given mass element this change can be described by (Bruenn

(1985))

$$k_BT\dot{s} = (\mu_e + \mu_p - \mu_n - \langle E_\nu\rangle_{emit}) \cdot (-\dot{y}_e) -$$
$$- \langle \Delta E_\nu\rangle_{scat}\, r_{scat} + \langle E_\nu\rangle_{abs}\, \dot{y}_e \; . \qquad (2.4.3.1)$$

where $\langle E_\nu\rangle_{emit}$ and $\langle E_\nu\rangle_{abs}$ are the average neutrino energy in emission and absorption processes, resepectively. r_{scat} is the rate of ν_e-e^--scattering and $\langle \Delta E_\nu\rangle_{scat}$ is energy transfer to the electrons by scattering. All other quantities have their usual meaning. Before neutrino-trapping ν-e-scattering and ν-absorptions are small. Therefore eq. (2.4.3.1) reduces to

$$k_BT\dot{s} = (\mu_e + \mu_p - \mu_n - \langle E_\nu\rangle_{emit}) \cdot (-\dot{y}_e) \; . \qquad (2.4.3.2)$$

For e^--captures on free protons we have $\langle E_\nu\rangle_{emit} \simeq \frac{5}{6}\,\mu_e$ (since $\mu_e \gg k_BT$) and consequently

$$k_BT\dot{s} \simeq (\tfrac{1}{6}\,\mu_e + \mu_p - \mu_n) \cdot (-\dot{y}_e) \; , \qquad (2.4.3.3)$$

which in general is negative.

For e^--captures on heavy nuclei part of the decay energy is stored in nuclear excitations and we get

$$\langle E_\nu\rangle_{emit} \simeq \tfrac{3}{5}(\mu_e + \mu_p - \mu_n - \Delta) \; , \qquad (2.4.3.4)$$

where Δ is the excitation energy.

Now we find

$$k_BT\dot{s} = (\tfrac{2}{3}(\mu_e + \mu_p - \mu_n) + \tfrac{3}{5}\,\Delta) \; , \qquad (2.4.3.5)$$

an always positive contribution.

After neutrino-trapping the change of the entropy is approximately

$$k_BT\dot{s} \simeq (\mu_e + \mu_p - \mu_\nu - \mu_n) \cdot (-\dot{y}_e) \; , \qquad (2.4.3.6)$$

which again is always postitive.

From numerical models one obtains a change in y_e of at most 0.1 and the difference of the chemical potentials is smaller than $5k_BT$. The increase of the entropy is therefore at most $0.5k_B$/nucleon and the entropy stays low as was first discussed by Bethe et al. (1979).

3. NUMERICAL METHODS FOR MODELLING STELLAR COLLAPSE

3.1. A simple analytic solution

As was demonstrated in the preceding section stellar collapse is approximately adiabatic and is well described by an EOS with an adiabatic index $\gamma \simeq 4/3$. If in addition it is assumed that the stellar structure corresponds to a polytrope of index n=3, the hydrodynamic equations can be solved analytically (Goldreich and Weber (1980)).

One finds that solutions exist with $u=r\cdot\dot{a}/a$ where u is the matter velocity in an Eulerian system of coordinates, r is the distance from the center of the star, and a(t) is a scale factor given by

$$a(t) = (\gamma p_c/\rho_c)^{\frac{1}{2}}/(\pi\gamma G\rho_c)^{\frac{1}{2}} = \rho_c^{-1/3}\left[\frac{\kappa}{\pi G}\right]^{\frac{1}{2}} . \qquad (3.1.1)$$

Here the index c denotes central values and $\kappa:=p\rho^{\gamma}$=constant.

The collapse is thus homologous and the velocity depends only on the entropy which determines κ. For zero total energy the radial structure is given by the Lane-Emden equation and one finds

$$M = \frac{4\pi}{3}\left[\frac{r_s}{a}\right]^3\left[\frac{\bar{\rho}}{\rho_c}\right]\left[\frac{\kappa}{\pi G}\right]^{3/2} , \qquad (3.1.2)$$

$$r_s/a = 6.879\ , \quad \bar{\rho}/\rho_c = 0.01846\ .$$

For a relativistic degenerate electron gas we have $\kappa\simeq1.4\mathrm{x}10^7 y_e^{4/3}$ (in cgs units) and therefore $M=5.72M_\odot y_e^2 \equiv M_{CH}$. This means that also for realistic equations of state about a Chandrasekhar-mass will collapse homologously up to nuclear matter density. Slight modifications will be due the exact EOS, non-zero total energy and general relativity, but also rotating models will evolve differently.

3.2. Spherical collapse: the one-dimensional problem

3.2.1. Lagrange coordinates, Newtonian gravity. For non-rotating models it is convenient to use the mass as a Lagrange coordinate in which case mass-conservation is simply expressed by

$$\rho(r) = \frac{3}{4\pi}\frac{dm}{dr^3} , \qquad (3.2.1.1)$$

the equation of motion becomes

$$\frac{du}{dt} = -4\pi r^2\frac{dp}{dm} - 4\pi Gm(r)\frac{dr}{dm} , \qquad (3.2.1.2)$$

and the entropy equation reads

$$\frac{ds(m,t)}{dt} = \text{source terms}. \qquad (3.2.1.3)$$

Of course, an EOS, rate-equations and ν-transport equations have to be added.

3.2.2. A simple discretisation scheme. In most astrophysical applications the differential equations (3.2.1.1) to (3.2.1.3) are approximated by a system of finite difference equations. A natural and simple way of doing this is to divide the star into N concentric mass-shells of mass Δm_i, $\Sigma\,\Delta m_i=M$, and density ρ_i. The thickness of a mass-shell is then given by $\Delta r_i^3=\Delta m\cdot 3/4\pi\rho_i$ (eq. (3.2.1.1)). Such schemes are numerically more stable if matter variables like pressure, density and entropy are attached to the center of a zone whereas radii and velocities should be taken at its outer boundary.

The boundary conditions at the center of the star are then $m_o=0$, $r_o=0$, $u_o=0$ and $p_1=p_c$, $\rho_1=\rho_c$, etc., and at the outer boundary $m_N=M$ and $r_N=R$. There is some ambiguity in how to choose outer boundaries for u, p, ρ and s, but these can partially be eliminated by putting the outer boundary sufficiently far away from the center where these quantities do not change much on the time scale of collapse.

The equation of motion, for example, now reads

$$u_i(t_j + \Delta t) - u_i(t_j) =$$

$$= \Delta t \cdot \left[-4\pi r_i^2 \frac{p_i - p_{i+1}}{m_{i-1} - m_i} - 4\pi G m_i \frac{r_{i-1} - r_i}{m_{i-1} - m_i} \right] \Bigg|_{t=?} . \qquad (3.2.2.1)$$

and the other equations are replaced in the same way.

There are several problems related to this approach. Firstly, the difference equations in general are only accurate to first order in $\Delta m/M$. Therefore many mass-zones (100 to 200) are needed. Secondly, it is not clear at which time $(t_j, t_j+\Delta t, \ldots)$ the terms on the right hand side of eq. (3.2.2.1) should be taken. Schemes that choose $t=t_j$, the old time step, are called "explicit", because at the new time step all quantities are determined from old values only. If, on the other hand, $t=t_j+\Delta t$ is used, the new functions have to be determined from a non-linear system of implicit equations and the schemes are therefore called "implicit". There is, of course, always the possibility to mix the two approaches leading to "semi-implicit" schemes. The advantage of an implicit method is that very different time scales can be treated but at each time step a large system of non-linear equations has to be solved. Explicit schemes, on the other hand, are simple and fast but Δt has to be smaller than the shortest time scale in the problem. Together with the requirement of small mass-zones, this means that the time step Δt can become very small ($\lesssim 10^{-6}$s!). One partially can overcome these difficulties by either redistributing the mass-zones once in a while ("re-zoning") or by solving the hydrodynamic equations explicitly but the transport equations implicitly, because most of the time the transport time step is much smaller than the others.

3.2.3. Treatment of shocks: pseudo-viscosity. Discontinuities in a shock front cannot be treated exactly by the discretization scheme described above. One possibility to solve this problem is to smear the shock over 2 to 3 mass zones by a so-called pseudo-viscosity. This should have the properties of a physical kinematic viscosity. Therefore the viscosity tensor should be trace-free, and it should be zero if the matter is not compressed or in homologous contraction. An expression which fulfils these condition is (Tscharnuter and Winkler (1979))

$$\underline{\underline{Q}} := \begin{cases} \ell^2 \rho \ \mathrm{div}(\underline{u}) [(\underline{\underline{\nabla u}}) - \frac{1}{3} \mathrm{div} \ (\underline{u}) \underline{\underline{e}} \\ \qquad\qquad \text{if div } (\underline{u}) < 0 \\ 0 \qquad \text{otherwise,} \end{cases} \tag{3.2.3.1}$$

where $\underline{\underline{e}}$ is the unit tensor.

In spherical symmetry eq. (3.2.3.1) reduces to

$$q \equiv Q_r^r = \begin{cases} 3\ell^2 \rho \frac{\partial (r^2 u)}{\partial r^3} \left[\frac{\partial u}{\partial r} - \frac{\partial (r^2 u)}{\partial r^3} \right] \\ \qquad \text{if } \frac{1}{r^2} \frac{\partial r^2 u}{\partial r} < 0 \\ 0 \qquad \text{otherwise} \end{cases} \tag{3.2.3.2}$$

and

$$Q_\vartheta^\vartheta = Q_\Phi^\Phi = 3\ell^2 \rho \ \frac{\partial r^2 u}{\partial r^3} \left[\frac{u}{r} - \frac{\partial (r^2 u)}{\partial r^3} \right] \tag{3.2.3.3}$$

In the equation of motion $\frac{\partial p}{\partial r}$ is now replaced by

$$\frac{\partial p}{\partial r} + Q^k{}_{r\,;k} = \frac{\partial p}{\partial r} + \frac{3}{r} \ \frac{\partial r^3 q}{\partial r^3}$$

and the entropy equation becomes

$$k_B T \dot{s} = \epsilon_q = - \frac{9}{2} \ \ell^2 \ \frac{\partial (r^2 u)}{\partial r^3} \ \left[\frac{\partial u}{\partial r} - \frac{\partial (r^2 u)}{\partial r^3} \right]^2 . \tag{3.2.3.4}$$

In these expressions ℓ is a constant of the order of $\alpha \cdot \Delta r$, $\alpha \simeq 1-2$.

3.2.4. <u>General relativistic hydrodynamics</u>. The general relativistic version of the hydrodynamic equations in spherical symmetry has been given by Misner and Sharp (1964, 1965) and a finite difference scheme similar to the one described here has been discussed by Van Riper (1979). They can be solved in very much the same way and cause no additional complications. It should be noted that near core-bounce relativistic corrections to the gravitational potential are of the order of a few percent and should not be neglected in numerical models.

3.3. Collapse with rotation: the two-dimensional case

3.3.1. <u>The basic equations</u>. For a rotating collapsing star we may assume rotational and equatorial symmetry. It is then convenient to use spherical polar coordinates: $\underline{r}=(r, \vartheta, \Phi)$, $\underline{v}=(u, v, w)$ and rotational symmetry implies $\partial/\partial\Phi=0$.

The Poisson equation now becomes

$$3 \frac{\partial}{\partial(r^3)} \left[r^2 \frac{\partial \Phi}{\partial r} \right] + \frac{1}{r^2 \sin\vartheta} \frac{\partial}{\partial \vartheta} \left[\sin\vartheta \frac{\partial \Phi}{\partial \vartheta} \right] = 4\pi G\rho, \quad (3.3.1.1)$$

the continuity equation reads

$$\frac{\partial \ln \rho}{\partial t} + \frac{u-u_g}{r} \frac{\partial \ln \rho}{\partial \ln r} + \frac{v}{r} \frac{\partial \ln \rho}{\partial \vartheta} + 3 \frac{\partial(r^2 u)}{\partial(r^3)} +$$

$$+ \frac{1}{r \sin\vartheta} \frac{\partial}{\partial \vartheta}(v \sin\vartheta) = 0 , \qquad (3.3.1.2)$$

the equations of motion are

$$\frac{\partial u}{\partial t} + (u-u_g) \frac{\partial u}{\partial r} + \frac{v}{r} \frac{\partial u}{\partial \vartheta} - \frac{v^2+w^2}{r} + \frac{p}{\rho} \frac{\partial \ln p}{\partial r} +$$

$$\frac{\partial \Phi}{\partial r} + \frac{1}{\rho} u_g = 0 \qquad (3.3.1.3)$$

$$\frac{\partial v}{\partial t} + (u-u_g) \frac{\partial v}{\partial r} + \frac{v}{r} \frac{\partial v}{\partial \vartheta} + \frac{vu}{r} - \frac{w^2}{r \tan\vartheta} +$$

$$+ \frac{p}{\rho r} \frac{\partial \ln p}{\partial \vartheta} + \frac{1}{r} \frac{\partial \Phi}{\partial \vartheta} + \frac{1}{\rho r} v_q = 0 \qquad (3.3.1.4)$$

$$\frac{\partial \Omega}{\partial t} + (u-u_g) \frac{\partial \Omega}{\partial r} + \frac{v}{r} \frac{\partial \Omega}{\partial \vartheta} + \frac{2\Omega}{r} \left[u + \frac{v}{\tan\vartheta} \right] +$$

$$+ \frac{1}{\rho (r \sin\vartheta)^2} w_q = 0 , \qquad (3.3.1.5)$$

and finally the entropy equation is

$$\frac{\partial \ln s}{\partial t} + \frac{u-u_g}{r} \frac{\partial \ln s}{\partial \ln r} + \frac{v}{r} \frac{\partial \ln s}{\partial \vartheta} - \frac{\dot{\epsilon}_q}{k_B T \cdot s} = 0 .$$

Here Ω is the angular frequency, $w=\Omega r \sin\vartheta$, u_q, v_q and w_q the momentum transfer by viscous forces, ϵ_q the (viscous, ...) energy generation, and $\partial/\partial t$ are derivatives with respect to the radially moving coordinate system, which has to be introduced in order to avoid resolution problems during collapse. Note that we now can no longer use Lagrange coordinates. This fact introduces several numerical problems which so far have only partially been solved. For example, mass and angular momentum conservation can be guaranteed globally in certain schemes but not locally.

3.3.2. <u>A "hybrid" scheme</u>. If deviations from spherical symmetry are expected to be small, i.e. if the collapsing stellar core is rotating

slowly, one may solve the hydrodynamic equations by using finite differences in radial direction and an expansion in Legendre polynomials $P_\ell(\cos\vartheta)$ in ϑ-direction. This method which was derived by Rozyczka et al. (1979) has been applied to the collapse problem by Müller et al. (1980) and Müller and Hillebrandt (1981). It has the advantage that the Poisson equation (3.3.1.1) separates and that it can easily be extended to three dimensions.
The density, for example, is expressed by

$$\ell n\rho(r,\vartheta,t) = a_o(r,t) + \sum_{\ell=2}^{\infty} [a_\ell(r,t) P_\ell(\cos\vartheta)] , \qquad (3.3.2.1)$$

and in the same way Φ, u, $\ell n\Omega$ and ℓns are expanded. Because of the symmetry conditions only even values of ℓ appear in these equations. Because v≡0 on the symmetry axis the expansion of v contains only odd values of ℓ and reads

$$v(r,\vartheta,t) = \sum_{\ell=1}^{\infty} [v_\ell(r,t)\ P_\ell{}^1(\cos\vartheta)] , \qquad (3.3.2.2)$$

where

$$P_\ell^1 \triangleq (1 - \cos^2\vartheta)^{1/2} \frac{dP_\ell}{d\cos\vartheta} = \sin\vartheta \frac{dP_\ell}{d\cos\vartheta}. \qquad (3.3.2.3)$$

In applications to stellar collapse it is usually sufficient to truncate the expansions at $\ell \simeq 12$. The time evolution is now given by:

$$\Delta a_\ell(r,t^n) = \Delta t \int \left.\frac{\partial \ell n\rho}{\partial t}\right|_{t=t^n} P_\ell(\cos\vartheta)\ d\,(\cos\vartheta) , \qquad (3.3.2.4)$$

$$a_\ell(r,t^{n+1}) = a_\ell(r,t^n) + \Delta a_\ell(r,t^n) . \qquad (3.3.2.5)$$

and similar for all other quantities, which means that we are using an explicit scheme.

Note that this method, though being simple, is not conservative and that the grid velocity u_g is a free parameter.

3.4. Neutrino transport schemes

3.4.1. Equilibrium diffusion. In order to derive simple transport equations for the neutrinos one may replace the Boltzmann equation by moment equations

$$M^n{}_\nu := {1/2} \int_{-1}^{1} I_\nu\ \mu^n\ d\mu . \qquad (3.4.1.1)$$

where $\mu = \cos\Theta$ and I_ν is the intensity of neutrinos of energy ϵ_ν. Then we obtain

$$M_\nu^o = J_\nu =: \frac{1}{4\pi} G_\nu \text{ (neutrino energy)} . \qquad (3.4.1.2)$$

$$M_\nu^1 = H_\nu =: \frac{1}{4\pi} F_\nu \text{ (neutrino energy flux)} \ . \qquad (3.4.1.3)$$

$$M_\nu^2 = K_\nu \text{ (neutrino pressure)}. \qquad (3.4.1.4)$$

The diffusion approximation follows if we use the closure conditions

$$M_\nu^3 \equiv 0 \ , \quad K_\nu = J_\nu/3 = p_\nu \ . \qquad (3.4.1.5)$$

and we get in Lagrange coordinates

$$\dot{G}_\nu + \frac{1}{r^2} \frac{\partial(r^2 F_\nu)}{\partial r} + \frac{\dot{\rho}}{3\rho} \frac{\partial G_\nu}{\partial \ln \epsilon_\nu} = c\Sigma_a^\nu G_\nu + q_\nu \ . \qquad (3.4.1.6)$$

$$\frac{1}{c} \dot{F}_\nu + \frac{c}{3} \frac{\partial G_\nu}{\partial r} - F_\nu \left[\frac{2v}{rc} + \frac{2}{c}\frac{\dot{\rho}}{\rho}\right] - \frac{v}{cr} \frac{\partial}{\partial \epsilon_\nu}(\epsilon_\nu F_\nu) =$$

$$= - \Sigma_t^\nu F_\nu \ . \qquad (3.4.1.7)$$

(see, e.g. Arnett (1977b)) where Σ_a^ν is the macroscopic absorption cross-section, q_ν is the spectral emissivity, and Σ_{t-1}^ν denotes the macroscopic transport cross-section and is equal to $(\lambda_\nu)^{-1}$.

If, in addition, we assume a thermal distribution of neutrinos we have

$$G_\nu = \text{const.} \ \epsilon_\nu^3 \ (1+e^{(\epsilon_\nu-\mu_\nu)/k_B T})^{-1} \ , \qquad (3.4.1.8)$$

and in the non-relativistic limit

$$F_\nu = - \frac{c}{3\Sigma_t^\nu} \frac{\partial G_\nu}{\partial r} \quad \text{and} \qquad (3.4.1.9)$$

$$\dot{G}_\nu - \frac{1}{r^2}\frac{\partial}{\partial r} \left[r^2 \frac{c}{3\Sigma_t^\nu} \frac{\partial G_\nu}{\partial r}\right] + \frac{\dot{\rho}}{3\rho} \frac{\partial G_\nu}{\partial \ln \epsilon_\nu} = -c\Sigma_a^\nu G_\nu + q_\nu \ .$$
$$(3.4.1.10)$$

By multiplying eq. (3.4.1.10) by ϵ_ν^{-1} and integrating over $d\epsilon_\nu$ we finally get

$$\frac{\partial}{\partial t} (\rho y_\nu) - \frac{1}{r^2} \frac{\partial}{\partial r} \left[r^2 \frac{c}{3\Sigma_t} \frac{\partial}{\partial r} (\rho y_\nu)\right] = \qquad (3.4.1.11)$$

$$= \text{sources and sinks},$$

which is the usual diffusion equation for the neutrino number density and the mean free path is given by

$$(\bar{\Sigma}_t)^{-1} = \int_0^\infty (\Sigma_t^\nu)^{-1} d\epsilon_\nu / \int_0^\infty d\epsilon_\nu \ . \qquad (3.4.1.12)$$

Since in thermal equilibrium $n_\nu \sim T^3 F_2 \left[\frac{\mu_\nu}{k_B T}\right]$ and $g_\nu \sim T^4 F_3 \left[\frac{\mu_\nu}{k_B T}\right]$, the neutrino energy flux can be computed once the particle flux is known.

There are two major problems, however. Firstly at densities below $10^{12}\,\mathrm{gcm^{-3}}$ neutrinos will not be in thermal equilibrium, and secondly the diffusion approximation will break down at the neutrino-sphere defined by $\int_R^{R_\nu} (\bar{\lambda})^{-1} dr = 2/3$, $\bar{\lambda} = (\bar{\Sigma}_t)^{-1}$. The second short-coming is usually circumvented by introducing a flux-limiter which is tailored in such a way that for small neutrino mean free path ($\bar{\lambda} \ll \Delta r$) the ordinary diffusion approximation results and that for $\bar{\lambda} \gg \Delta r$ the free streaming limit is obtained. The first problem, on the other hand, requires non-equilibrium transport models.

3.4.3 A simple non-equilibrium transport scheme. One can retain the simplicity of the diffusion approximation in a non-equilibrium transport model if one assumes that the neutrino distribution function remains a Fermi distribution but allows for neutrino temperatures different from the matter temperature (Arnett, 1982). We then have

$$\rho y_\nu \sim T_\nu^3\, F_2\, (\eta_\nu)\ , \tag{3.4.3.1}$$

$$g_\nu \sim T_\nu^4\, F_3\, (\eta_\nu)\ , \tag{3.4.3.2}$$

where $\eta_\nu = \mu_\nu / k_B T_\nu$, and F_2, F_3 are relativistic Fermi-integrals. Then:

$$\dot{y}_\nu = (\text{emission} + \text{absorption}) - \frac{1}{\rho}\, \underline{\nabla} \cdot \underline{F}_y\ , \tag{3.4.3.3}$$

$$\dot{g}_\nu = (\text{emission} + \text{absorption}) - \underline{\nabla} \cdot \underline{F}_g - \frac{g_\nu}{3} \frac{\dot{V}}{V}\ , \tag{3.4.3.4}$$

where

$$\underline{F}_y = -\frac{c}{3}\ \lambda_\eta \rho y_\nu\ \underline{\nabla}\eta_\nu - \frac{c}{3}\ \lambda_T \rho y_\nu \frac{3}{T_\nu}\ \underline{\nabla} T_\nu\ , \tag{3.4.3.5}$$

$$\underline{F}_g = -\frac{c}{3}\ \ell_\eta g_\nu\ \underline{\nabla}\eta_\nu - \frac{c}{3}\ \ell_T g_\nu \frac{4}{T_\nu}\ \underline{\nabla} T_\nu\ , \tag{3.4.3.6}$$

and

$$\lambda_\eta = \int (\Sigma_t^\nu)^{-1} \frac{\partial G_\nu}{\partial \eta_\nu}\, d\epsilon_\nu \Big/ \int \frac{\partial G_\nu}{\partial \eta_\nu}\, d\epsilon_\nu\ , \tag{3.4.3.7}$$

$$\lambda_T = \int (\Sigma_t^\nu)^{-1} \frac{\partial G_\nu}{\partial T_\nu}\, d\epsilon_\nu \Big/ \int \frac{\partial G_\nu}{\partial T_\nu}\, d\epsilon_\nu\ , \tag{3.4.3.8}$$

and ℓ_η and ℓ_T are defined in a similar way.

This simple scheme has the advantage that all quantities of interest can be expressed in terms of relativistic Fermi-integrals. It has the disadvantage, however, that the neutrinos may not always have a Fermi-distribution and that it becomes numerically unstable if the neutrinos are strongly degenerate.

3.4.3. Multigroup flux-limited diffusion. This more general method has been used by several people (see, e.g., Arnett (1977b), Bowers and Wilson (1982), Bruenn (1985)). The transport equations in diffusion approximation are replaced by equations for a finite number of neutrino energy groups. In this method also the spectral form of G_ν is determined, but numerically it is much more complicated, although certainly more accurate. Once again a flux-limiter has to be introduced causing uncertainties in the neutrino spectrum near the ν-sphere, because the diffusion approximation is still employed.

4. Results of Numerical Simulations

4.1. General Considerations

The outcome of computer simulations of stellar collapse will depend on the pre-collapse stellar model, the adopted microphysics input data and the numerical methods. It is important to note that so far only in one case has a simulation been performed that uses the same stellar model and the same input physics but different numerical methods, and the results do not agree (Hillebrandt (1982), Bruenn (1985))!

Generally speaking, the California-group (Wilson et al. 1985) use their own equation of state which differs from all the others and multi-group diffusion for the neutrinos; the Munich-group (Hillebrandt and Wolff, 1985) use a Hartree-Fock EOS but only quasi-equilibrium neutrino transport; and finally Bruenn (1985) tried different EOS, including the Munich-EOS and multi-group diffusion, but has so far computed only one particular stellar model.

There are, however, several features that are the same in all computations. During the collapse phase the entropy stays low ($s<2k_B/N$), and, therefore, all models collapse to nuclear matter density, if rotation is neglected. In addition, in agreement with the analytic considerations presented earlier, about a Chandrasekhar-mass collapses homologously. Since M_{CH} is proportional to Y_e^2 it will decrease during collapse, and the question is by how much. The answer depends on the entropy of the initial model, on e^--capture rates, the ν-transport scheme and the EOS, and turns out to be different in most computations. There is, on the other hand, general agreement that there exists a sonic point near M_{CH}.

At core-bounce the central density of the star is only slightly higher than nuclear matter density ($\gtrsim 3\times10^{14}\,gcm^{-3}$). The inner core, defined by the condition that its velocity is subsonic, is stopped on a sound-travel time ($\lesssim 1ms$) once in the center of the star nuclei are dissolved into a homogeneous nuclear fluid and the EOS becomes stiff ($\gamma\simeq 2.5-3$). Since the outer material is still falling with supersonic velocity, a shock must form near the sonic point which is at a radius of about 20km. The unshocked inner core cannot expand against the ram-pressure of the infalling outer layers and achieves a hydrostatic equilibrium soon after bounce. From energy-conservation one can estimate the energy that is put into the shockfront (Brown et al., 1981) and finds typically

$$E_{shock} \simeq E_B^{ic} \simeq (4\text{-}8)\times10^{51}\text{erg} \quad , \tag{4.1.1}$$

where E_B^{ic} is the binding energy of the unshocked inner core.

There are several uncertainties, however, among which the stiffness of the EOS just below and above nuclear matter density seems to be the most important at present. But also e^--captures and ν-transport can change these numbers considerably. The shock-propagation phase is characterized by strong damping due to nuclear photodissociations, which cost about 8×10^{18}erg per gram, and neutrino losses including processes such as $e^+e^-\rightarrow\nu\bar{\nu}$. Neglecting neutrino losses for a moment, we see from equation (4.1.1) that at most $0.5M_\odot$ of heavy nuclei can be dissociated by the shock, and that a necessary condition for successful propagation is

$$M_{\text{"Fe"}} - M^{ic} \lesssim 0.5M_\odot \quad . \tag{4.1.2}$$

Since numerical computations predict $M^{ic}\lesssim0.7M_\odot$ (from $Y_e\lesssim0.38$) it follows immediately that only stellar models with iron-cores less than $1.2M_\odot$ can lead to explosions, which relates to He-core masses between 2 and 2.4 $M_\odot$ or main-sequence masses around 8 to 10 $M_\odot$. This is, of course, a necessary condition only, and additional uncertainties enter from the way neutrino transport in the shockfront is computed, from the EOS, and possibly from convective instabilities.

4.2. Low mass stellar cores, prompt explosions

So far only in two simulations has a prompt explosion caused by a hydro-dynamical shock wave been found. Hillebrandt (1982) computed the collapse of a $10M_\odot$ model with a $2.7M_\odot$ He-core after core-H burning (Weaver and Woosley, 1980) and obtained a very weak explosion. This model, however, was computed with a Boltzmann-gas EOS and without neutrino transport. The final lepton concentration in the inner core turned out to be 0.4, which is probably too high. It is therefore not surprising that Bowers and Wilson (1982) and Bruenn (1985) did not find prompt explosions for the same initial model. Wilson (1985b), on the other hand, could obtain an explosion if he modified his EOS in such a way that it was somewhat stiffer below and somewhat softer beyond nuclear matter density. So the conclusion is that this particular model will at best and within the uncertainties of the input physics lead to a marginal explosion, but it is obvious that the final lepton concentration at core-bounce is crucial for success (see Bruenn (1985) and also table I).

More recently Nomoto (1984) has evolved stellar models with even smaller He-core masses to the onset of collapse and one of his models has been used in a core-collapse computation by Hillebrandt et al. (1984). Some properties of this particular model are summarized in table II.

TABLE I. Properties of a $10M_\odot$ model at core-bounce

	y_ℓ^f	$M_{HC}(M_\odot)$
Hillebrandt (1982)	0.40	0.81
Bowers and Wilson (1982)	0.38	0.65
Bruenn (1985)	0.31-0.33	0.59

y_ℓ^f : final lepton concentration at core bounce.
M_{HC} : mass of the homologous inner core.

Note that the Bowers-Wilson EOS must be very "soft" because of the low value of M_{HC}.

TABLE II. Properties of an approximately $9M_\odot$ star

	initially	at core-bounce
$M_{\text{"Fe"}}$	$0.7M_\odot$	$1.1M_\odot$
S_c/k_BN	0.9	1.1
y_e^c	0.41	0.35

Due to the rather low lepton concentration at core-bounce, which was caused by electron-captures during the early stages of contraction just behind the oxygen diffagration front, the mass of the homologous core of this model was $0.68M_\odot$ only and the incident shock had an energy of slightly less than 7×10^{51}erg.

The results of this computation, which is displayed also in fig. 4 and 5, can be summarized as follows. The explosion energy was 2×10^{51}erg, the mass of the remaining neutron star was about $1.25M_\odot$ and the chemical composition of the ejecta was found to resemble very much what is observed in the Crab nebula. So, if these results were to be confirmed we would be led to the following conclusions:

1. Stars with O-Ne-Mg cores do collapse if the matter is degenerate.
2. These are stars with main sequence masses around $(9\pm1)M_\odot$.
3. They explode and leave neutron stars behind and the neutron star masses will be around $(1.3\pm0.1)M_\odot$.
4. This will also be true for O-Ne-Mg white dwarfs, if their mass exceeds the Chandrasekhar limit.

The last conclusion may help to explain peculiar sub-luminous type I supernovae (see, e.g., Panagia et al. (1985)).

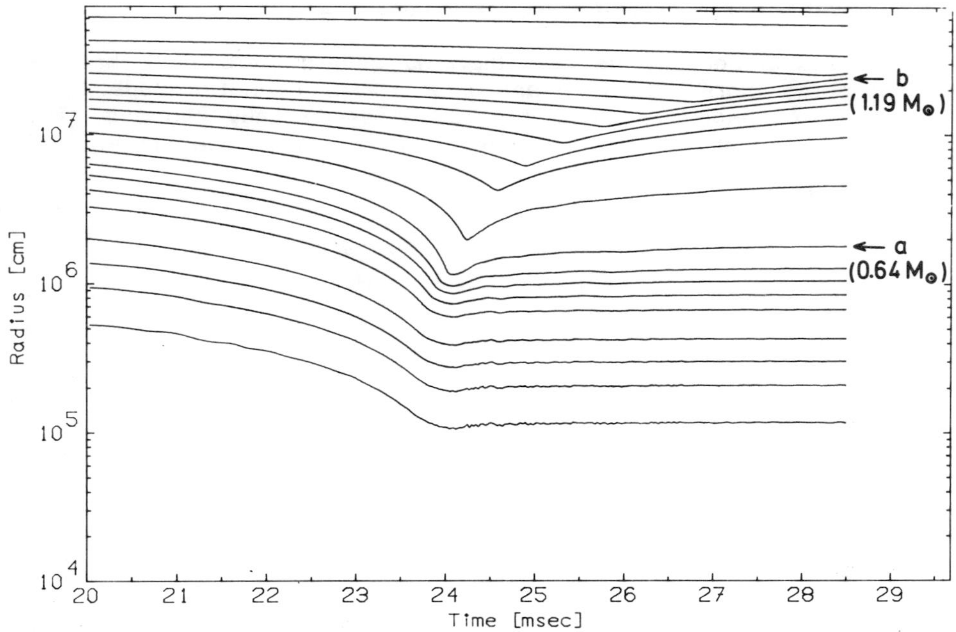

Figure 4 and figure 5: Radius versus time for selected mass zones and snap-shots of density versus mass of a $9M_\odot$ supernova model (from Hillebrandt et al. (1984)).

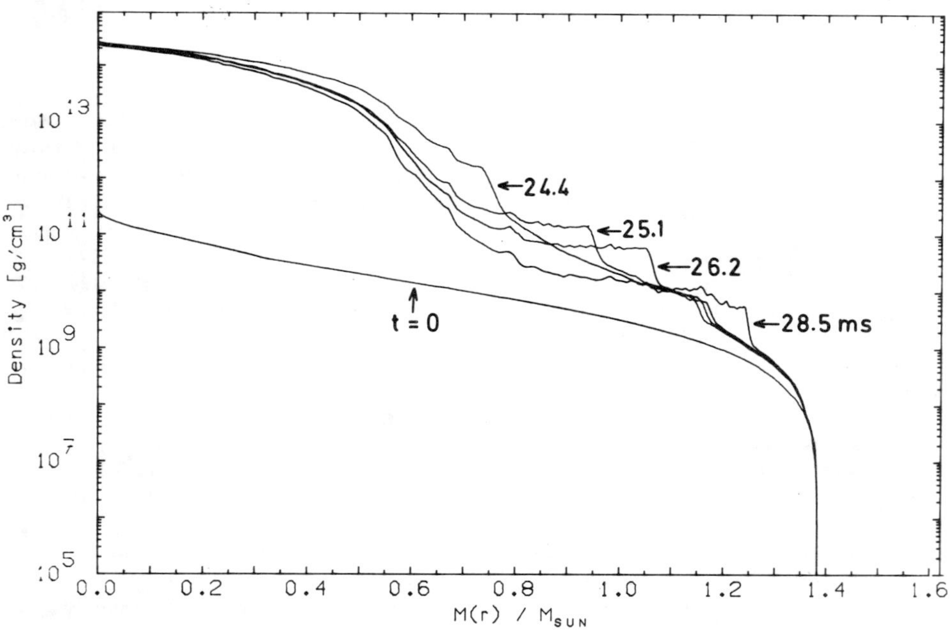

4.3. Massive stellar cores, delayed explosions

Among all researchers in the field there is general agreement that stars with iron cores more massive than about $1.4M_{\odot}$ cannot explode by the core-bounce mechanism, unless the EOS is dramatically changed. Table III (from Wilson et al., 1985) relates initial stellar masses to iron core masses for different nuclear physics input data.

TABLE III.

$M_{\mu s}$ ($M_{\odot}$)	$^{12}C(\alpha,\gamma)^{16}O$ from Fowler et al. (1975)	$^{12}C(\alpha,\gamma)^{16}O$ *2.5	$^{12}C(\alpha,\gamma)^{16}O$ *3
10	1.4*	---	---
11	0.2	---	---
12	---	---	1.31
15	1.56*	---	1.33
20	1.36	---	1.90
25	1.37*	1.63	2.2
50	1.79	---	---
100	1.85	---	---

*: without revised e^--capture rates.

A typical example of a computation of such a more massive stellar core is shown in figures 6 and 7.

The question is then: Why <u>should</u> some more massive stars explode? One argument has already been presented before. We have found that stars up to $10M_{\odot}$ and maybe even up to $15M_{\odot}$ contribute little to the enrichment of the Galaxy with heavy elements and stars around $(25\pm5)M_{\odot}$ are believed to do this job. Moreover, the very narrow mass-range discussed earlier may not contain sufficiently many stars to account for the observed type II supernova rate. Finally, supernova remnants such as Cas A or Puppis A show large overabundances of oxygen indicating progenitor masses of at least around $20M_{\odot}$.

All these arguments have led to an intensive search for alternative explosion mechanisms. Wilson (1985a) (see also Bethe and Wilson (1985) and Wilson et al. (1985)) has found that a few hundred milliseconds after core-bounce energy transport by neutrinos may revive a stalling shock. At that time the edge of the unshocked inner core ($\simeq 0.7M_{\odot}$) will be at a radius of about 20km and the density and temperature there will be about $10^{12}gcm^{-3}$ and 2 to 3 MeV, respectively. The shock has changed to an almost standing accretion shock at $1.4M_{\odot}$ and a radius of about 500km, the temperature and density just behind the front being $10^{7}gcm^{-3}$ and 0.8MeV, respectively. The shocked matter is irradiated by a neutrino flux originating from the neutrino-sphere at 30

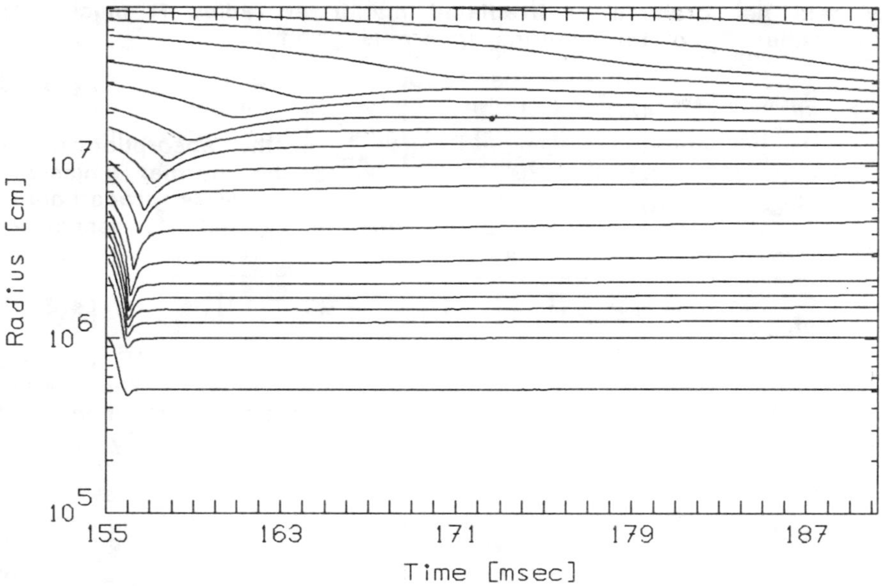

Figure 6: Radius versus time for selected mass zones of a $20M_{\odot}$ stellar model. Due to the large iron core no prompt explosion was observed.

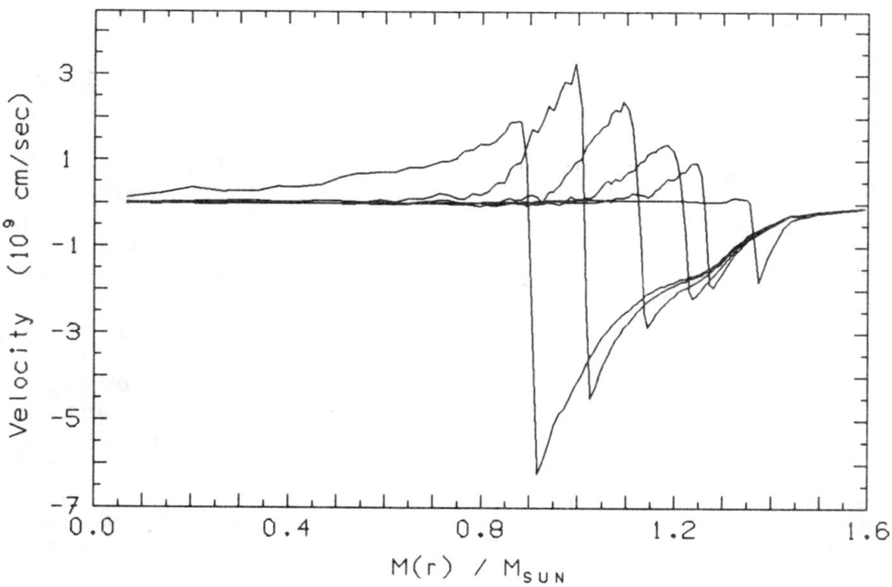

Figure 7: Snap-shots of velocities of a $20M_{\odot}$ model. Due to photo-dissociation of heavy nuclei the shock stagnates at about $1.3M_{\odot}$.

to 70 km. Typical neutrino energies will range from 3 to 5 MeV. The energy gain per gramme of irradiated matter at radius R much larger than the radius R_ν of the neutrino-sphere is given by

$$\dot{E}_{gain} = \kappa_{abs} \frac{L_\nu}{4\pi R^2}, \tag{4.3.1}$$

where L_ν is the neutrino luminosity and κ_{abs} is the absorption opacity. Each gramme of matter will also loose energy by thermal emission of neutrinos, $E_{loss} \sim T^4 \kappa_{emiss}$. Since the neutrino luminosity is proportional to $T_\nu^4 R_\nu^2$, and κ_{abs} and κ_{emiss} are proportional to T_ν^2 and T^2, respectively, the net energy gain can be written as

$$\frac{dQ}{dt} = \frac{7}{16} \, ac \, \kappa_{abs} T_\nu^4 \left[\frac{1}{4} (R_\nu/R)^2 - (T/T_\nu)^6\right], \tag{4.3.2}$$

where temperatures are in MeV, $a=1 \cdot 37\times10^{26}$ erg/c cm^3 MeV^4 and $\kappa_{abs} \simeq 6\times10^{20} <(\epsilon_\nu/m_e c^2)^2> cm^2 g^{-1}$ (Lattimer and Burrows, 1984). The maximum matter temperature that can be obtained from this heating mechanism can be determined from the equilibrium condition dQ/dt=0. At 10 times the radius of the ν-sphere T_{max} is about $0.37T_\nu$ and thus about 1MeV. The heating time scale can also be estimated and one finds a few hundred ms. It is, however, clear that the quantity of key importance is the neutrino temperature at the neutrino-sphere which, due to uncertainties in the neutrino transport schemes, is not very accurately determined. Small changes in T_ν, on the other hand, may change the net-heating considerably.

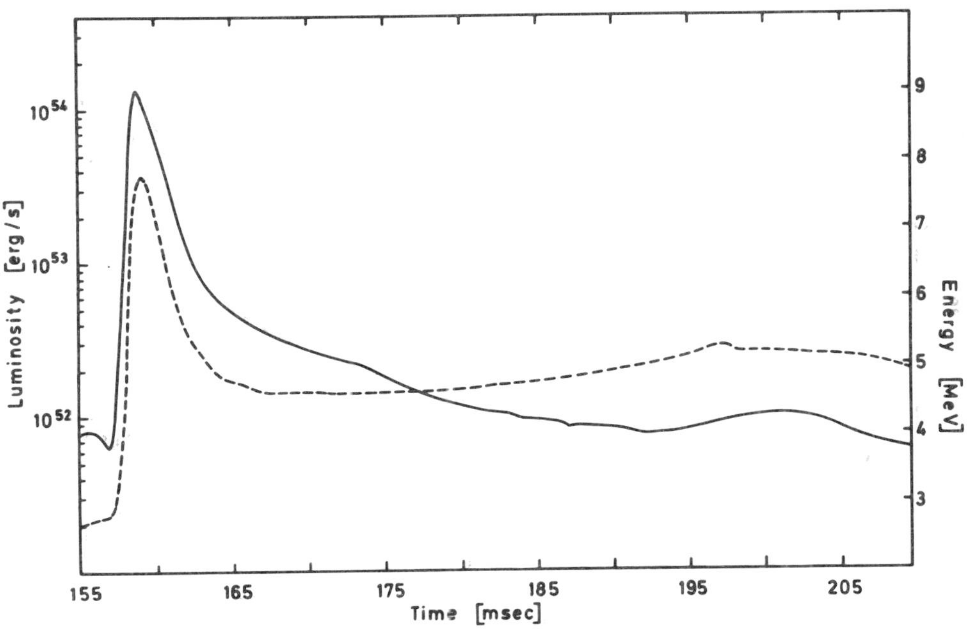

Figure 8: Neutrino luminosity (solid curve) and average neutrino energy (dashed curve) versus time for a $20M_\odot$ model.

Neutrino-luminosities can be estimated in a straight-forward way. Inside the ν-sphere but outside the unshocked core, electron captures on free protons and positron captures on free neutrons will produce neutrinos and anti-neutrinos and their energy in erg per cm^3 and second is approximately given by (Takahashi et al., 1978)

$$Q_{\nu_e} \simeq Q_{\bar{\nu}_e} \simeq 2.4\times10^{28} x_n \, T_{MeV}^6 \, \rho_{10} \, , \tag{4.3.3}$$

where we have assumed an equal number of electrons and positrons and neutrons and protons, respectively. ρ_{10} is the density in $10^{10}gcm^{-3}$. Integrating over the shocked matter inside the ν-sphere we obtain for the neutrino (and anti-neutrino) luminosity

$$L_{\nu_e} \simeq L_{\bar{\nu}_e} \simeq (2-4)\times10^{52} erg\ s^{-1} \, , \tag{4.3.4}$$

in good agreement with numerical results (see fig. 8). Other processes such as neutrinos from e^-e^+-pairs and those diffusing out of the unshocked core contribute little to the total luminosity on time-scales less than one second.

Wilson (1985a) and Wilson et al. (1985) have recently computed the hydrodynamic evolution of several stellar models for approximately 1s after core-bounce and found explosions in all cases considered. Mostly, however, the explosion energy was low, $\lesssim 4\times10^{50}$erg, so typical type II supernova light curves cannot be fitted by these models. Only for rather massive stars ($M \gtrsim 25M_\odot$) explosive oxygen-burning added enough energy to the explosion to explain typical outbursts ($\simeq 10^{51}$erg). Hillebrandt (1985) has also computed a model of $20M_\odot$ of Weaver et al. (1985) which had a core-structure similar to the $15M_\odot$ model considered by Wilson et al. (1985) and did not find the desired explosion (see fig. 9). The main difference between the computations turned out to be the position and the temperature of the ν-sphere, which can be caused by either differences in the equation of state or in the neutrino transport models, and the results seem to be quite sensitive to both. A summary of the Wilson et al. (1985) computations is given in table IV.

TABLE IV. Properties of delayed explosion models.

$M/M_\odot$	$M_{\text{"Fe"}}/M_\odot$	E_{exp}	$M_B/M_\odot$	$M_G/M_\odot$
12	1.31	3.8	1.35	1.26
15	1.33	3.5	1.42	1.31
25	2.2	10	2.44	1.96

E_{exp}: explosion energy in 10^{50}erg.
M_B, M_G: baryon mass and gravitating mass of the compact remnant, respectively.

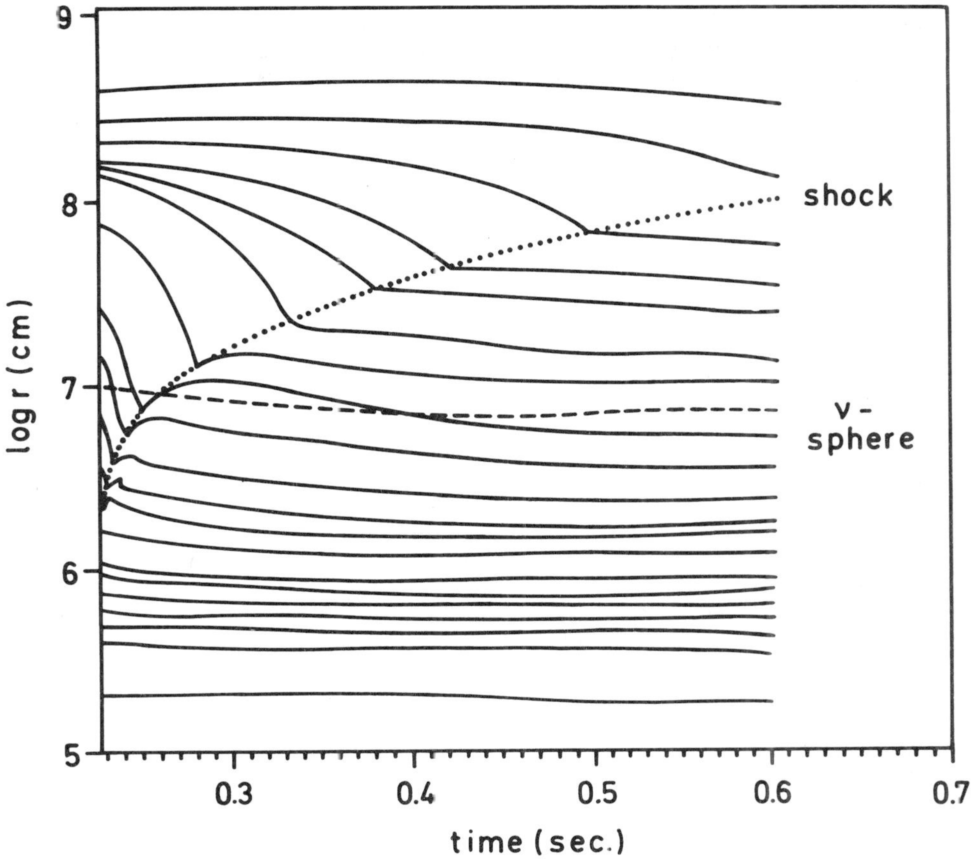

Figure 9: Radius versus time for selected mass zones of a $20M_\odot$ model (from Hillebrandt (1985)). The position of the shock front and of the neutrino sphere are also shown. The time axis is shifted to coincide with the $15M_\odot$ model of Wilson et al. (1985) at core-bounce. Both models have a very similar core structure. No indication of a delayed explosion is observed probably because a stiffer equation of state was used in my computations and the radius of the neutrino sphere stayed at around 70 km.

As an interesting consequence of their computations one may speculate that supernova explosions leaving behind a black hole rather than a neutron star are also possible because for very massive iron cores most of the energy comes from nuclear burning and the remnant mass may exceed that of a stable neutron star.

4.4. Collapse with rotation

Rotation may alter the general picture outlined so far considerably. For a rigidly rotating stellar core the critical adiabatic index is given by

$$\gamma_c = \frac{4}{3} - \frac{2}{9}\frac{J\Omega^2}{|E_{pot}|} < \frac{4}{3}\ , \tag{4.4.1}$$

where J is the moment of inertia, so the collapse may not proceed to nuclear matter density. Non-spherical velocity fields will lead to non-spherical shocks which may introduce large-scale circulations and thus a more efficient neutrino transport. Finally, rapidly rotating cores will be unstable to triaxial deformations or even to fragmentation (Hoyle, 1946), but a numerical treatment of those latter effects requires 3-dimensional rather than 2-dimensional hydro-codes. From equation (4.4.1) it is obvious, that even slow rotation will be extremely important if the equation of state predicts an adiabatic index very close to 4/3. Collapsing rotating stellar cores have been modelled by Le Blanc and Wilson (1970), Müller et al. (1980), Müller and Hillebrandt (1981), Tohline et al. (1980) and recently by Symbalisty (1984). So far the conclusion was that no prompt explosions triggered by rotation alone have been found. "Slowly" rotating models, i.e. $|E_{rot}/E_G| \lesssim 1-3\%$, collapsed directly to nuclear matter density, but this may change if stiffer equations of state are used. In simulations of more rapidly rotating cores, i.e. $|E_{rot}/E_G| \gtrsim 5\%$, on the other hand, the collapse ended before nuclear matter density was reached, and high-velocity jets of $10^{-4}M_\odot$ may escape from the core (Symbalisty, 1984). Only a combination of rapid rotation and an extremely strong magnetic field ($B \simeq 2\times10^{15}$ Gauss!) may lead to an explosion powered by a strong polar jet (Symbalisty, 1984)

5. SUMMARY

We have shown that stars more massive than about $8M_\odot$ on the main sequence will collapse and form neutron stars of masses of at least about $1.3M_\odot$. The upper mass limit for neutron star formation is not yet known but may be somewhere between $25M_\odot$ and $50M_\odot$, if the delayed explosion mechanism turns out to be correct. Some of those stars, more likely at the low mass end, give rise to type II supernova explosions with Crab-like composition of the ejecta. Some very massive stars ($M_{MS} \gtrsim 25M_\odot$?) may form a black hole and a type II (?) supernova. Most type II supernovae will not enrich the galactic matter with heavy elements and, due to a revised $^{12}C(\alpha,\gamma)^{16}O$ reaction rate (Kettner et al., 1982), massive stars can no longer explain the abundance of carbon.

All those conclusions are, of course, preliminary and problems related to rotation, neutrino transport and the equation of state need further investigation.

References

Arnett, W.D. 1977a, Astrophys. J. Suppl. 35, 145
——————— 1977b, Astrophys. J. 218, 815
Arnett, W.D. 1982, private communication
Barranco, M., and Buchler, J.R. 1980, Phys. Rev. C22, 1729
Bethe, H.A., and Wilson, J.R. 1985, Astrophys. J. 295, 14
Bethe, H.A., Brown, G.E., Applegate, J., and Lattimer, J.M. 1979, Nucl. Phys. A324, 487
Bonche, P., and Vautherin, D. 1981, Nucl. Phys. A372, 496
Bowers, R.B., and Wilson, J.R. 1982, Astrophys. J. 263, 366
Brown, G.E., Bethe, H.A., and Baym, G. 1981, Nucl. Phys. A375, 481
Bruenn, S.W. 1985, Astrophys. J. Suppl. 58, 771
Castor, J.I. 1972, Astrophys. J. 178, 779
El Eid, M.F., and Hillebrandt, W. 1980, Astron. Astrophys. Suppl. 42, 215
Fowler, W.A., Caughlan, G.R., and Zimmermann, B.A. 1975, Ann. Rev. Astron. Astrophys. 13, 69
Fuller, G.M., Fowler, W.A., and Newman, M.J. 1982, Astrophys. J. 252, 715
Goldreich, P., and Weber, S.V. 1980, Astrophys. J. 238, 991
Hillebrandt, W. 1982, Astron. Astrophys. 110, L3
——————— 1985, in Proc. 35d Workshop on Nuclear Astrophysics; W. Hillebrandt, E. Müller, F.-K. Thielemann, eds., MPA 199, p.59
Hillebrandt, W., and Müller, E. 1984, in "Neutrino Physics and Astrophysics", K. Kleinknecht and E.A. Paschos, eds., World Sci. Publ., p.229
Hillebrandt, W., and Wolff, R.G. 1985, in "Nucleosynthesis: Challenges and New Developments", W.D. Arnett and J.W. Truran, eds. Univ. Chicago Press, p.131
Hillebrandt, W., Nomoto, K., and Wolff, R.G. 1984, Astron. Astrophys. 133, 175
Hoyle, F. 1946, Mon. Not. R. Astr. Soc. 106, 343
Lamb, D.Q., Lattimer, J.M., Pethick, C.J., and Ravenhall, D.G. 1978, Phys. Rev. Lett. 41, 1623
——————— 1983, Nucl. Phys. A411, 449
Lattimer, J.M., and Burrows, A. 1984, in "Problems of Collapse and Numerical Relativity", D. Bancel and W. Signore, eds. NATO-ASI C134, Reidel, p. 147
Le Blanc, J.M., and Wilson, J.R. 1970, Astrophys. J. 161, 541
Mathews, G.J., Bloom, S.D., and Hausman, R.F. 1983, Phys. Rev. C28, 1367
Mazurek, T.J., Lattimer, J.M., and Brown, G.E. 1979, Astrophys. J. 229, 713
Misner, C.W., and Sharp, D.H. 1964, Phys. Rev. B136, 571
——————— 1965, Phys. Rev. Lett. 15, 279
Müller, E., and Hillebrandt, W. 1981, Astron. Astrophys. 103, 358

Müller, E., Rozyczka, M., and Hillebrandt, W. 1980, Astron. Astrophys. 81, 288

Nomoto, K. 1984, Astrophys. J. 277, 791

Ogasawara, R., and Sato, K. 1982, Prog. Theor. Phys. 68, 222

Panagia, N. Sramek, R.A., and Weiler, K.W. 1985, Astrophys. J. Lett., in press

Rozyczka, M., Tscharnuter, W.M., Winkler, K.-H., and Yorke, H.W. 1980, Astron. Astrophys. 83, 118

Symbalisty, E. 1984, Astrophys. J. 285, 729

Takahashi, K., El Eid, M.F., and Hillebrandt, W. 1978, Astron. Astrophys. 103, 358

Tohline, J.E., Schombert, J.M., and Boss, A.P. 1980, Space Sci. Rev. 27, 555

Tscharnuter, W.M., and Winkler, K.-H. 1979, Comp. Phys. Comm. 18, 171

Tubbs, D.L. 1979, Astrophys. J. 231, 846

Tubbs, D.L., and Schramm, D.N. 1975, Astrophys. J. 201, 467

Van Riper, K.A. 1979, Astrophys. J. 232, 558

Weaver, T.A., and Woosley, S.E. 1980, B.A.A.S., 11, 724

Weaver, T.A., Woosley, S.E., and Fuller, G.M. 1985, in "Numerical Astrophysics", J. Centrella, J. Le Blanc and R. Bowers, eds., Jones and Bartlett, p.374

Wilson, J.R. 1985a, in "Numerical Astrophysics", l.c., p.422

Wilson, J.R. 1985b, private communication

Wilson, J.R., Mayle, R., Woosley, S.E., and Weaver, T.A. 1985, Ann. New York Acad. Sci., in press

Wolff, R.G. 1985, Ph. D. thesis, Munich

Woosley, S.E., Weaver, T.A., and Taam, R.E. 1980, in "Typ I Supernovae", J.C. Wheeler, ed., Univ. Texas, p.96

Yueh, W.R., and Buchler, J.R. 1976, Astrophys. Space Sci. 41, 221

WHITE DWARFS VS. NEUTRON STARS: PROPERTIES, CONSTRAINTS, AND ANALOGIES

Virginia Trimble
Astronomy Program
University of Maryland
College Park MD 20742 USA

and

Department of Physics
University of California
Irvine CA 92717 USA

ABSTRACT. A review is given of theoretical expectations for, and observational constraints on, fundamental properties of white dwarfs and neutron stars, including masses, rotation periods, moments of inertia, magnetic fields, space velocities, formation processes, birthrates, and compositions, particularly for objects like pulsars and cataclysmic variables with high energy activity in their vicinity. Perhaps surprisingly NS models are much more tightly constrained by the data than are WD ones, even though uncertainties in the equation of state of dense nuclear matter leave much more flexibility in the former. In addition, useful analogies can occasionally be drawn between the two classes, the rotation periods of single and binary white dwarfs, for instance, adding weight to the hypothesis that rapidly-rotating pulsars are typically the products of interacting binary systems.

1. MASSES

1.1. Neutron Stars

Stars supported by degenerate neutron pressure can exist only in a range of masses limited at the lower end by the danger of the neutrons decaying back to ordinary matter and at the upper end by the danger of further collapse to black hole configurations. The lower limit is a few tenths of a solar mass and is unlikely to be approached in any currently-contemplated scenario for neutron star formation. The upper limit, however, is both rather close to measured NS masses and quite uncertain, owing to the wide range of possible equations of state for dense nuclear matter.

Table I presents many (not all) of the possible EoS's, with the maximum non-rotating neutron star (gravitational) mass permitted by each, the ratio of mass in amu to gravitational mass (the difference is the gravitational binding energy), and the maximum moment of inertia for a non-rotating star. The equations of state are listed, roughly, from stiff (lots of hard-core repulsion at small particle separations) to soft. The maximum mass and moment of inertia for rotating stars exceed those shown by about 15% for soft EoS's to 30% for hard ones. The largest possible moment of inertia always occurs for a mass less than the maximum.

F. Pacini (ed.), High Energy Phenomena Around Collapsed Stars, 105–120.

TABLE I

Masses and Moments of Inertia of Neutron Stars Constructed Using a Representative Sample of Equations of State for Dense Nuclear Matter

Equation of State	M^{max}_{grav} (M_o)	M_{amu}/M_{gr} (M_{gr} = 1.4)	$I^{max}_{non-rot}$ (10^{45} g-cm^2)
			STIFF
L	2.7	----	4.85
N	2.58	1.14	4.03
I	2.45	----	3.30
O	2.39	1.14	3.13
M	1.956	1.12	3.06
C(140, 3)	1.89	----	----
C	1.853	1.17	1.65
MBJ I-H	1.83	----	1.57
MBJ V-N	1.75	----	1.38
Moszkowski	1.74	----	1.26
E	1.73	1.215	1.254
MBJ III-H	1.73	----	1.32
F	1.7	----	0.95
C(90, 3)	1.69	----	----
AM 5	1.657	----	1.14
A	1.655	1.21	1.10
AM 9	1.653	----	1.15
D	1.65	1.19	1.306
MBJ V-H	1.59	----	1.17
CHH	1.50	----	0.93
B	1.414	1.34	0.66
C(140, 2.5)	1.36	----	----
G	1.36	impossible	0.605
non-rel Fermi	0.9	"	0.6
C(140, 2)	0.835	"	----
H (TOV)	0.71	"	0.55
			SOFT

Equations of state labeled by a single letter are those considered by Arnett and Bowers (1978). More complicated designations are the EoS's discussed by Mølvik and Østgaard (1985), apart from the four with names like C(140,3), which are the ones of many considered by Baron et al. (1985) that permitted a prompt shock following core collapse in a massive star to eject enough mass and energy to make a Type II supernova.

There are no data that directly constrain masses of single neutron stars. The subject badly needs a brand new idea. Among non-interacting binaries, we have two measured values, 1.42 $\pm$ .03 M_o for the pulsar 1913+16 and 1.40 $\pm$ .03 M_o for its companion, presumed also to be a neutron star (Weisberg and Taylor 1984). These numbers come from a complete analysis of changes in pulse period as both the pulsar and its orbit lose energy and are somewhat dependent upon General Relativity being the right weak-field theory of gravity and upon a further (perhaps) approximation called the quadrupole formula (Trimble 1982). But the system is gasless and, in general, the determination very clean.

Things are much less tidy among interacting (X-ray) binaries, but there are also many more of them. At least eight systems have mass determinations, most achieved by combining velocity data for the companion star with Doppler shifts in X-ray pulsation periods of the NS's (Kelley and Rappaport 1981; de Loore et al. 1982; Hutchings 1984; Kelley et al. 1984). The eight objects and their masses, roughly from small to large, are SMC X-1 (0.5-1.5), Cen X-3 (0.4-1.7), Her X-1 (0.5-2), 4U1538 (0.7-3.5), A0535 (1.6 $\pm$ .2), LMC X-4 (1.6+1,-0.5), 4U0900 (1.6-2.6), Vela X-1 (1.74 +.15, -.25). All numbers are in solar masses and include 2σ error bars, apart from LMC and Vela, which are 1σ. There is then a clean gap in masses until we reach the $\gtrsim$6 and $\geq$ 10 M_o for Cyg X-1 and LMC X-3, generally advertized as black holes.

The data are thus consistent with all measured neutron stars having masses in a narrow range around 1.5 M_o, but also consistent with a much wider spread. In either case, at least a few values are clearly large enough to rule out EoS's softer than B in Table I. This includes two of the four that yielded prompt shock ejection (Baron et al. 1985). From the point of view of neutron star formation, it is worth remembering that to form a 1.4 M_o NS requires the collapse of from 1.6 to 1.9 M_o of material, depending on the equation of state.

Masses restrict the EoS much more severely if, as theoretical considerations suggest, neutron stars form near 1.4 M_o in binaries and then accrete up to 1.8-2.0 M_o over their X-ray lifetimes (D.Q. Lamb 1987). One could then rule out all the EoS's softer than about MJB V-N.

1.2. White Dwarfs

Stars supported by degenerate electron pressure are possible for masses between zero and an upper limit that depends largely on composition (Chandrasekhar 1935). For non-rotating, non-magnetic stars, there is a unique equation of state, and so unique values of the limiting (Chandrasekhar) mass: 1.43 M_o for helium cores, 1.39 for C and O, 1.3 for Ne and Mg, and 1.12 for Fe, each $\pm$ 0.02 or so (Hamada and Salpeter 1961; Ibáñez C. 1983). Rotation, especially differential rotation, can increase these to near 3 M_o (Ostriker and Bodenheimer 1968). Adding a thin H atmosphere will affect the radius (Sect. 3.2) but not the mass.

White dwarfs in visual binaries provide the cleanest mass determinations. Values for eight stars in six systems(references in Trimble 1986) have been published: Sir B (1.05), Sanduleak-Pesch (0.8 + 0.4), G 107-70AB (0.67 + 0.25), Procyon B (0.63), Stein 2051B (0.5) and 40 Eri B (0.43). The average is 0.59 M_o. Three spectroscopic, eclipsing systems (UU Sge, V471 Tau, and AA Dor) include white dwarfs that have probably been through common-envelope stripping, but are not now strongly interacting. The masses are 1.1, 0.79, and 0.25 M_o (Patterson 1984; Ritter 1984; Sion et al. 1984). None of these relatively well determined masses is at any risk of exceeding theoretically-permitted values.

Mass determinations in interacting binaries (cataclysmic variables) are both more numerous and more error-prone. Ritter (1984) and Patterson (1984) give partially-overlapping tables, from which the average for several dozen systems is 0.75 M_o, with ¼ as large as 1.0 M_o. The minimum is $\leq$ 0.4 M_o (T Leo) and the maximum $\sim$ 1.4 M_o (RU Peg). Thus, CV masses seem to have a rather wider range and a higher average than those in wider systems. Livio and Soker (1984) have concluded that the higher average is not just an observational selection effect.

Masses for single WD's can be derived by combining a model mass-radius relation with any of three spectroscopic constraints — a surface gravity (M/R^2), a radius from the Barnes-Evans relation (plus parallax), or a gravitational redshift (M/R). There is general agreement that WD masses are rather tightly peaked around a median value, the one-sigma dispersion being only 0.1-0.15 M_o (Greenstein 1984), but less agreement on what the median value is.

The Barnes-Evans value is the smallest, 0.50-0.55 M_o (Moffett et al. 1978). Measurements of gravitational redshift (Greenstein and Trimble 1967; Trimble and Greenstein 1972; Wegner 1974) give the largest values, 0.75-0.80 M_o for DA's and about 0.6 M_o for other types. Some of the spectroscopic determinations roughly concur, with Shipman (1979) and Guseinov et al. (1983, 1984) finding 0.75 M_o for DA's and 0.5-0.6 for the helium atmosphere types. Given these relatively large values, white dwarfs make a significant contribution to the local mass density, reducing the amount of "missing mass" in the galactic plane; and a typical 1-2 M_o population I star seems to be losing about half its mass during pre-WD evolution.

Other atmospheric analyses (Wickramasinghe et al. 1983; Weideman and Koester 1983, 1984; Oke et al. 1984) yield 0.55-0.60 M_o as the mean for all spectral types, including helium-rich atmospheres, the pulsationally variable ZZ Ceti stars, and members of Magellanic Cloud clusters (Weidemann 1984). All methods, naturally, give the right answer when applied to Sir B and 40 Eri B. Koester (1984) has considered differences between the Kiel and Soviet scales and remains on the Kiel side.

Clearly there is a problem. It cannot be resolved by combining any two of the spectroscopic indicators and eliminating the interior models. For instance, 8 Hyads have a mean log g = 7.96 in the Kiel system and a mean gravitational redshift of 56 km/s. Together, these imply an average mass of 2.34 M_o and an average radius of 0.027 R_o. The implication is that the difficulty is somewhere in our understanding of the atmospheres. In any case, all the average values are perfectly possible in terms of stellar evolution, and all are well under the Chandrasekhar limit for any composition on up to pure U^{238}.

2. ANGULAR VELOCITIES AND ROTATION PERIODS

2.1. White Dwarfs

The traditional envelope back readily yields a minimum rotation period, $P = (4\pi^2R^3/GM)^{\frac{1}{2}} = 12.5$sec for a 0.65 M_o, 10^{-2} R_o white dwarf. More precise calculations, including differential rotation, can reduce this to about 6.5 s at the equator and 0.5 s near the poles (Ostriker and Bodenheimer 1968; Ostriker 1968). No observations press uncomfortably upon these limits.

Single WD's are uniformly slow rotators. The largest spectroscopic v sin i's are 40-60 km/s, and many stars yield only limits, some as low as 5-10 km/s (Pilachowski and Milkey 1984; Milkey and Pilachowski 1985). The corresponding rotation periods are 20 min up to several hours. Among the Magnetic white dwarfs, absence of changes in field strength and activity implies either perfectly symmetric field geometries or, more likely, rotation periods that range upward from 100 min to many years (Angel et al. 1980). Finally, the pulsating WD's with hydrogen, helium, and oxygen-rich atmospheres can all be modeled by non-radial modes of stars in which rotation makes no perceptible contribution to the energetics (Cox 1984).

Among close binary white dwarfs, three classes can be distinguished. First, the AM Her stars, with strong magnetic fields, show evidence of only a single periodicity of order hours, indicating rotation is locked to orbit motion and thus slow (Liebert and Stockman 1985). Second, the DQ Her stars or intermediate polars (though their fields may or may not be of intermediate strength!) show a range of rotation periods from 33 to 18690 s (AE Aqr to TV Col). The average is near 800 s (Lamb and Patterson 1983) shorter than for single WD's but not approaching the model limits. The WD's are spun up or down to equilibrium by material accreted from their companions. Finally, many dwarf novae and some nova-like variables show 10-50 s coherent oscillations. These are not stable enough to be themselves rotation periods, but one of several models invokes non-radial modes of white dwarfs with dynamically important rotation (Papaloizou and Pringle 1978).

Apparently, then, WD rotation anywhere near break-up speed occurs, if at all, only in interacting binary systems. If neutron stars behave analogously, we would conclude that pulsars with periods less than about 10 msec must have passed through a binary phase.

2.2. Neutron Stars

The envelope back here yields $P = (4\pi^2R^3/GM)^{\frac{1}{2}}$ ms as the minimum possible rotation period for a 10^6 cm, 1.5 M_o neutron star. More careful calculations, in this case, increase the limit slightly (Friedman et al. 1984) instabilities of the m = 3 and 4 modes setting in at T/W = 0.12 rather than unity. Differential rotation is not thought to occur, but the limiting rotation period does depend somewhat on equation of state (Sect. 1) since softer EoS's yield smaller radii at a given mass, and so smaller kinetic energy at a given angular velocity. For the most massive star possible with each EoS, the range is 0.4-0.9 ms, and 0.6-1.8 ms for a 1.4 M_o neutron star.

Among single pulsars, we see periods of 1.55 to 4000 ms, most being in the range 1-2 s. Both statistical (Chevalier 1985) and energetic (Trimble and Rees) arguments have been advanced that few pulsars start out with periods close to break-up. I am not entirely convinced. Though non-interacting binary NS's (binary pulsars) are rare, their period range is similar to single NS's, 6.1 to 1066 ms (Stokes et al. 1985).

The interacting (X-ray) binary NS's display the widest range of rotation periods (Bradt and McClintock 1983, and a few later determinations). At last count, the direct measurements (from X-ray pulsations) included four systems with low mass companions and periods between 1.24 (HZ Her) and 120 or 227 s (GX 1+4); eight with Be star companions and periods from 0.069 (0535 in the LMC) to 835 s (X Per); 10 neutron stars with massive OB companions, whose periods range from 0.7 (SMC X-1) to 696 s (WRA 977); and three without optical ID's at 0.067, 9.3, and 191 s.

Finally, a handful of systems, including GX 5+1, Sco X-1, and Cyg X-2 (van der Klis et al. 1985) display quasi-periodic, intensity-dependent oscillations at 20-40 Hz and an excess of low frequency noise, reminiscent of the dwarf nova oscillations mentioned in 2.1. The analogy then immediately suggested applying any or all of the numerous models proposed for the CV phenomenon. Rapid rotation by the compact star is a feature of many of these, including a widely discussed model (Alpar et al. 1985; van den Klis et al. 1985; F.K. Lamb 1987) in which the quasi-period we see is a beat frequency between rotation and blobs moving in an accretion disc.

Data on neutron stars thus indicate rapid rotation for a few objects, both single and binary. There are, however, strong theoretical reasons to suppose that all or most of the rapid rotators have, in fact, been spun up by binary accretion (van den Heuvel 1984, 1987). In any case, the shortest pulsar period of 1.55 ms cannot be accomodated by NS's obeying the stiffest of the equations of state in Table I and serves to rule them out.

3. MOMENTS OF INERTIA AND RADII

3.1. Neutron Stars

Each of the equations of state tabulated in Sect. 1 implies a unique radius, R, and moment of inertia, I, at each mass. Smallest stable radius coincides with maximum central density and M; maximum I occurs at somewhat lower central density. The range of minimum R's is 7-14 km; the range of R's at 1.4 M_o is 7-18 km. I^{max} varies from 0.6 to 4.8 X 10^{45} g-cm^2 and I at 1.4 M_o from 0.6 to 3 X 10^{45} g-cm^2. Largest values all occur for the stiffest EoS's.

Liang (1986) has presented statistical data on X-ray bursters that favor radii near 10 km and so rule out the softest EoS's. In addition, we have some information on two individual neutron stars. First, Fujimoto and Taam (1985) have modeled the continuum and line fluxes from the burster MXB 1636-536, attributing the 4.1 keV line to electron-positron annihilation, modified by both gravitational and transverse Doppler redshifts. Two resulting equations in M and R then yield R = 10.3 km (range 9.1-11.3) and M = 1.45 M_o (range 1.28-1.65). The large ends of the two ranges go together, since gravitational redshift scales as M/R. Taken at face value, these R(M) combinations rule out equations of state "B" and softer (also AM5) and "M" and harder. In combination with the most stringent mass limit from Sect. 1.1, this would restrict the real EoS to a very narrow range around "C".

Second, the need to replenish short-lived electrons in the Crab Nebula sets a lower limit to I of its pulsar, the available energy being at most $dE/dt = I\omega\dot{\omega} = 4.6 \times 10^{-7}$ erg/s. The minimum required is 1-2 X 10^{38} erg/s (Woltjer 1971; Trimble and Rees 1970), and I_{min}, therefore, $0.2\text{-}0.4 \times 10^{45}$ g-cm^2, possible for all equations of state. A recent measurement of the rate at which the nebular radio continuum flux is declining (Aller and Reynolds 1985) enables us to do slightly better. The decline rate is only about 60% of what one would expect if adiabatic losses in magnetic field and relativistic electron energy were not replenished,

$$\frac{dL/dt}{L} = \frac{7}{4}\,\frac{dU_e/dt}{U_e} - \frac{3}{4}\,\frac{dV/dt}{V}$$

assuming equipartition between field and particle energies. Since the second term is known to be 0.00268 per year from the measured nebular expansion and (dL/dt)/L = -0.00167/yr, evidently there is partial replenishment of the lower-energy electrons that produce radio synchrotron (as well as complete replacement of the high-energy ones that radiate X-ray and optical photons), amounting to roughly another 2×10^{38} erg/s. Then our best estimate for I becomes $0.6\text{-}0.8 \times 10^{45}$ g-cm^2, assuming 100% efficiency for field and electron energy production.

This rules out only the same very soft EoS's that fail to make NS's as large as 1.4 M_o. Allowing for less than 100% efficiency (e.g. energy "wasted" in accelerating protons) would make I_{min} of NP0532 significantly more restrictive. The adiabatic energy losses do not, of course, disappear from the universe. The energy is converted to bulk kinetic motion, accelerating the present expansion, sweeping up new gas at the edges, or both. To sweep up previously-stationary gas as dense as the Murdin and Clark (1981) halo ($N_H \sim 7$ cm^{-3}) would deplete the nebular energy content much faster than any pulsar could replace it, and we should expect to see the Crab disappear completely about JD 2459301 (28 May 2021).

3.2. White Dwarfs

White dwarf models, as noted in Sect. 1.2, are (almost) unique functions of interior mean molecular weight. The "almost" includes the effects of rotation, magnetic fields, non-degenerate hydrogen atmospheres, and small remaining uncertainties in the physics. Unpublished models by Winget, Lamb, and Van Horn, for instance, apparently have somewhat larger radii than the corresponding Hamada-Salpeter (1961) models. The first three effects also increase radii, though to change measured surface gravities by as much as 0.1 in the log requires rotation speeds or field strengths nearly 20% of the maximum stable values, more than an order of magnitude larger than anything ever observed (Sect. 2.1 and 4.1). A non-degenerate hydrogen envelope of 0.01 M_o would produce a similar 0.1 decrease in log g, but is again more than we expect to see, even if the limits on H envelope thickness that come from models of the ZZ Ceti pulsational variables can be relaxed to 10^{-4} M_o (Cox 1985). Ignoring these minor effects, one expects an average $0^o.75$ M_o WD to have R = 0.0085 R_o (or 0.0126 R_o for 0.6 M_o, and a C-O interior in either case) and I = $3X10^{50}$ g-cm^2.

Data constraining I do not seem to exist. It is, obviously a parameter in models where magnetic stars are spun up or down by accretion, but other factors in these models are more uncertain than the moments of inertia. A few independent radii come from stars with well-determined parallaxes and colors. Where there are discrepancies, they are in the direction of the measured radii slightly exceeding the Hamada-Salpeter model ones. A similar conclusion can be reached by comparing radii determined by the Kiel group using more than one spectroscopic technique (Koester et al. 1979; Weidemann and Koester 1984). Finally, the difference between Barnes-Evans masses for average WD's and those arrived at by other methods (Sect. 1.2) could be resolved if real stars are a bit larger than the Hamada-Salpeter ones. This does, however, increase the discrepancy between surface gravity and gravitational redshift results, since z/g is proportional to R, and the redshifts already look too large for the surface gravities. As in Sect. 1.2, we can conclude that there is a problem, but in our atmospheric models, not the interior ones.

4. MAGNETIC FIELDS

4.1. White Dwarfs

Considerations of pressure balance and energy equipartition would permit interior white dwarf fields up to 10^{12} G. Such large fields would slightly raise the Chandrasekhar limiting mass and increase the radius at a given mass. We have no strong observational evidence for or against them. Among single WD's, 95% or more have surface fields $\leq$ 1-$2X10^5$ G. The limit comes from the absence of quadratic Zeeman shifting of H Balmer lines (Trimble 1971) and the real value could be much smaller. In contrast, a few percent of white dwarfs have magnetic fields, revealed by circular polarization and gross line shifts, in the range $10^{6.5-8.5}$G (Angel 1978; Wunner et al. 1985). These few span nearly the same range of surface temperatures (mean 9000 K) as the general run of DA's (mean 14000 K) and are thus neither younger nor older than average. This is consistent with predicted time scale = 10^9 yr for the decay of WD fields (Chanmugam and Gabriel 1972).

Among the close binaries (mostly CV's of one sort or another) we again see a bimodal distribution, a few stars with B near 10^7 G (magnetic moments of 10^{33-34} G-cm^2) among the AM Her systems, and a much larger number (by 20:1, according to Patterson 1984) with magnetic moments $\leq 10^{30}$ G-cm^2 (King et al. 1985) among the other classes of CV's. As with the single stars, these must reflect the fields with which the stars were born.

The strongest white dwarf fields cannot have been generated by dynamo processes (Fontaine et al. 1973) and must represent frozen-in flux from earlier evolutionary phases. Since BR^2 is nearly the same for WD's and NS's, analogy would suggest flux-freezing for the latter as well, though various kinds of in-situ flux creation are possible for them

(Blandford et al. 1973).

5. SPACE VELOCITIES

White dwarfs represent an old disc population, having a velocity dispersion of 40-50 km/s (Sion et al. 1978), possibly somewhat correlated with progenitor mass, and hence age (Guseinov et al. 1984). Their scale height is correspondingly large (a few hundred parsecs), indicating that they have come to terms with the galactic gravitational potential and were either born with their largish velocities or at least acquired them more than a few epicyclic periods ($\gtrsim 10^8$ yr) ago, via encounters with giant molecular clouds or other massive objects (e.g. Villumsen and Binney 1985)

Neutron stars (meaning pulsars and, probably, X-ray binaries), on the other hand, have a velocity distribution (nearly 200 km/s) that is disproportionately large for their scale height of 350 pc (Lyne 1982). One concludes that they haven't been doing it very long, the velocities having been acquired within the last few million years. Recoil from asymmetric explosions, disruption of close binary systems, or reaction to asymmetric radiation processes have all been suggested. In any event, we expect to see most of the objects going away from the galactic plane. This seems to be the case (Lyne 1982).

6. FORMATION PROCESSES

White dwarfs and neutron stars both come from main sequence precursors considerably more massive than themselves, otherwise the former would be a good deal more massive and the latter a good deal more numerous than they are. The three main questions are (a) how much material does a typical star of main sequence mass M shed, (b) what mechanisms produce the losses, and (c) how do the first two depend on composition, rotation, magnetic field, binaricity, and other properties of the stars?

All the answers are more or less incomplete. But the standard synthesis (Weidemann and Koester 1983; Iben 1985; Morris and Zuckerman 1985) has put the WD/NS progenitor mass cut at 8-10 M_o and included enough mass shedding to turn 1-2 M_o stars into 0.5-0.6 M_o WD's and OB main sequence stars into Wolf-Rayets before core collapse occurs and dissipates everything else outside the Fe core. Implications are that we should see large amounts of neutral material outside most planetary nebulae (which themselves comprise only a few tenths of a solar mass) and evidence for extensive circumstellar shells around Type II supernovae. We do, at least occasionally (Amnuel et al. 1984 on PNe; Niemala et al 1985 on SN 1983k).

Bertelli et al. (1985) have recently suggested a modified scheme in which a new set of stellar evolution calculations both lowers the WD/NS progenitor mass cut to about 6 M_o and predicts lower mass-loss rates that will let a 1 M_o star form at least a 0.6 M_o WD and will leave typical OB stars with enough hydrogen-rich envelope to make a Type II supernova light curve. An assortment of other problems with Cepheid masses, absence of very bright stars in the LMC, and statistics of carbon stars also disappears.

Observations, both direct (Prylyser et al. 1985 on OB star winds in the Magellanic Clouds) and indirect (Shklovskii 1984 on the absence of Type II SNe in irregular galaxies), tell us that there are real differences among stars with the same mass but different compositions (etc.). Some mechanism problems remain. Among the small stars, most loss must occur gradually in winds, since PNe are relatively small. One major question is whether stellar radiation carries enough momentum to drive these without invoking more complex hydrodynamical processes (Holzer and MacGregor 1985).

Among the larger stars, at least those that make SNII's of normal brightness, several solar masses must be ejected in an explosion immediately following core collapse. Recent discussions have focussed on two drivers — a prompt shock, created when the core reaches nuclear density, becomes stiff, and bounces (Bethe et al. 1980) and the same shock, but stalled, delayed, and re-energized by neutrino heating until it has enough power to eject the overlying layers (Wilson 1985). In order for a prompt shock to get out (Baron et al. 1985), the collapsing core mass must be $\lesssim 1.2\ M_o$. Building a 1.4 M_o final NS may then be difficult. Also the equation of state must resemble one of the four in Table I labeled C(140;3) etc. Several of these are soft enough to be ruled out by measured neutron star masses (Sect. 1.1) or radii (Sect. 2.1). The delayed shock, if it works, would be more attractive, permitting a stiffer EoS and allowing material trapped inside the shock when it stalls to fall back down and build a heftier NS. In some cases a black hole might even result (Woosley 1984), to which there is no objection, as long as it doesn't happen too often!

There exist also several more complex scenarios for formation of neutron stars, involving mass transfer in close binaries, electron capture in ONeMg white dwarfs, merging of WD's, and so forth (Saio and Nomoto 1985 and references therein).

7. BIRTHRATES IN THE MILKY WAY

7.1. Neutron Stars

An already excessively voluminous literature has established that the pulsar birthrate, SNR formation rate, and SNII rate are the same to within the uncertainties of the determinations at one per 30-60 yr. The birthrate of neutron stars in binaries is probably comparable (Blair and Candy 1984), as one would expect if roughly half of all massive stars are in close binaries that are disrupted only by the second supernova (Radhakrishnan 1987). Clearly these systems cannot all spend $\geq 10^5$ yr emitting 10^{38} erg/s in X-rays, or the galaxy would be much brighter than it is.

At times, it has seemed a bit difficult to provide enough progenitors from the local Initial Mass Function (van Buren 1985) if only stars more massive than 8-10 M_o could ignite carbon peacefully and so eventually core-collapse to neutron stars. Incorporation of semi-convection and the new $C^{12}(\alpha,\gamma)O^{16}$ rate (Fowler 1985) into evolutionary calculations for intermediate mass stars (Bertelli et al. 1985) has lowered the non-degenerate carbon ignition limit to about 6 M_o, alleviating this problem. Blaauw (1985) also recently concluded that some such change was necessary from statistics of nearby OB associations and pulsars.

7.2. White Dwarfs

For single stars, one expects that the white dwarf birthrate should slightly exceed the planetary nebula (progenitor) formation rate, because only WD's more massive than 0.55 M_o can ionize PNe (Schoenberner 1983). The data (discussed more thoroughly by Trimble 1986) indicate a slight difference between the rates, but in the wrong direction. Green's (1984) latest white dwarf birthrate, derived from the bright DA stars in the Palomar-Green survey, is 5-8 X 10^{-13} WD/yr-pc^3. Estimates of the PN formation rate, on the other hand, range from 1 to 6 X 10^{-12} PN/yr-pc^3 (Drilling and Schoenberner 1985; Amnuel et al. 1984; Mallik 1985), the lower rates obtaining only if one assumes the largest of several possible PN distance scales. The discrepancy is marginally worrying, particularly given that increasing the mean WD mass would slow down cooling and make it all worse.

Among the interacting binary white dwarfs (CV's) there may also be a glitch in the direction of progenitors exceeding products. Patterson (1984) estimates that the CV birthrate (from numbers of eclipsing binary PN nuclei etc.) is about 2% of the total WD birthrate, while the space density is only 10^{-3} of the WD value. He concludes that CV's die in 10^{8-9} yr, perhaps by merging. Alternatively, they may spend much of their time too faint to get counted in most samples. The recovery of Nova 1670 (CK Vul) at M_R = +10.4 (Shara et al. 1985) favors the latter possibility.

8. COMPOSITIONS

8.1. White Dwarfs

Evolutionary calculations for single stars predict lower limits to the core masses capable of igniting particular fuels and so, in turn, a correlation between WD mass and composition (Iben 1985). One expects He white dwarfs up to about 0.4 M_o, CO from 0.4 to 1.1 M_o, and ONeMg for the most massive objects. White dwarf spectroscopy is not terribly informative. As a result of gravitational settling (Schatzman 1945) most show atmospheres of essentially pure hydrogen or helium. Small amounts of heavier elements seen in a few stars probably reflect accretion of interstellar gas, apart from carbon, which can be dredged up from the interior (Wegner et al. 1983), implying that stars with strong carbon lines belong to the middle mass range. There is also evidence for significant interior oxygen. The very hot, pulsating PG 1159 stars are most readily modeled with oxygen contributing about half the mass and an oxygen ionization zone just under the photosphere (Starrfield et al. 1984). The rather high O/C ratio is consistent with the new, larger cross-section for $C^{12}(\alpha,\gamma)O^{16}$ appropriate to stellar helium-burning conditions (Fowler 1985).

The best probes of WD interior composition occur in interacting binary systems. All three of the expected sorts apparently exist. The gas being shed by the low-mass white dwarf in AM CVn is unusually helium-rich (Nather 1985), while two different groups of novae show ejecta with excess CNO and excess heavier elements (Truran 1985), implying material scooped up from the underlying star and partially burned. The techniques of mass determination in cataclysmic variables are not currently equal to the task of deciding whether the latter group is systematically the more massive.

8.2 Neutron Stars

Neutron stars are generally reputed to consist of about 99% neutrons, though we cannot expect to have direct spectroscopic evidence for this. In one very speculative model, however, Cyg X-3 constitutes an exception, being made of stable, strange quark matter (Witten 1984; Barnhill et al. 1985; Shaw et al. 1985)!

The motivation is a bit complex. First, gamma rays above 10^{15} eV, pulsed at the 4.8 hr X-ray period, are coming from a region within 2° of the Cyg X-3 position. There are good reasons to believe both the flux and the identification. It has been seen by several groups (Samorski and Stamm 1983; Lloyd-Evans et al. 1985; and others reporting at the 18th International Cosmic Ray Conference, La Jolla CA, August 1985); the periodicity and a better position also turn up in 10^{12} eV gamma ray data from many air show arrays (Cawley et al. 1985 and references therein); other neutron stars of known periods and positions are beginning to turn up at slightly lower flux levels (Baltrusaitis et al. 1985 on Her X-1; Protheroe et al. 1984 on Vela X-1; Protheroe and Clay 1985 on LMC X-4; Chadwick et al. 1985 on 4U 0115+63 and Psr 1953+29 = 2CG 065); and there exist reasonably conventional models for producing even such high energy gamma rays (Vestrand and Eichler 1982; Hillas 1984; Kazanas and Ellison 1985).

Second, there have been several reports from groups operating underground detectors (designed to look for proton decay) of pulsed muons coming from the general vicinity of Cyg X-3 (Marshak et al. 1985a,b,; Battistoni et al. 1985). The air showers seen at Kiel (Samorski and Stamm 1983) are apparently also muon-rich. Other proton decay experiment groups reporting at the 18th Cosmic Ray Conference indicated either very weakly positive (Frejus Tunnel, Kolar Gold Field) or inconclusive (IMB, Baxan, HPW) results (Sreekantan 1985). There are several reasons to be more cautious in accepting the muon flux and its identification. The several reports are not quite consistent about the peak phase, flux, and position, or how flux changes with altitude; one has apparently seen a second, longer periodicity (Marshak et al. 1985b) that appears in the X-ray data only occasionally (Priedhorsky and Terrell 1986); and, finally, we are forced into some rather contrived models to account for them.

Muons don't live long enough to be coming to us as primary particles from the source. They must be secondaries (atmospheric, not rock ones, judging from the Kiel mu-rich showers; though the Akeno group reported that they see the 4.8 hr period only in mu-poor shows), produced when some long-lived neutral particle from the source reaches us. Neutrons can't get here; photons should not create a flux of muon secondaries nearly as numerous as themselves; and neutrinos (Stecker et al. 1985) or photinos (Stenger 1985), though not impossible, must be remarkably copious relative to the gammas to do the job.

It is against this background that theorists have tried to come up with some new sort of long-lived, neutral particle having a baryon-like cross-section for muon production, and arrived at the di-Λ or H particle, a tightly bound $\Lambda\Lambda$ pair that can, perhaps, only decay to a pair of protons by a double-weak process, giving it a lifetime at rest of months to years (enough to get here from Cyg X-3 at sufficiently high relativistic speed). If the di-lambda is this stable, then strange quark matter can be completely stable in a strong gravitational field. From this follows the possibility of a truly strange composition for Cyg X-3, some of whose surface material then gets stripped off in charged chunks and accelerated, eventually leaving the di-Λ's as part of the neutral beam that reaches us. If any reader has hung on this far and managed to believe it all simultaneously, there is a very nice bridge crossing the East River between Brooklyn and Manhattan that I would like to sell you.

But the situation is not completely hopeless. The Kolar Gold Field air shower array is currently increasing its collecting area and muon-sensitivity, and, over the next couple of years, the Frejus Tunnel and Irvine-Michigan-Brookhaven proton-decay experiments should accumulate enough additional events to be able to confirm or rule out a pulsed flux of muons from the direction of Cyg X-3.

ACKNOWLEDGEMENTS

I am grateful to Dr. Franco Pacini for the invitation to participate in the Advanced Study Institute and to NATO for partial support of my travel expenses.

REFERENCES

Aller, H.D. & S.P. Reynolds 1985. Astrophys. J. 293, L78
Alpar, M.A. & J. Shaham 1985. Nature 316, 239
Amnuel, P.R. et al. 1984. Astrophys. Space Sci. 107, 19
Angel, J.R.P. 1978. Ann. Rev. Astron. Astrophys. 16, 487
Angel, J.R.P., J.D. Landstreet & E. Borra 1980. Ap.J. Sup. 45, 457
Arnett, W.D. & R.L. Bowers 1977. Astrophys. J. Suppl. 33, 415
Baltrusaitis, R.M., G.L. Cassiday, R. Cooper et al. 1985. Ap.J. 293, L69
Barnhill, M.V. et al. 1985. Nature 317, 409
Baron, E., J. Cooperstein & S. Kahana 1985. Phys. Rev. Lett. 55, 126
Battistoni, G. et al. 1985. Phys. Lett. B155, 465
Bertelli, G., A.G. Bressan & C. Chiosi 1985. Astron. Astrophys. 150, 33
Bethe, H.A., J. Applegate & G.E. Brown 1980. Astrophys. J. 241, 343
Blaauw, A. 1985. in W. Boland & H. van Woerden (eds.) <u>Formation and Evolution of Massive Stars and Stellar Associations</u> (Reidel)
Blair, D.G. & B.N. Candy 1984. Unpublished preprint
Blandford, R.D., J.H. Applegate & L. Hernquist 1983. MNRAS 204, 1025
Bradt, H.V.D. & J.F. McClintock 1983. Ann. Rev. Astron. Ap. 21, 13
van Buren, D. 1985. Astrophys. J. 294, 567
Cawley, M.F. et al. 1985. Astrophys. J. 296, 185
Chadwick, P.M. et al. 1985. Nature 317, 236
Chandrasekhar, S. 1935. Mon. Not. Royal Astron. Soc. 95, 207
Chanmugam, G. & M. Gabriel 1972. Astron. Astrophys. 16, 149

Chevalier, R.A. 1985. in V. Radhakrishnan & G. Srinivasan (eds.) Supernovae and Their Remnants (Bangalore: Indian Academy of Sciences)
Cox, J.P. 1984. Nature 311, 509
..... 1985. Private communication
Drilling, J.S. & D. Schoenberner 1985. Astron. Astrophys. 146, L23
Fontaine, G., J.H. Thomas & H. Van Horn 1973. Astrophys. J. 184, 911
Fowler, W.A. 1985. in W.D. Arnett & J.W. Truran (eds) Nucleosynthesis (Univ. Chicago Press) p. 8
Friedman. J.L., J.R. Ipser & L. Parker 1984. Nature 312, 255
Fujimoto, M.Y. & R.E. Taam 1985. Preprint. 'The Mass and Radius of a Neutron Star in X-Ray Burster MXB 1636-536'
Green, R. 1984. Talk at conference on Low Luminosity Stars, Caltech, October 1984
Greenstein, J.L. 1984. Astrophys. J. 276, 602
Greenstein, J.L. & V. Trimble 1967. Astrophys. J. 149, 283
Guseinov, O.H., H.I. Novruzova & Yu.S. Rustamov 1983. Astrophys. Space Sci. 96, 1
..... 1984. Astrophys. Space Sci. 97, 309
Hamada, T. & E.E. Salpeter 1961. Astrophys. J. 134, 683
van den Heuvel, E.P.J. 1984. J. Astrophys. Astron. 5, 209
..... 1987. This volume.
Hillas, A.M. 1984. Nature 312, 50
Holzer, T.E. & K.B. MacGregor 1985. in Morris & Zuckerman (1985) p. 229
Hutchings, J.B. 1984. Publ. Astron. Soc. Pac. 96, 312
Ibáñez C., J.M. 1983. Rev. Mex. Astron. Astrofis. 8, 7
Iben, I. 1985. Q.J. Royal Astron. Soc. 26, 1
Kazanas, D. & D.C. Ellison 1985. Preprint 'On the Origin of Ultra High Energy Gamma-Rays from Cygnus X-3 and Related Objects' Submitted to Nature
Kelley, R.L. & S. Rappaport 1981. in W. Sieber & R. Wielebinski (eds.) IAU Symp. No. 95, Pulsars (Dordrecht: D. Reidel) p. 353
Kelley, R.L. et al. 1984. Astrophys. J. 264, 568
King, A.R., J. Frank & H. Ritter 1985. Mon. Not. Royal Astron. Soc. 213, 181
van der Klis, M. et al. 1985. Nature 316, 225
Koester, D. 1984. Astrophys. Space Sci. 100, 471
Koester, D., H. Schultz & V. Weidemann 1979. Astron. Astrophys. 76, 262
Lamb, D.Q. 1987. This volume.
Lamb, D.Q. & J. Patterson 1983. in G. Shaviv & M. Livio (eds). Cataclysmic Variables and Related Objects (Dordrecht: Reidel) p. 229
Lamb, F.K. 1987. This volume.
Liang, E.P.T. 1987. This volume.
Liebert, J. & H.S. Stockman 1985. in D.Q. Lamb & J. Patterson (edd.) Cataclysmic Variables and Low Mass X-ray Binaries (Reidel) p. 151
Livio, M. & N. Soker 1984. Mon. Not. Royal Astron. Soc. 208, 613
Lloyd-Evans, J. et al. 1983. Nature 305, 784
de Loore, C. 1982. ESO Messenger No. 27, p 14
Lyne, A.G. 1982. in M.J. Rees & R.J. Stoneham (eds.) Supernovae: A Survey of Current Research (Dordrecht: D. Reidel) p. 405
Mallik, D.C.V. 1985. Astrophys. Lett. 24, 173
Marshak, M.L. 1985a. Phys. Rev. Lett. 54, 2079

Marshak, M.L. et al. 1985b. Preprint DOE/ER/40105-522 'Time Distributions for Underground Muons from Cygnus X-3'. Submitted to Phys. Rev. Lett.
Milkey, R.W. & C.A. Pilachowski 1985. Publ. Astron. Soc. Pac. 97, 634
Moffett, J., T.G. Barnes & D.S. Evans 1978. Astron. J. 83, 820
Mølvik, T. & E. Østgaard 1985. Nucl. Phys. A237, 239
Morris, M. & B. Zuckerman (eds.) 1985. Mass Loss from Red Giants (Reidel)
Murdin, P. & D.H. Clark 1981. Nature 294, 543
Nather, R.E. 1985. in J.E. Pringle & R.A. Wade (eds.) Interacting Binary Stars (Dordrecht: D. Reidel)
Niemela, V.S., M.T. Ruíz & M.M. Phillips 1985. Astrophys. J. 289, 52
Oke, J.B., V. Weidemann & D. Koester 1984. Astrophys. J. 281,267
Ostriker, J.P. 1968. Nature 217, 1227
Ostriker, J.P. & P. Bodenheimer 1968. Astrophys. J. 151, 1089
Papaloizou, J. & J.E. Pringle 1978. MNRAS 182, 423
Patterson, J. 1984. Astrophys. J. Suppl. 54, 443
Pilachowski, C.A. & R.W. Milkey 1984. Publ. Astron. Soc. Pac. 96, 821
Priedhorsky, W. & J. Terrell 1986. Astrophys. J. (1 February)
Protheroe, R.J. & R.W. Clay 1985. Nature 315, 205
Protheroe, R.J., R.W. Clay & P.R. Gerhardy 1984. Astrophys. J. 280, L47
Pylyser, E., C. Doom & C. de Loore 1985. Astron. Astrophys. 148, 379
Radhakrishnan, V. 1987. This volume.
Ritter, H. 1984. Astron. Astrophys. Suppl. 57, 585
Saio, H. & K. Nomoto 1985. Astron. Astrophys. 150, L21
Samorski, M. &. W. Stamm 1983. Astrophys. J. 268, L17
Schatzman, E. 1945. Ann. d'Astrophys. 8, 143
Schoenberner, D. 1983. Astrophys. J. 272, 708
Shara, M. et al. 1985. Astrophys. J. 294, 271
Shaw, G.L., G. Benford & D.L. Silverman 1985. Preprint UCI 85-14, 'Proposed Test of a Strange Matter Pulsar by Observations of the Neutrino Flux from Cygnus X-3'
Sion, E.M., O.L. Lupis & K.N. Young 1978. Publ. Astron. Soc. Pac. 90, 154
Sion, E.M., F. Wesemael & E.F. Guinan 1984. Astrophys. J. 279, 758
Sreekantan, B.V. 1985. Colloquium at U. Maryland, October
Starrfield, S. et al. 1984. Astrophys. J. 281, 800
Stecker, F.W., A.K. Harding & J.J. Bernard 1985. Nature 316, 418
Stenger, J.V. 1985. Nature 317, 411
Stokes, G.H., J.H. Taylor, & R.J. Dewey 1985. Astrophys. J. 294, L21
Taam, R. & E.P.J. van den Heuvel 1985. Preprint, 'Magnetic Field Decay and the Origin of Neutron Star Binaries'
Trimble, V. 1971. Nature 231, 124
..... 1982. Nature 297, 357
..... 1986. Q.J. Royal Astron. Soc. (March)
Trimble, V. & J.L. Greenstein 1972. Astrophys. J. 177, 441
Trimble, V. & M.J. Rees 1970. Astrophys. Lett. 5, 93
Truran, J. 1985. in J. Pringle & R. Wade (eds.) Interacting Binary Stars (Dordrecht: Reidel)
Vestrand, W.T. & D. Eichler 1982. Astrophys. J. 261, 251
Villumsen, J.V. & J.J. Binney 1985. Astrophys. J. 295, 388
Wegner, G. 1974. Mon. Not. Royal Astron. Soc. 166, 271
Wegner, G. et al. 1983. Astrophys. J. 275, 240
Weidemann, V. 1984. Astron. Astrophys. 134, L1

Weidemann, V. & D. Koester 1983. Astron. Astrophys. 121, 77
..... 1984. Astron. Astrophys. 132, 195
Weisberg, J.M. & J.H. Taylor 1984. Phys. Rev. Lett. 52, 1348
Wickramasinge, D.T. et al. 1983. Mon. Not. Royal Astron. Soc. 203, 887
Wilson, J.R. 1985. Work in progress
Witten, E. 1984. Phys. Rev. D30, 272
Woltjer, L. 1971. in R.D. Davies & F.G. Smith (eds.) IAU Symp. No. 46, The Crab Nebula (Dordrecht: Reidel) p. 389
Woosley, S.E. 1984. Colloquium at U. Maryland, October
Wunner, G., W. Roesner, H. Herold & H. Ruder 1985. Astron. Astrophys. 149, 102

OBSERVATIONS OF PULSARS

A.G. Lyne
University of Manchester
Nuffield Radio Astronomy Laboratories
Jodrell Bank, Macclesfield,
Cheshire SK11 9DL

1. INTRODUCTION

Pulsars were discovered at Cambridge in 1968 by Jocelyn Bell, Anthony Hewish and their collaborators (Hewish et al. 1968) and provided the first observational evidence of the existence of neutron stars, a form of collapsed star first discussed by Baade and Zwicky (1934). They seem to possess extremely strong magnetic fields, among the strongest anywhere in the Universe, and hence provide a unique opportunity for the study of a number of unusual electrodynamic processes. As the end-point of stellar evolution for at least some massive stars, their observation provides significant constraints on theories of this evolution. The impulsive and highly linearly polarized nature of the radiation makes them extremely useful probes of the interstellar medium, revealing information upon the distribution of free electron densities, free electron turbulence and interstellar magnetic fields through the Galaxy. They have also yielded information on General Relativity and given the first observational evidence for the existence of gravitational radiation.

Over 400 radio pulsars are now known, all but one in our own Galaxy. In this review, I will discuss the properties of pulsars in three main parts. The first will describe the observed properties of the pulsed emission and their implications for models of the radiation mechanism. Secondly, I will cover the observed variations in periods, including the secular decrease, discontinuous changes and those due to binary motion. Finally, I will discuss the Galactic distribution of pulsars as revealed by the major surveys, and how their properties evolve as a function of time. I do not propose to describe the invaluable information they have given on the interstellar medium in these lectures as this region is rather remote from any collapsed objects, the main topic of this meeting.

F. Pacini (ed.), High Energy Phenomena Around Collapsed Stars, 121–143.

2. THE EMISSION CHARACTERISTICS

2.1 The Periods

The 420 known radio pulsars in the Galaxy exhibit a wide range of properties. In particular the range of repetition periods of the pulses covers over three orders of magnitude. The distribution of these periods is shown in Figure 1 and some of the most notable of these are listed in Table 1, together with a few comments.

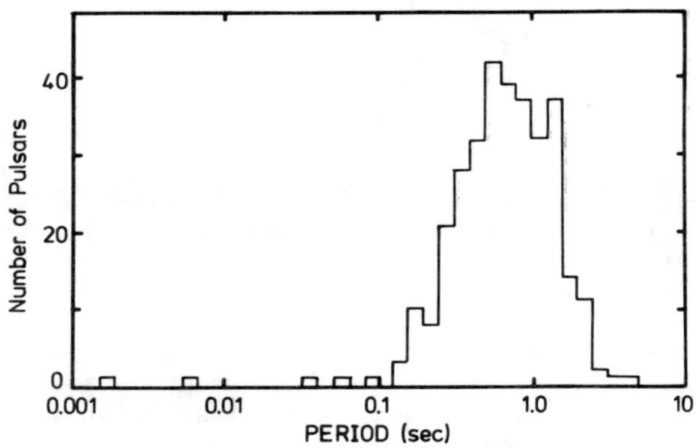

Figure 1. The period distribution of pulsars

Table 1

Period	Pulsar	Comment
1.6 ms	PSR 1937+21	The Millisecond Pulsar
6 ms	PSR 1953+29	In a 120 day binary orbit
33 ms	PSR 0531+21	The Crab Pulsar. Also displays optical, X-ray and γ-ray pulses
59 ms	PSR 1913+16	In a 7.7 hour binary orbit
89 ms	PSR 0833-45	The Vela Pulsar. Also displays optical and γ-ray pulses
0.67 s		Median period
4.3 s	PSR 1845-19	Maximum period

Shortly after the discovery of the Crab pulsar in 1968, it became clear that the only explanation for the repetition period was that the radiation originated from a rotating neutron star (Gold, 1968). Indeed Pacini (1967) had described some of the properties of a magnetised neutron star before the detection of the first pulsar and few people now doubt that we are observing such objects in pulsars. As I will describe, there is a wealth of observational material which supports this interpretation.

2.2 The Period Derivatives

Pacini (1967) described how a rotating neutron star of radius R_o and rotating with angular velocity Ω would experience a braking torque of order $2B_o^2R_o^6\Omega^3/3c^3$ where B_o is the value of the magnetic field at the surface. This will cause the period P to increase, i.e. the period derivative $\dot{P}$ to be positive, a phenomenon seen in all known pulsars. Indeed from this argument using the measured values of P and $\dot{P}$ one can estimate the magnitude of the surface magnetic field as

$$B_o \approx \left[\frac{3c^3}{8\pi^2} \quad \frac{P\dot{P}I}{R_o^6}\right]^{\frac{1}{2}}$$

where I is the moment of inertia of the neutron star. One can also deduce an age for the pulsar, the characteristic age τ, assuming that the pulsar was originally spinning much more rapidly than it is now and that the effective magnetic field has remained constant. Then

$$\tau = \frac{P}{2\dot{P}}$$

Assuming a moment of inertia $I=10^{45}$ gm cm^2 and a stellar radius of 10^6 cm, we can calculate values of B_o and τ for a number of pulsars as shown in Table 2. This clearly implies that all the magnetic fields are extremely strong, far stronger than we can experience in the laboratory on Earth.

Table 2

Pulsar	P (s)	$\dot{P} \times 10^{15}$ (s s^{-1})	B_o (gauss)	τ (year)
1509-58	0.150	1540	1.5×10^{13}	1540
Crab	0.033	422	4×10^{12}	1230
Median	0.67	2.3	$\sim 10^{12}$	$\sim 5 \times 10^6$
1913+16	0.059	0.009	2×10^{10}	10^8
1937+21	0.0016	0.00011	4×10^8	5×10^8

2.3 The Integrated Profiles

Figure 2a shows the integrated profiles of a number of pulsars. The widths and shapes of the profiles vary greatly from pulsar to pulsar but a few generalisations can be usefully made. The mean equivalent width of the pulses is about 10° of longitude (where 360° represents the whole pulsar period). Often two or more components are seen, sometimes in a symmetrical formation. The profiles are on the whole extremely stable and no secular variations have so far been detected.

It is natural to identify these profiles with the shape of a pencil

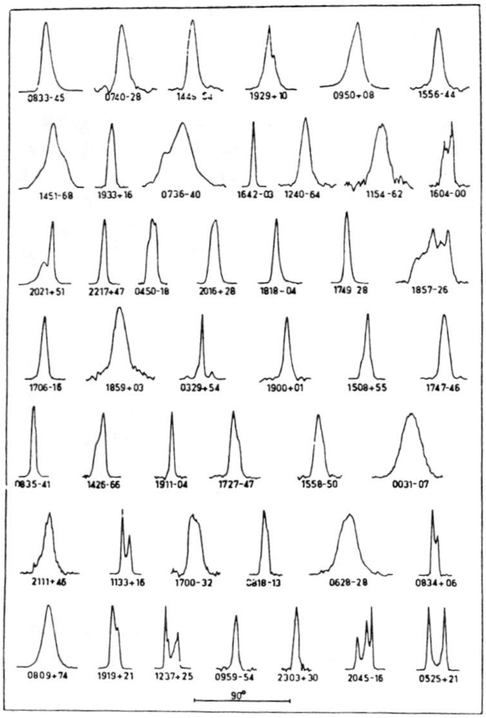

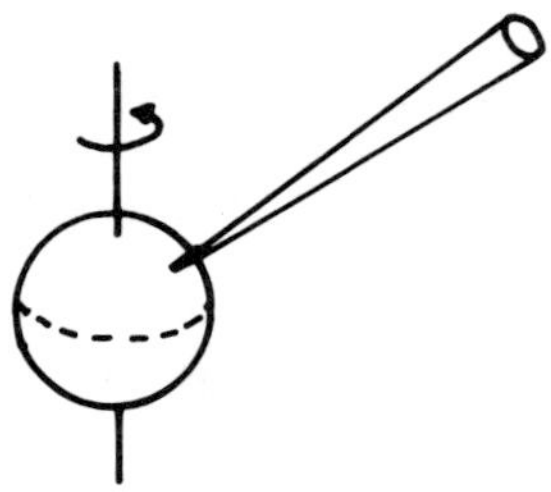

Figure 2b. The pulsar 'light-house' beam

Figure 2a. The integrated profiles of 45 pulsars (Manchester and Taylor 1977)

beam fixed to the surface of the neutron star (see Figure 2b). The symmetry of the double profiles suggests that in some cases the beam takes the form of a hollow cone.

2.4 Single pulses

While the integrated profiles are very stable in form, individual pulses are very erratic both in shape and intensity. In general they consist of sub-pulses within the integrated profiles, these sub-pulses being bursts of radiation of only a few degrees of longitude in width. Within these sub-pulses, one sees micropulses, which are fine pulses having widths of substantially less than one degree, typically of 10 -100 μs duration.

It is generally believed that the integrated profile and sub-pulses result from the rotation of the beam as described above, while micropulses represent a temporal modulation of the emission. If so, their duration puts a stringent limit on the size of the emitting source and hence its brightness temperature, as discussed in the next section.

2.5 The Pulse Emission Characteristics

For a "median" pulsar at a frequency of around 400 MHz, the observed flux density $S \sim 20$ mJy, its distance $d \sim 2$ kpc and the integrated pulse width $w \sim 10^{o}$. Assuming that the beam is circular in form and is beamed orthogonally to the rotation axis, then one can estimate the total radio luminosity of such a pulsar as

$$L_r \doteq \pi^3 d^3 \frac{w}{360} S\Delta\nu = 5 \times 10^{27} \text{ erg s}^{-1}$$

where $\Delta\nu \sim 400$ MHz is the bandwidth of the radiation.

This can be compared with the rate of loss of rotational energy from the star from the observed slowdown rate:

$$\dot{E} = I \,\Omega\, \dot{\Omega} = -4\pi \frac{I\dot{P}}{P^3} \sim 3 \times 10^{32} \text{ erg s}^{-1}$$

for a moment of inertia of 10^{45} gm cm^2. Clearly most of the energy is lost in forms of energy other than the radio pulses which account for only 10^{-5} of the total rotational energy loss. Most of the energy is probably lost in the form of energetic particles or in the form of electromagnetic dipole radiation at the pulsar rotation frequency.

If the micropulses truely represent the "elemental" emission, then we have a limit on the size of the emitting region of less than about $l = 30 \times 10^{-6}$ light seconds or about 10^6 cm. One can then calculate the specific intensity of the region as about 2×10^9 erg cm^{-2} s^{-1} Hz^{-1} ster^{-1}. Using the Rayleigh-Jeans approximation this leads to a brightness temperature of about 10^{29}K! Clearly a coherent emission process of some sort is required, and about the only ones we know involve either the bunching of charged particles or some form of maser process.

2.6 The magnetic pole model

Pulsars are notable for the high degree of linear polarisation displayed by many of them, sometimes reaching very close to 100%. They also display the characteristic "S" form of position angle variation through the pulse profile. Some examples are displayed in Figure 3. The amount of circular polarisation is usually much lower, but often shows an abrupt change in the middle of the profile from one sense to another.

Such polarisation observations provide the main support for the widely accepted magnetic pole model for the emission from pulsars (Radhakrishnan and Cook 1969). In this model, charged particles are drawn from the surface of the star by strong electrodynamic effects and the radiation is emitted by the particles as they move along the bundle of 'open' field lines above one of the magnetic poles. The plane of linear polarisation of the emitted radiation is either parallel or orthogonal to the local magnetic field lines, depending upon the details of the emission process. The emission probably arises from curvature

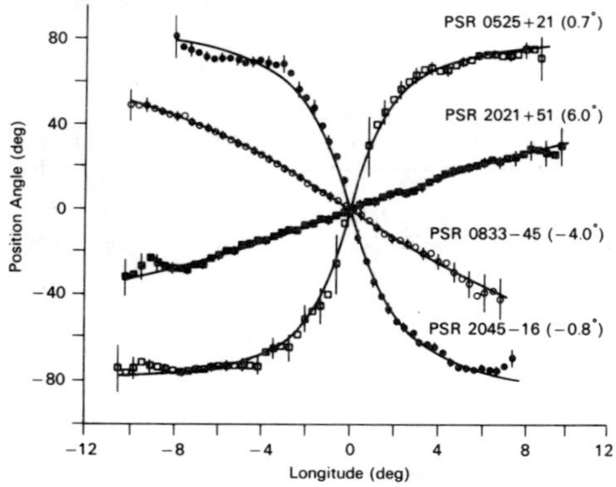

Figure 3. The variation of the position angle of linear polarisation across the pulses of 4 pulsars. (Manchester and Taylor, 1977).

radiation as bunches of charged particles moving along the field lines accelerate normal to their direction of motion because of the field curvature. The radiation will be beamed along the field lines. The outer annulus of open field lines will have greatest curvature and might be expected to radiate most strongly. We can then interpret the single and double component morphologies seen in many pulsars in terms of different impact parameters between the lines of sight from the Earth and the magnetic pole. (Figure 4). There is also considerable evidence that the very centre of the polar region is a strong emitter, giving rise sometimes to a symmetrical triple pulse structure. High frequency radiation is expected to be emitted near to the surface of the star, lower frequencies further out where the bundle of open field lines will have flared more, giving rise to wider profiles. This effect can be seen in most pulsars and a classic example is given in Figure 5.

On the whole this model describes the observations of polarisation and shapes remarkably well, although it is quite clear that the intensity of emission across the polar region is often far from symmetrical, while the position angle variation usually is.

There has recently been conjecture as to whether the beam is circular or whether it is more elongated in pulsar rotational latitude than longitude. If it is elongated, this will mean that we have observed a greater proportion of the pulsar population than we previously believed and hence reduce our estimates of the number of pulsars in the Galaxy and of their formation rate. The main argument in favour of elongation (Narayan and Vivekanand 1983) is based upon the predominance of pulsars with small swings in position angle of the linear polarisation through the pulse, implying a large impact parameter of the line of sight relative to the magnetic pole. However in view of the extreme "lumpiness" of the emission region, I believe that in many of these cases we may be seeing only a portion of the emitting region in an individual pulsar, making such arguments rather uncertain. In my view the case for elongated beams is unproven.

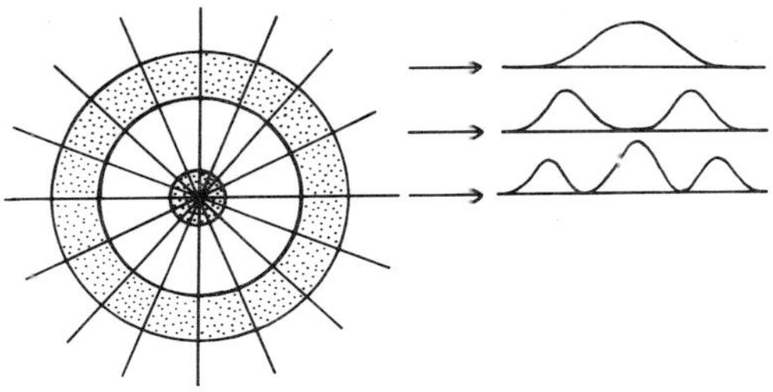

Figure 4. A schematic view of a magnetic pole, showing how different line of sight impact parameters lead to various pulse morphologies.

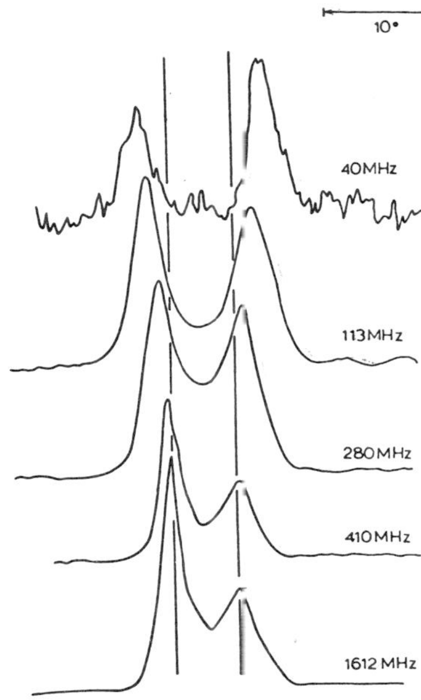

Figure 5. The flaring of the pulse profiles of PSR 1133+16 at low frequencies.

2.7 Pulse Nulling, Mode Changing and Sub-pulse Drifting

Three other phenomena displayed by pulsars must be accounted for by any comprehensive model of the emission from pulsars. The most dramatic is that of pulse nulling, in which emission from the pulsar ceases abruptly for between a few periods and several thousand periods and then, just as suddenly, starts emitting again. During the null state, there is usually no detectable radio emission from the pulsar. An associated phenomenon is that of mode-changing in which the form of the integrated profile changes abruptly. No satisfactory explanation for these effects exists.

Sub-pulse drifting is the phenomenon of a regular drifting of sub-pulses through the integrated profile. This is commonly interpreted in terms of hot-spots moving around the emission core, driven by electro-dynamic forces. Sometimes pulse nulling occurs within the middle of a sequence of drifting. The nulling then clearly causes a perturbation in the drift rate which then takes some time to recover to its initial value.

3. PULSE TIMING MEASUREMENTS

A pulsar, being essentially an isolated freely rotating 'dead' body with a well-defined 'tick', offers the possibility of providing an excellent clock. In this section we discuss how well this promise is fulfilled and to what uses we can put such clocks. The departures from the ideal also give us substantial information about the internal structure of pulsars themselves.

3.1 Technical Aspects

A measured arrival time of a pulse on Earth, t, has to be corrected to an inertial reference frame before we can compare it with other times. This frame is conveniently chosen to be the barycentre of the Solar System, and the arrival time in this frame, t_b, is calculated as

$$t_b = t + \frac{\underline{r}_o . \underline{s}}{c} + \Delta t_e$$

where $\underline{r}_o$ is the position of the observatory relative to the barycentre, $\underline{s}$ is the unit vector towards the source and Δt_e is a clock correction to account for transverse Doppler and gravitational redshift effects, amounting to a maximum of about 1.5 milliseconds.

One can compare these arrival times with a model by calculating the phase of the pulse as

$$\phi = \phi_o + \Omega_o(t_b - t_o) + \tfrac{1}{2}\dot{\Omega}_o(t_b - t_o)^2$$

where ϕ_o, Ω_o and $\dot{\Omega}_o$ are the pulsation phase, angular frequency and first derivative at epoch t_o.

For a good model, $\phi - \phi_o \approx 2\pi n$ where n is an integer, so that the quantity $(\phi - \phi_o - 2\pi n)/\Omega$ is known as the residual and is a measure of the accuracy of the model. Least squares fitting procedures are then used to adjust the model in order to minimise the residuals.

An error in the source position $\underline{s}$ will give rise to an annual sinusoid in the residuals. For example, a one arcsecond position error will give rise to a sinusoidal residual having an amplitude of about 2 milliseconds. The measurement of such residuals incidentally allows the determination of the accurate positions of pulsars and even, when measured over a long enough period of time, the determination of proper motions (Figure 6). There will be discussion of the results of such measurements in section 4.

3.2 Period Derivatives

The measured period derivatives $\dot{P}$ are all positive - due to loss of rotational energy by one mechanism or another. For loss by magnetic dipole radiation at the rotation rate of the pulsar, the expected braking force is proportional to Ω^3 so that we can write $\dot{\Omega} = -K\Omega^n$ where K is a constant and n=3 is a quantity known as the braking index. For loss of

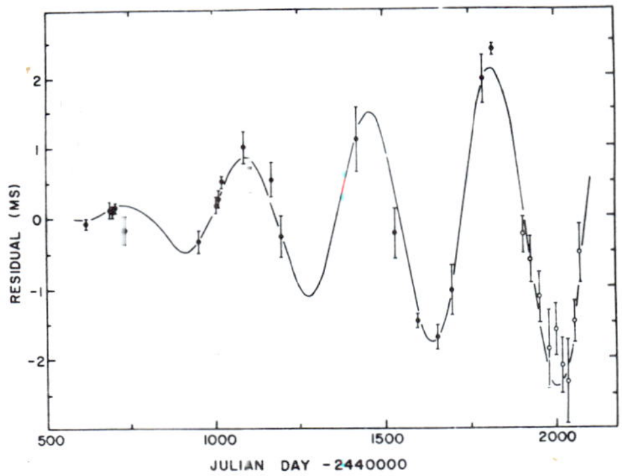

Figure 6. The timing residuals for PSR 1133+16 showing the effects of proper motion. (Manchester and Taylor 1977).

energy in the form of charged particles from an aligned dipolar field, it also turns out that n=3 (Goldreich and Julian 1969). Now, to find out what the true value of n is for a pulsar, we can in principle calculate n from the measurement of the second derivative:

$$n = \frac{\Omega\ddot{\Omega}}{\dot{\Omega}^2}$$

However, in practice, only two significant values of $\ddot{\Omega}$ have been obtained, mainly because measurements are limited by intrinsic fluctuations in the rotation rate which will be discussed in the next section. These measurements are for the Crab pulsar (n = 2.51, Groth 1975) and for PSR 1509-58 (n = 2.7, Manchester et al. 1985). Both of these are somewhat less than the expected value of 3. There are a number of possible reasons for this, the most likely being that the field is distorted from its dipolar form by a stellar wind.

The characteristic age for a pulsar with a braking index of n may be calculated assuming a high initial rotation rate and a constant magnetic field as

$$\tau_c = \frac{\Omega}{(n-1)\dot{\Omega}} = \frac{P}{(n-1)\dot{P}} = \frac{P}{2\dot{P}} \quad \text{(for n=3)}$$

For the Crab pulsar, this characteristic age turns out to be 1240 years, compared to the known age of 925 years. The modest discrepancy may be the result of either a slow rotation rate at birth or some magnetic field evolution since then.

3.3 Period Instabilities

The variations described above are stable and predictable. However some pulsars show unpredictable changes of two kinds, either discrete events with an abrupt decrease in period, known as "glitches", or continuous noise-like variations known as timing noise.

Glitches occur in a small number of pulsars and consist of a perturbation in the period and its first derivative, followed by a relaxation on timescales which vary from pulsar to pulsar between days and years. The Vela pulsar has suffered seven major glitches since 1969. Figure 7a shows a number of these glitches in period over an eight year data span. Table 3 lists the major glitch events so far observed in all pulsars.

Table 3. The parameters of the major glitches

Pulsar	Number of events	$\Delta\Omega/\Omega$	$\Delta\dot{\Omega}/\dot{\Omega}$
Vela	7 at 2-3 year intervals	2-3 x 10^{-6}	8 x 10^{-3}
Crab	2	10^{-8},4 x 10^{-8}	2 x 10^{-3}
1641-45	1	2 x 10^{-7}	-
0525+21	1	10^{-9}	5 x 10^{-3}

These events provide clear evidence for two or more components within the neutron star which are rather poorly coupled to one another.

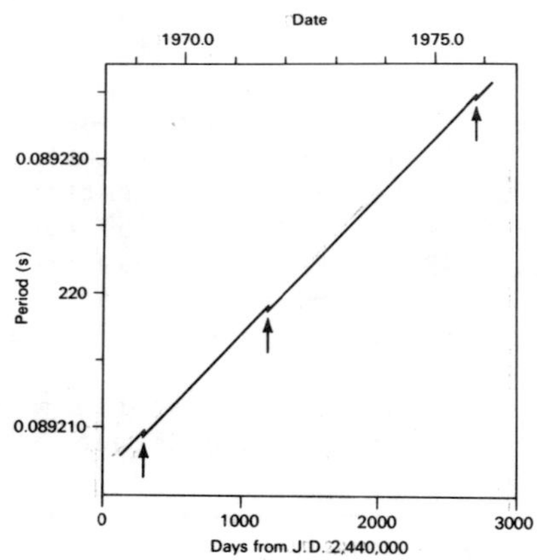

Figure 7a. The period of the Vela pulsar showing three of the glitches (Manchester and Taylor 1977)

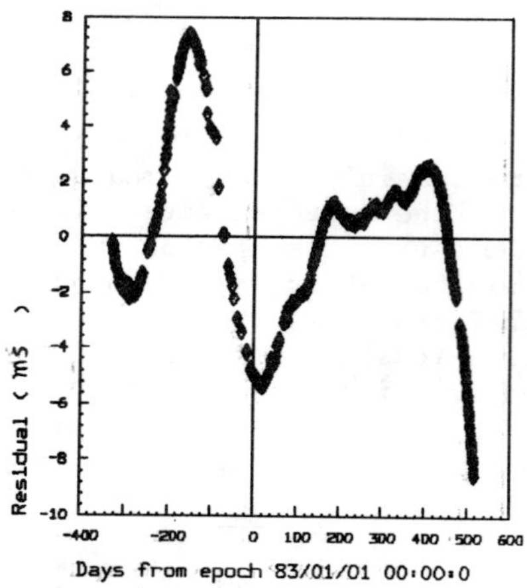

Figure 7b. Timing residuals for the Crab pulsar

The simplest models consist of a solid crust with a superfluid interior. Since the crust is being slowed down, the interior will be rotating more rapidly than the crust. Catastrophic variations in the coupling of momentum between the two, such as might arise of the pinning and unpinning of vortices (e.g. Ruderman 1976), may give rise to the observed effects. Alternatively the initial glitch may be the result of a "Starquake" in the rigid solid crust (Baym et al. 1969): As the pulsar slows down, the equilibrium oblateness will decrease and so therefore will the moment of inertia. Possibly this change occurs only when the rigidity of the crust is unable to sustain the stresses which build up. We will then see a sudden increase in the rotation rate due to the decrease in moment of inertia, followed by a relaxation as rotational equilibrium between the crust and fluid core is restored through the weak frictional coupling forces between the two. However this cannot explain the frequency and magnitude of the glitches in the Vela pulsar. The physics of these processes will be discussed later by David Pines.

Timing noise is characterised by a continuous unpredictable phase wandering. Again it is most predominant in the Crab pulsar (Figure 7b) and other large period derivative objects. The dependence of the timing noise upon the derivative was first established by Cordes and Helfand (1980). This can be understood qualitatively in terms of the two component model described briefly above. The greater the period derivative, the greater will be the difference in rotation rates of the fluid and the crust, and the variable coupling between the two will cause a greater fluctuation in the rotation rate of the crust.

The greatest stability is seen in the millisecond pulsar PSR 1937+21 which has a very small period derivative of only about 10^{-19}. Observations of this pulsar over about two years (Davis et al. 1984) show rms residuals of less than a microsecond (Figure 8). This indicates a

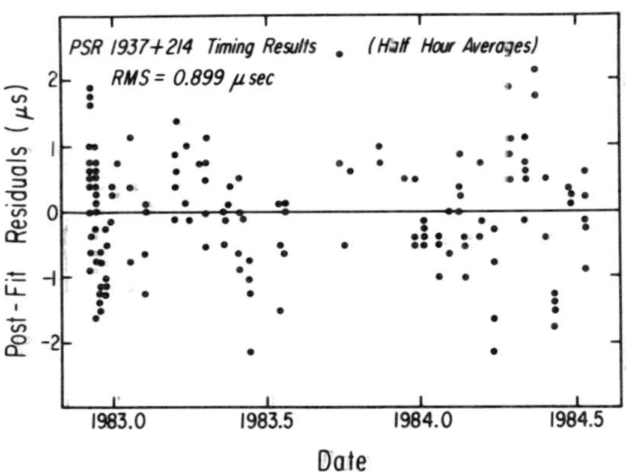

Figure 8. Timing residuals for the millisecond pulsar PSR 1937+21.

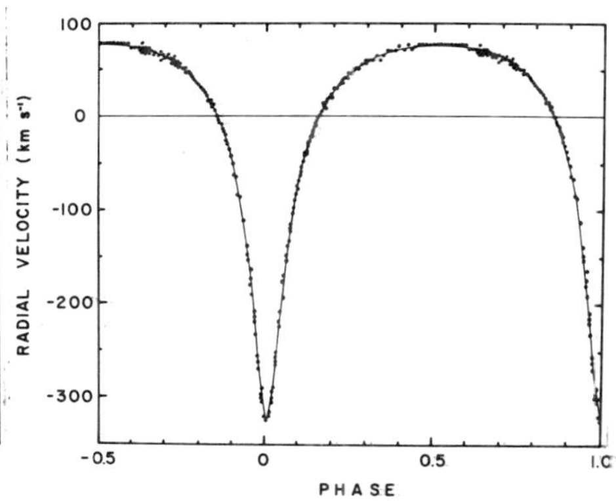

Figure 9. The velocity curve of the binary pulsar PSR 1913+16. (Manchester and Taylor 1977)

frequency stability of about 1 part in 10^{14} and is comparable with the stabilities of the best banks of atomic clocks. More precise measurements over longer timescales will certainly reveal significant residuals, but it will not then be clear which is at fault, the pulsar or the terrestrial time standard. Improved time standards or further millisecond pulsars are required to act as referees.

3.4 Binary Pulsars

The first pulsar to be discovered in a binary system, PSR 1913+16, was detected by Hulse and Taylor (1974) at Arecibo. This pulsar is in a highly eccentric orbit (e = 0.617), has an orbital period of just under 8 hours and $a\sin i \sim 1\ R_\odot$, where a is the orbit semi-major axis and i is the orbital inclination to the plane of the sky. No occultation of the pulsar was observed and both objects appeared to act gravitationally as point objects, compatible with the absence of any optical identification. The velocity curve from the earlier observations is shown in figure 9 and from this one can calculate the mass function of the system,

$$f(m_p) = \frac{m_c{}^3\sin^3 i}{(m_p + m_c)^2} = 0.132\ M_\odot$$

Classically, one cannot completely solve such a system without measurements on the orbit transverse to the line of sight, i.e. involving resolution of the orbit which in this case is only about 10^{-6} arcsec. However, the large eccentricity and mildly relativistic velocity has subsequently allowed a complete solution. Firstly, the advance of the longitude of periastron of 4.2^o/year showed that the total mass of the system was

$$m_p + m_c = 2.83\ M_\odot.$$

Secondly the observation of transverse doppler and gravitational redshift effects were detected giving a further function of m_p and m_c. These three equations can be solved explicitly to provide solutions (Weisberg and Taylor 1984):

$$m_p = 1.43 \pm 0.07\ M_\odot$$
$$m_c = 1.40 \mp 0.07\ M_\odot$$
$$\text{and}\quad i = 47^o \mp 3^o$$

These masses are quite consistent with the system consisting of two neutron stars. No pulsations have been detected from the companion, perhaps because it is not emitting pulsed radiation, the beam does not intersect the line of sight to the Earth or, possibly, the radio luminosity is just too small to be detected at such a large distance (see Section 4.4).

With the orbit now precisely determined it is possible to predict from general relativistic theory the rate at which the orbital period will change due to loss of energy from the system in the form of gravitational radiation. The predicted value of

$$\dot{P}_b = -2.38 \times 10^{-12}\ ss^{-1}$$

agrees with observation within the errors of measurement at about the 1 percent level at present confirming the general relativistic theory with unprecedented precision.

The main properties of the system which have allowed such studies are the high eccentricity, high orbital velocity and a very stable pulsar clock. This stability in the case of PSR 1913+16 clearly arises from the very small value of the period derivative of the pulsar, $\dot{P} \sim 10^{-17}$ ss^{-1}. None of the four other pulsars so far detected in binary systems has provided such a good set of circumstances (Table 4) and, apart from possibly PSR 2303+46 (Stokes, Taylor and Dewey 1985), we are unlikely to determine the masses from timing measurements alone.

Table 4 Binary Pulsars

Pulsar	P_b (days)	e	P_{pulse} (msecs)
1913+16	0.32	0.617	59
0655+64	1.03	$\leq 10^{-5}$	196
2303+46	12.34	0.66	1070
1953+29	117.0	<0.01	6.1
0820+02	1232.0	0.012	864

There is no chance of resolving the orbits with interferometry, although there is a method which has been employed which involves the use of interstellar scintillation to determine the transverse velocity of the pulsar in its orbit. Figure 10 shows how the rate of interstellar scintillation varies as the pulsar PSR 0655+64 moves in its circular orbit. The large variation in speed shows that the inclination of the orbit is very high and puts an upper limit on the mass of the companion of about 0.7 $M_{\odot}$ (Lyne 1984).

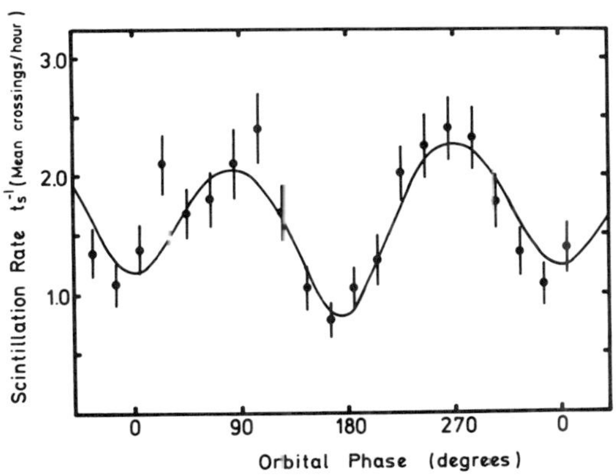

Figure 10. The variation of scintillation speed around the orbit of PSR 0655+64.

4. THE GALACTIC DISTRIBUTION AND EVOLUTION OF PULSARS

In this section I am going to discuss what is known about the galactic distribution, evolution and origin of pulsars and what we can learn of the formation event. Much of the work is the result of a recent analysis with Dick Manchester and Joe Taylor (Lyne, Manchester and Taylor 1985), while other parts come from studies at Jodrell Bank concerning the measurements of proper motions (Lyne, Anderson and Salter 1981).

4.1 The Observed Distribution

There are now 420 known pulsars, most of which have been discovered in the four main surveys carried out at Arecibo, Jodrell Bank, Molonglo and Greenbank, all at frequencies of about 400 MHz (Hulse and Taylor 1974, 1975; Davies, Lyne and Seiradakis 1972, 1973; Large and Vaughan 1971; Manchester et al. 1977 and Damashek et al.1978). Figure 11 shows a plot of the 316 pulsars detected in the four surveys in galactic

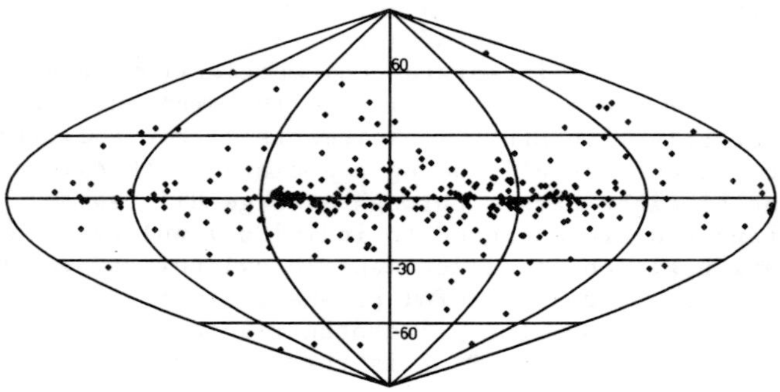

Figure 11. The distribution of observed pulsars on the sky.

coordinates and in this the concentration of pulsars along the galactic plane is clear. However, although there is now fairly good coverage of the whole sky, the depth of coverage is not completely uniform. For instance the high density of pulsars around longitude 50° is due to the very deep Arecibo survey which covered only a very limited area of sky close to the galactic plane. The low density of pulsars towards the centre of the Galaxy is due, at least in part, to the reduced receiver sensitivities caused by the very high galactic background noise in this direction.

4.2 The distance scale

In order to study the distribution of pulsars through the Galaxy, it is necessary to know their distances. Fortunately the dispersion measure provides an approximate indicator when combined with a model of the galactic distribution of electrons. Also, independent distance estimates can be obtained for a number of pulsars. The main independent distance estimates are obtained from the absorption of the 21 cm radiation from the pulsar by interstellar HI when combined with a model for the differential rotation of the Galaxy. Such observations have been carried out for about 30 objects. See for example Graham et al. (1974), Ables and Manchester (1976) and Booth and Lyne (1976).

These observations have been used, together with a certain amount of other information, to develop and to calibrate a model for the distribution of the electron density, n_e, through the Galaxy. In the model most recently adopted, n_e has the form

$$n_e = 0.025 + 0.015 \exp(-|z|/70\text{pc}) \frac{2}{(1+R/10)} + 0.28\alpha_{GN} \text{ pc cm}^{-3}$$

where z is the distance above the galactic plane and R is the galactocentric radius (Lyne, Manchester and Taylor 1985). The last term represents a contribution which arises when a line of sight passes through the region of the Gum Nebula.

To determine the distance to a pulsar using this model, we move out from the Sun along the line of sight to the pulsar until we reach the distance at which the integrated column electron density is equal to the dispersion measure of the pulsar. The distances so obtained agree satisfactorily with other measurements within a factor of about 2, the main discrepancies usually being associated with circumstances in which the line of sight passes through a large HII region.

Once distances d are assigned to individual pulsars, one can then calculate their position in galactocentric cylindrical coordinates and derive three astrophysically useful parameters; the distance from the galactic plane, $z = d\sin b$, the galactocentric radius, R, and a radio luminosity, $L = S_{400}d^2$, where S_{400} is the observed flux density at 400 MHz.

Figure 12 shows the distribution of pulsars projected onto the galactic plane and shows the strong selection effect caused by the

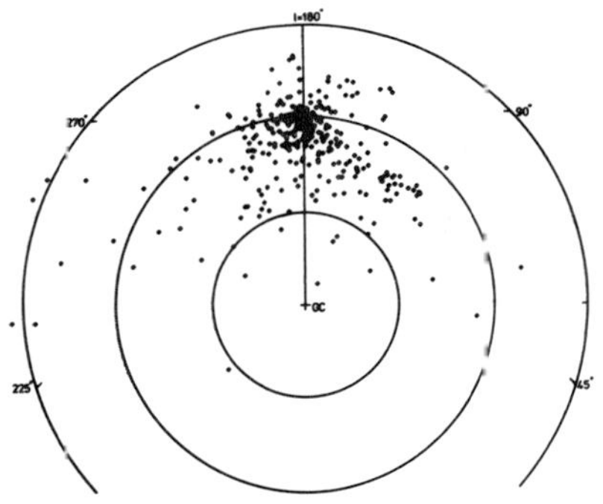

Figure 12. The distribution of pulsars projected onto the galactic plane.

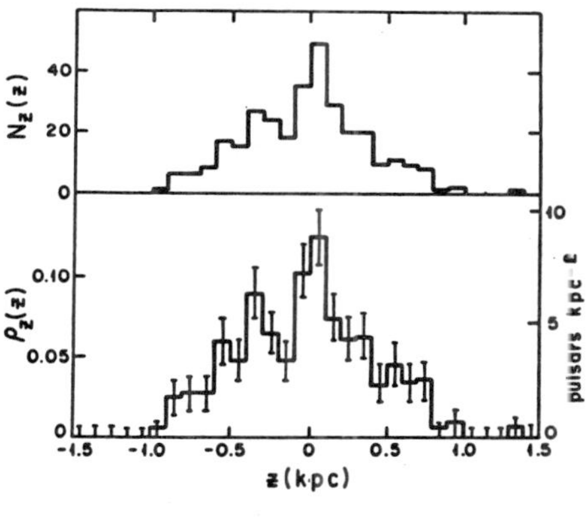

Figure 13. The observed (top) and derived (bottom) distribution in z.

inverse square law. The observed distributions in the three parameters are dominated by the three main selection effects at play in the data:

1) The inverse square law
2) The areas of sky covered by each survey

and 3) The changing receiver sensitivity due to galactic background radiation.

The analysis procedures used have to correct for these effects and allow us to determine unbiassed estimates of the true distributions of z, R and L. The top halves of figures 13-15 show the observed distributions in the three parameters.

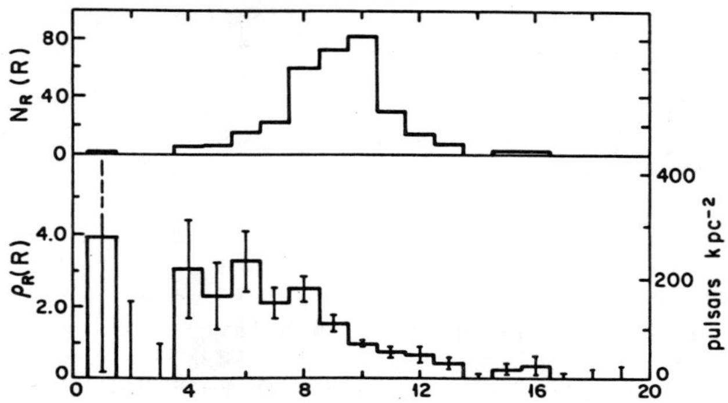

Figure 14. The observed (top) and derived (bottom) distributions in R.

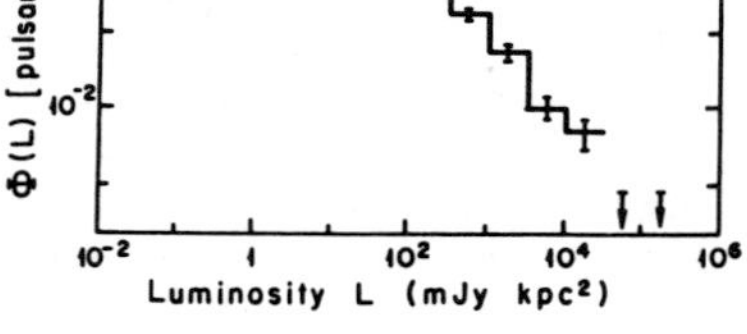

Figure 15. The observed (top) and derived (bottom) distributions in L.

4.2 The Derived Distributions

The analysis procedure used by Lyne, Manchester and Taylor (1985) will not be described here in any detail. In effect the procedure consists of calculating the volume of the Galaxy searched for pulsars having luminosity L and lying between z and z+dz and R and R+dR.

From the number of pulsars observed and the volume searched it is possible to calculate a space density for a particular z, R and L. Figures 13, 14 and 15 show the distributions derived for each of the three quantities. The z distribution shows pulsars to have an approximately exponential distribution having a scale height of about 400 pc, much greater than that of most galactic population I species. The reasons for this will be discussed in the next section. The radial distribution shows an increasing density of pulsars towards the galactic centre, decreasing towards the outer parts of the Galaxy. This distribution shows a marked resemblance to that of supernova remnants and HII regions and probably also of massive stars. It is reassuring that the distribution varies reasonably continuously through the position of the Sun (R = 10 kpc) showing that the powerful selection effects responsible for the sharp peak in the observed distribution have been substantially removed by the analysis. The apparent fall in density within about 4 kpc of the galactic centre is probably not real. A recent survey at 1400 MHz (Clifton and Lyne 1985) has shown that many pulsars in this region of the Galaxy have probably been missed at the lower frequencies around 400 MHz because of strong turbulence in the interstellar medium causing multipath scattering of the pulsar radiation and the smearing out of the pulses so that periodicity can no longer be detected.

The distribution in luminosity at the position of the Sun shows a galactic surface density of pulsars which increases towards low luminosities but flattens or even turns over at luminosities of less than about 1 mJy kpc^2. Integrating under this distribution we find that there are about 70 ± 15 observable pulsars per square kpc at the position of the Sun. In this context, observable means having a luminosity in the direction of the Earth of greater than 0.3 mJy kpc^2. The true density of pulsars is certainly substantially greater than this because the majority are probably not beamed towards the Earth. This incompleteness or beaming factor may be as large as five, as discussed in Section 4.4.

4.3 Pulsar Velocities

The scale height of about 400 pc which we see in z-distance is substantially greater than that of most massive stars which are likely to be the progenitors of pulsars, and also than that of supernova remnants, both of which have scale heights of about 70 pc. Gunn and Ostriker (1970) provided the explanation for this as pulsars being runaway stars which have moved large distances since their birth. This can be seen explicitly from proper motion measurements obtained from timing observations (e.g. Manchester, Taylor and Van 1974, Helfand et al. 1980) and more recently from direct interferometry at Jodrell Bank (Lyne, Anderson and Salter 1981). Some early timing measurements of proper motion have been shown to suffer from the effects of timing noise and must be treated with caution.

In Figure 16 the significant proper motions have been converted to transverse velocities using pulsar distance estimates and are shown as a function of galactic longitude and galactic z-distance. It is clear that, with two exceptions, all the pulsars lying outside a

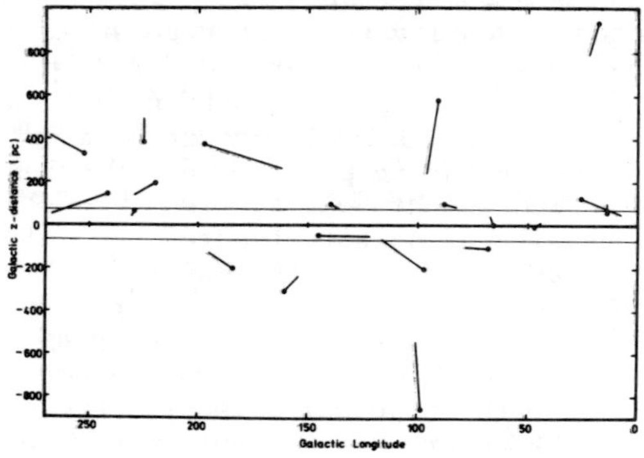

Figure 16. The velocities of pulsars relative to the galactic plane. The filled circles represent the pulsars and the lines leading to them the velocities in amplitude and direction. The largest velocity is about 400 km s^{-1}.

"progenitor" layer of width of about 70 pc above and below the galactic plane are migrating away from the plane. The two exceptions are PSR 1237+25 and PSR 0943+10 which both lie at high galactic latitude so that the observed transverse motion is likely to be dominated by the velocity component parallel to the galactic plane and not that normal to the plane. Thus these two objects may also be moving away from the plane as well.

The mean transverse velocity of pulsars is about 150 km s^{-1}, showing that they do indeed have high space velocities, having a mean value of about 200 km s^{-1}. We thus have the picture of pulsars being born close to the plane from progenitors having a population I form of distribution and being given a "kick" of a few hundred km s^{-1}, an order of magnitude greater than the random motions within such populations.

4.4 The Galactic Population of Pulsars

The total number of observable pulsars is obtained by integrating the distribution in R through the galactic disc and gives a value of 70×10^3. Note that the rather poorly determined density near the galactic centre probably does not affect the accuracy of this value much because of the relatively small volume of space involved. Because of beaming effects as discussed earlier, we can expect to observe only a small fraction, f, of all the active pulsars in the Galaxy, i.e. those whose radiation beams cross the Earth during a rotation of the pulsar. If the radiation beams have the same width in latitude as in longitude (and for most polar cap models this is the case) then f ~ 0.2. For most relativistic beaming models it seems that a similar value of f is also appropriate. Certainly, it is difficult to see how a value of f much less than this can come about. Adopting a value for f of 0.2, then the total number of active pulsars in the Galaxy is

$N_G = (350 \pm 150) \times 10^3$ for $L > 0.3$ mJy kpc^2.

It should be noted that although only about 4% of the observed pulsars have luminosities between 0.3 and 3 mJy kpc^2, they represent between 80 and 90 percent of the galactic population. This highlights the fact that the pulsars observed from the Earth are mostly a small minority of very high luminosity objects which have a mean luminosity of about 100 times the mean luminosity of the galactic population (see Figure 15).

We see therefore that the estimates of the galactic population rely heavily upon the statistical measurements of only a small number of low luminosity pulsars which lie within a few hundred parsecs of the Sun. They are too close for HI absorption measurements to determine their distance and unfortunately we have had to rely upon the dispersion measure as described in section 2. However, both Lyne (1974) and Hall (1980) have concluded that the local mean electron density is similar to that elsewhere in the Galaxy and provided that this is the case the population estimates will not be greatly in error. In fact if Hall (1980) is correct and the electrons are more closely confined to the plane than previous models have suggested and the population figures will have to be revised upwards. The measurement of the trigonometric parallax on PSR 1929+10 (Salter, Lyne and Anderson 1979), one of the low luminosity pulsars, indicates a mean electron density of 0.069 cm^{-3} and lends some support to Hall's model.

4.5 The lifetimes and birthrate of pulsars

In order to determine the birthrate of pulsars which is required to sustain the galactic population, it is necessary to investigate the evolution and "lifetimes" of pulsars.

Let us consider again the proper motion observations and the migration from the plane. Clearly it is possible to extrapolate a pulsar path back onto the galactic plane and thus estimate an "age" since the initial ejection from the progenitor population. There will be errors in such estimates due to the uncertainty of z at birth, the unknown component of velocity along the line of sight (and hence the precise form of the track on the sky) and the errors in the proper motion measurements themselves. In figure 17, these "kinetic" ages are compared with the "characteristic" ages, $t_c = P/2\dot{P}$, obtained from an assumption of magnetic braking arising from a constant magnetic field. The form of this plot confirms the doubts expressed by Lyne, Ritchings and Smith (1975) and Taylor and Manchester (1977) that the characteristic age is not a good estimate for ages of greater than a few million years. Figure 17 is consistent with the effective magnetic dipole moment decaying on a timescale, t_B, of between 4 and 10 million years. This was first proposed by Gunn and Ostriker (1970) but the experimental evidence in its favour has been weak up to now. In their model, the surface magnetic field decays exponentially with time and it is possible to calculate the true age, t, from the characteristic age, t_c. Thus

$$B = B_o \exp(-t/t_B)$$
$$t = \tfrac{1}{2} t_B \ln(1 + 2t_c/t_B)$$

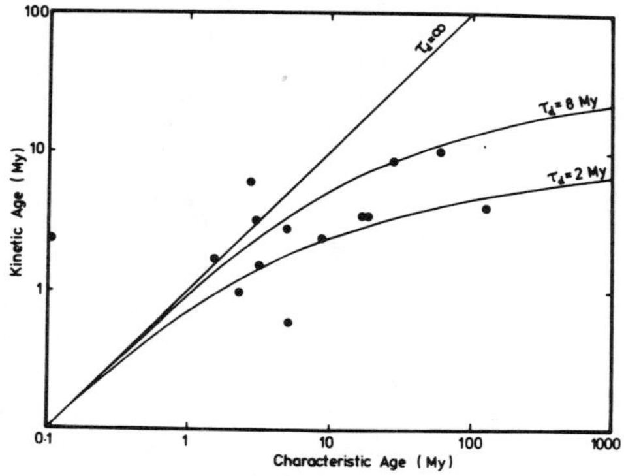

Figure 17. The kinetic and characteristic ages of pulsars. The lines represent the expected paths for the magnetic field decay time indicated.

Gunn and Ostriker also suggested that the luminosity depended upon the square of the magnetic field. Their simple model fits the data remarkably well. For instance in figure 18 we see the distribution of characteristic ages and the expected distribution of a flux limited sample for a magnetic field decay time of 9 million years. Figure 19, showing the luminosity as a function of characteristic age, lends some further support to the model where the data are reasonable consistent with a luminosity decay time of 4.6 million years, i.e. a magnetic field decay time of 9 million years.

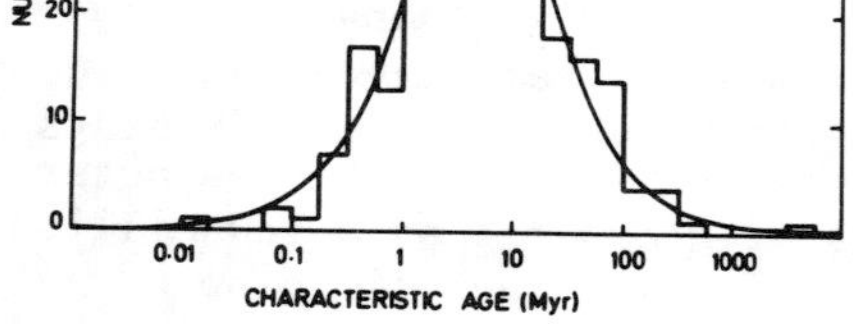

Figure 18. The distribution of the characteristic ages of pulsars.

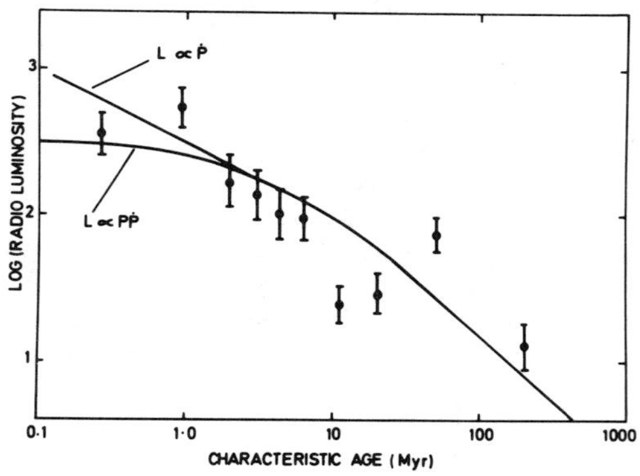

Figure 19. The mean luminosities of pulsars as a function of characteristic age.

In this model therefore pulsars move from one luminosity bin to the next (figure 15 bottom) in 4.6 My. Then the number required to be born in 4.6 million years in order to sustain the observed density near the Sun is

$$\sum_{L=L_{min}}^{\infty} N(L) - N(L+1) = N(L_{min})\ \mathrm{kpc}^{-2} = 22\ \mathrm{kpc}^{-2}$$

leading to a minimum birthrate of 4±1 $\mathrm{kpc}^{-2}\ \mathrm{My}^{-1}$. Integrating through the Galaxy and applying a 0.2 beaming factor gives a required birthrate of 1 every 46 years. Taking into account the various uncertainties we believe the birthrate must lie in the range of 1 every 30 to 1 every 120 years.

Many other estimates of the birthrate (e.g. Davies, Lyne and Seiradakis 1977; Taylor and Manchester 1977) have used the concept of pulsar "lifetime", the pulsar dying suddenly after a certain age, the death probably being associated with the phenomenon of pulse nulling (Ritchings 1976). These lifetimes have then been estimated from the ages of the observed pulsars, which as we have seen are the super-luminous minority and which we have now seen to be relatively young. This has caused the underestimation of the age of the pulsar population as a whole and an overestimation of the birthrate in the past.

4.5 Discussion and Conclusions

The main factors pertinent to the origin and evolution of pulsars are as follows:-

1) The galactic radial distribution is the same as that of massive stars and supernova remnants.
2) The scale height of pulsars in z is about 400pc, and most are moving away from the galactic plane.
3) The scale height of the progenitors is <100 pc, implying a population I origin.
4) The number of active pulsars in the Galaxy is about 350,000.
5) The magnetic fields decay on a timescale of about 9 My.
6) The radio luminosity is roughly proportional to B^2 and decays on a timescale of 4.6 My.
7) Ejection velocities from the birth event are about 200 km s^{-1}.
8) Only 5 pulsars out of 420 known are in binary systems.
9) A birthrate of 1 pulsar every 30-120 years is required to sustain the observed population.

This birthrate is just about compatible with recent estimates of the rates of supernovæ from historical and extragalactic evidence, but rather higher than those derived from supernova remnants (Tammann 1982, Caswell and Lerche 1979, Clark and Stephenson 1977).

Although it is now generally accepted that pulsars are formed in the supernovae of massive stars, the direct experimental evidence for this is still not very convincing. The main evidence is the almost incontrovertible association of the pulsars PSR 0531+21 and PSR 0833-45 with the Crab Nebula and Vela supernova remnants respectively because of the positional agreements and similarity of ages.

Supernova events also provide a theoretically acceptable formation mechanism as well as an explanation for the rather high space velocities of pulsars and the small number in binary systems. The similarity in the galactic radial distributions of pulsars and supernova remnants also provides some weak circumstantial evidence, although there are many other populations of galactic objects which have the same distributions. It seems that if supernovae are solely responsible for pulsar production, most supernovae must produce an active radio pulsar. Many supernova remnants are observed which do not contain observable pulsars. This is not in conflict with the conclusion above because the distances to most remnants are such that pulsars with luminosities typical of the galactic population would be too weak to detect.

From the evidence as it stands, we must still be prepared to question whether all pulsars are born in supernovae and to consider instead whether they may be born in rather less violent and less obvious events which are nevertheless violent enough to provide high space velocities and disrupt most binary systems.

REFERENCES

Ables, J.G. and Manchester, R.N., 1976. Astr.Astrophys., 50, 177.

Baade, W. and Zwicky, F., 1934. Proc.Nat.Acad.Sci., 20, 254.

Baym, G.C., Pethick, D., Pines, D. and Ruderman, M., 1969. Nature 224, 872.

Booth, R.S. and Lyne, A.G., 1976. Mon.Not.R.astr.Soc., 174, 53P.

Caswell, J.L. and Lerche, I., 1979. Mon.Not.R.astr.Soc., 187, 201.

Clark, D.H. and Stephenson, F.R., 1977. Mon.Not.R.astr.Soc., 179, 87P.

Clifton, T.R. and Lyne, A.G., 1985. Nature, submitted.

Cordes, J.M. and Helfand, D.J., 1980. Astrophys.J., 239, 640.

Damashek, M., Taylor, J.H. and Hulse, R.A., 1978. Astrophys.J., 225, L31.

Davies, J.G., Lyne, A.G. and Seiradakis, J.H., 1972. Nature, 240, 229.

Davies, J.G., Lyne, A.G. and Seiradakis, J.H., 1973. Nature, 244, 84.

Davies, J.G., Lyne, A.G. and Seiradakis, J.H., 1977. Mon.Not.R.astr.Soc. 179, 635.

Davis, M.M., Taylor, J.H., Weisberg, J.M. and Backer, D.S., 1984. Proceedings of NRAO workshop on Millisecond Pulsars, p12.

Gold, T., 1968. Nature, 218, 731.

Goldreich, P. and Julian, W.H., 1969. Astrophys.J., 157, 869.

Graham, D.A., Mebold, U., Hesse, K.H., Hills, D.L. and Wielebinski, R., 1974. Astr.Astrophys., 37, 405.

Groth, E.J., 1975. Astrophys.J.Suppl.Ser. 29, 431.

Gunn, J.E. and Ostriker, J.P., 1970. Astrophys.J., 160, 979.

Hall, A.N., 1980. Mon.Not.R.astr.Soc., 191, 751.

Helfand, D.J., Taylor, J.H., Backus, P.R. and Cordes, J.M., 1980. Astrophys.J., 237, 206.

Hewish, A., Bell, S.J., Pilkington, J.D.H., Scott, P.F. and Collins, R.A., 1968. Nature 217, 709.

Hulse, R.A. and Taylor, J.H., 1975. Astrophys.J.(Lett.)., 195, L51.

Large, M.I. and Vaughan, A.E., 1971. Mon.Not.R.astr.Soc., 151, 277.

Lyne, A.G., 1974. Galactic Radioastronomy, I.A.U. Symposium No.60., Reidel, p87.
Lyne, A.G., 1984. Nature, 310, 300.
Lyne, A.G., Anderson, B. and Salter, M.J., 1982. Mon.Not.R.astr.Soc., 201, 503.
Lyne, A.G., Manchester, R.N. and Taylor, J.H., 1985. Mon.Not.R.astr.Soc., 213, 613.
Lyne, A.G., Ritchings, R.T. and Smith, F.G., 1975. Mon.Not.R.astr.Soc., 171, 579.
Manchester, R.N., Lyne, A.G., Taylor, J.H., Durdin, J.M., Large, M.I. and Little, A.G., 1978. Mon.Not.R.astr.Soc., 185, 409.
Manchester, R.N., Taylor, J.H. and Van, Y.Y., 1974. Astrophys.J., 189, L119.
Manchester, R.N. and Taylor, J.H., 1977. Pulsars, W.H. Freeman and Co. San Francisco.
Narayan, R. and Vivekanand, M., 1983. Astr.Astrophys., 122, 45.
Pacini, F., 1967. Nature, 216, 567.
Radhakrishnan, V. and Cooke, D.J., 1969. Astrophys.Lett., 3, 225.
Ritchings, R.T., 1976. Mon.Not.R.astr.Soc., 176, 249.
Ruderman, M., 1976. Astrophys.J., 203, 213.
Salter, M.J., Lyne, A.G. and Anderson, B., 1979. Nature 280, 477.
Stokes, G.H., Raylor, J.H. and Dewey, R.J., 1985. Astrophys.J., 294, L21.
Tammann, 1982. In: Supernovae: A survey of current research, p371. Eds. Rees, M.J. and Stoneham, R.J., Reidel, Dordrecht, Holland.
Taylor, J.H. and Manchester, R.N., 1977. Astrophys.J., 215, 885.
Weisberg, J.M. and Taylor, J.H., 1984. Phys.Rev.Lett., 52, 1348.

ENERGETIC RADIATION FROM MAGNETIZED NEUTRON STARS

Malvin A. Ruderman*
Department of Physics, Columbia University

1. INTRODUCTION

Intense beams of very energetic radiation have been observed from at least two rather different families of neutron stars:

i) Very young radiopulsars such as the Crab and Vela pulsars;
ii) Much older neutron stars in binaries identified mainly by the X-ray emission associated with accretion from their companions.

The isolated radiopulsars are much more abundant: About one of each five or so supernova explosions leaves behind a highly magnetized rapidly spinning pulsar (Helfand 1982) imbedded in and powering an X-ray emitting nebula. The present pulsar birthrate appears to be of order one every century so that the total population of neutron stars in our Galaxy which have been formed in this way $\sim 10^8$. Because such neutron stars turn off as radiopulsars after 10^7 years, only about 10^5 are presently observable. Of these over 400 have now been detected. Many, perhaps most, of these were probably formed from binaries which broke apart as a consequence of the explosion which formed the neutron star.

Neutron stars which are still in binaries are a very much smaller population. They are formed in relatively rare events when a supernova explosion does not disrupt the binary, perhaps in mild explosions when certain White Dwarfs accrete enough mass from a companion to allow electron capture to begin in their cores.

Among all the observed radiopulsars only 10^{-2} are in binaries. These neutron stars have weaker magnetic fields and longer spin-down times than those of typical isolated radiopulsars, and they probably will have a longer life as observable radiopulsars. The birthrate of such neutron stars in binaries would then be of order, 10^{-2} - 10^{-3}

*Research supported in part by the National Science Foundation under grant NSF PHY 80-23721.

F. Pacini (ed.), High Energy Phenomena Around Collapsed Stars, 145–191.

that of isolated neutron stars, leading to an expected Galactic population $\sim 10^5 - 10^6$. A birthrate of neutron stars in binaries of at least 10^{-4} yr and a total Galactic population in excess of 10^6 is inferred from an estimated 10^5 year lifetime of 20 X-ray pulsars presently in strongly radiating accreting binaries. Over 30 neutron stars with close companions have definitely been observed as accreting X-ray binaries, and perhaps 300 have been seen through such X-rays and other kinds of emission. (If, as is commonly suggested, γ-ray burst sources are a nearby population of old neutron stars in weakly accreting binaries, another few hundred have been observed.) Thus despite the great disparity in population the number of observed and possibly of observable binary and isolated neutron stars may be quite comparable.

Typical radiopulsars have been observed only through their coherent microwave emission, but this usually accounts for only a very small fraction of their measured spin-down energy loss rate. A rich variety of much more energetic and more intense electromagnetic and particle emission has been observed, or inferred, from certain pulsars which have both large magnetic fields and short periods. For (isolated) radiopulsars these include:

1) γ-rays in the MeV-GeV range from the young (10^3 yr) Crab and (10^4 yr) Vela pulsars;
2) (pulsed) X-rays from several pulsars including the Crab pulsar and its very close relation (LMC 0540-693);
3) strong optical emission from the Crab pulsar and LMC 0540-693, and weak optical emission from the Vela pulsar;
4) 10^{12} eV γ-rays from the Crab and probably also the Vela pulsar;
5) 10^{15} eV γ-rays from the Crab which may have been observed at least occasionally; this would imply the creation of a nucleon beam of even greater energy;
6) very intense $e^{\pm}$ winds of average energy $\langle E \rangle \sim 10^{12}$ eV inferred for the Crab pulsar by modelers both of the pulsar EM radiation and of the synchrotron radiation from the surrounding nebula which is presumed to be supported by injection from the pulsar. All pulsars which have a spin down energy loss rate exceeding 10^{34} ergs s^{-1} have an observed X-ray synchrotron halo (except for the distant very low magnetic field "millisecond pulsar") (Helfand 1982).

Members of the second family of (generally older) neutron stars in binaries are identified by X-ray emission from companion fed accretion. Claimed observations of much more energetic emission from such systems include:

1) 10^{12} eV γ-rays from Her X-1, 4U0115+53, and Cyg X-3.
2) 10^{15} eV γ-rays from Her X-1, Vela X-1, LMC X-4, and Cyg X-3.

There exist other energetic radiation emitters whose sources are unidentified, but for which neutron stars are likely candidates:

1) unknown sources of 10 keV - 10 MeV "γ-ray bursts". The bursts are frequent, isotropic, and rarely repeating from

any one direction so that old neutron stars (possibly in binaries) may be the only sufficiently abundant (exotic) source candidates;

2) about 20 unidentified Galactic 10^2 MeV - GeV (Cos B) sources, even more intense than the Crab but not associated with younger supernovae (or with observed radiopulsars);

3) X-ray binaries such as Cyg X-3 which are the sources of superenergetic γ-rays and possibly other kinds of emission, but for which blackholes or White Dwarfs have not been excluded as the collapsed star which is almost certainly part of the binary system.

4) strange sources of energetic emission, such as SS433 or Geminga.

In these lectures we shall consider first, and in most detail, the problems of the origin of the non-radio emission from the young isolated radiopulsars, effects from the evolution of neutron star properties (especially magnetic fields), and how these might lead to older neutron stars in accreting binaries becoming sources of superenergetic radiation. We shall begin in the next section with the non radio-emission from the Crab and Vela neutron stars. This will be the theme of Sections 2-12. The following Sections 13-17 can be read quite independently of the earlier ones.

2. EXTREME RELATIVISTIC PARTICLE WINDS FROM YOUNG PULSARS

While there is no compelling argument that isolated, high magnetic field, rapidly spinning neutron stars must be sources of a very strong, extreme relativistic $e^{\pm}$ winds, attempts to understand both the observed synchrotron emission from surrounding nebulae and the near environment of pulsars are strongly supportive of these neutron stars being a source of such winds. We consider first some features of the Crab pulsar and its nebula which bear on this hypothesis.

The Crab pulsar has a period (P) $\sim$ 0.33 s. In almost all estimates and models for energy loss mechanisms for spinning down neutron stars the measured spin down rate of the Crab pulsar implies a surface dipole magnetic field component $B_s \sim 4\times10^{12}$ G. If such a spinning magnetic dipole is inserted into a strongly ionized plasma of electron density n_e whose plasma frequency

$$\omega_p = \left(\frac{4\pi n_e e^2}{m}\right)^{1/2} = 5\times10^4\ n_e^{\ 1/2}\ s^{-1} \qquad (2.1)$$

exceeds the electromagnetic radiation frequency of the spinning dipole,

$$\Omega(\text{Crab}) \sim 200\ s^{-1}\ , \qquad (2.2)$$

then an empty (except for 33 Hz EM radiation) expanding bubble would form around the spinning dipole. For a plasma like the interstellar medium, $n_e \sim 10^{-1}$ cm^{-3} and the "overdense" condition for pushing out the plasma would be easily met. The expansion velocity of the expanding bubble would be determined by the rate at which EM energy is pumped into it ($\sim 10^{39}$ ergs s^{-1} for the present Crab pulsar) and the inertia of the mass which must be accelerated as it is pushed away by the expanding bubble. For a Crab-like magnetic dipole inserted into a typical interstellar region, the resulting bubble would have a radius r_B growing with time (t) as

$$r_B(t) \sim 2\times 10^{18} \left(\frac{t}{10^3 \text{ yrs}}\right)^{3/5} \text{cm} , \tag{2.3}$$

at a velocity

$$\frac{d}{dt} r_B(t) \sim 2\times 10^{8}\left(\frac{10^3 \text{ yr}}{t}\right)^{2/5} \text{cm s}^{-1} , \tag{2.4}$$

and containing a characteristic magnetic field

$$B_B(t) \sim 10^{-3} \left(\frac{10^3 \text{ yrs}}{t}\right)^{3/5} \text{G} . \tag{2.5}$$

The t = 931 yr old Crab Nebula has, indeed, r_B, dr_B/dt, and B_B very similar to those of Eqs. (2.3), (2.4), and (2.5) (although its previous expansion velocity must have been somewhat smaller, not greater than its present one). However, instead of an empty volume (except for the 33 Hz EM radiation) it is filled with extreme relativistic $e^-(e^+)$ whose intense synchrotron radiation in $B \sim 10^{-3}$ G is a substantial fraction of the total energy rate injected by the pulsar into the nebula around it.

The synchrotron X-ray and optical emission from the Crab Nebula together with the growing nebulae expansion kinetic energy imply an injection of $e^-(e^+)$ at a rate

$$\dot{N}(e^-) \sim 10^{39} \text{ s}^{-1} \tag{2.6}$$

with

$$\langle E(e^-)\rangle \sim 10^{12} \text{ eV} \tag{2.7}$$

The electric current from the huge electron flow of Eq. (2.6) must be almost exactly balanced in the nebula by one of opposite sign, since for the net current (J),

$$J \lesssim B_B\, cr_B \sim 10^{26}\ (\text{cgs}) \sim 10^{-4} e\, \dot{N}(e^-)\ . \qquad (2.8)$$

Thus it is difficult to model the nebula unless the net local current on a scale of 10^{-4} r_B is negligible; e^- could not be accelerated out into the nebula from a region around the star and return to be reaccelerated along a very different path. Moreover it is difficult to see how the acceleration of such a net current as big as that of Eq. 2.6 would not generate a magnetic field greatly exceeding that of the accelerator, a relationship which generally cannot be maintained. If such an $\dot{N}(e)$ passed into and out of the spinning Crab neutron star, the resulting $\vec{B}_s \times \vec{J}$ torque would spin down the star in less than a day. Even if $\dot{N}(e^-)$ passed only through the stellar magnetosphere (extending out to a "corotation radius" $c/\Omega \sim 10^8$ cm) the electron current's magnetic field,

$$B(e^-) \gtrsim \frac{\dot{N}e}{c^2}\Omega \sim 10^{11}\ G\ , \qquad (2.9)$$

would greatly exceed that from the star ($B \sim 10^7$ G at $r \sim c\Omega^{-1}$). Therefore if e^- are injected and accelerated into the nebula by the pulsar, which is certainly the power source, an almost equal number of positively charged particles must flow with them. However there is no mechanism for lifting a flux of 10^{39} s^{-1} of positively charged ions from the neutron star surface to locally cancel the electric current of a similar e^- flux: an electric field which pulls out one will push back the other. However, as we shall discuss in more detail later, the near environment of a sufficiently high B_s, short period, neutron star is expected, for purely theoretical reasons, to be a copious source of $e^\pm$ pairs. Moreover that flux, $\dot{N}(e^\pm)$ is many orders of magnitude greater than the net current which may flow through the star or in its magnetosphere. Therefore, it is extremely suggestive that $e^\pm$ pair production near the Crab neutron star is the source of particles which are accelerated to extreme relativistic energies and injected into the nebula. The rather successful model for the nebula emission of Kennel and Coroniti (1984) begins with just this assumoption. The accelerator itself could not be a stationary (in the corotating frame of the spinning pulsar) electric field since this would accelerate 10^{39} s^{-1} of e^- and e^+ in opposite directions and thus give an unacceptably huge magnetic field. Exactly equal acceleration of e^+ and e^- so that no net electric current results, would be achieved from the copious injection of $e^\pm$ pairs into the strong magnetic dipole radiation of an unaligned pulsar. Pairs in the slowly varying ($eB/mc \gg \Omega$) electromagnetic field will be strongly accelerated until the electromagnetic power is mainly transferred to that of an extreme relativistic $e^\pm$ wind.

A quite different argument in support of very strong $e^\pm$ pair creation around the Crab pulsar follows from detailed modeling of the possible sources of the observed pulsar optical and X-ray radiation (cf. Section 9). Incoherent synchrotron (or curvature) radiation is simply too inefficient to give the measured intensities unless the number of radiating $e^-(e^+)$ exceeds 10^{37} s^{-1}.

For the Crab pulsar and LMC 0540-693 Eq. (2.6) implies that the spinning neutron star produces $e^\pm$ at the rate

$$\dot{N}_{CRAB}(e^\pm) \sim 10^{39}\ s^{-1} \tag{2.10}$$

A similar analysis for the synchrotron X-ray nebulosity around the Vela pulsar gives

$$\dot{N}_{VELA}(e^\pm) \sim 10^{36}\ s^{-1} \tag{2.11}$$

Since X-ray synchrotron halos have been observed around pulsars whose total energy loss rate exceeds 10^{34} ergs s^{-1}, it is extremely suggestive that intense extreme relativistic $e^\pm$ winds which carry away most of the neutron star spin down energy are a common property of such pulsars.

We turn next to the ways in which these $e^\pm$ pairs may be created.

3. PAIR PRODUCTION MECHANISMS

Production of $e^\pm$ pairs in the magnetosphere of a young pulsar would be caused by the materialization of high energy γ-rays. This can be accomplished in several ways. The most important are:

1) the conversion $\gamma \rightarrow e^+ + e^-$ in a very strong magnetic field (Sturrock 1971). Although the mean free path for such conversion is finite as long as $p_\perp$, the component of photon momentum perpendicular to B, satisfies $p_\perp > 2m_e c$, it is generally longer than a pulsar magnetosphere radius unless $B_{12}\, p_\perp/mc \gtrsim 1$. This pair creation mechanism is expected to be important only in the very strong ($B_{12} \sim 1$) magnetic field within several radii of the stellar surface.

2) the collision of a high energy γ-ray with another γ-ray or X-ray ($\gamma + \gamma' \rightarrow e^- + e^+$). This process has a maximum cross section of about one third the Thomson cross section, which is approached in the range between somewhat above and several times the center of mass threshold energy of $2m_e c^2$.

3) The conversion process (2) when a superenergetic $10^{12} - 10^{13}$ eV γ-ray materializes as the result of a collision with an optical or IR photon.

The necessary energetic γ-rays can be radiated (even, in part, by the very $e^\pm$ pairs they create) in several ways when appropriate accelerators exist.

a) Curvature radiation from an $e^-(e^+)$ moving along a curved magnetic field line. An electron with a Lorentz factor γ_e

moving on a trajectory with curvature radius s_c emits photons with characteristic energy

$$\hbar\omega \sim \gamma_e^3 \, \hbar c/s_c \, . \tag{3.1}$$

The electron energy loss rate to such radiation.

$$\dot{E}_{rad} \sim \frac{e^2 c \gamma_e^4}{s_c^2} \tag{3.2}$$

b) Inverse Compton scattering of extreme relativistic $e^-(e^+)$ by intense soft photon fluxes ($e + \gamma \rightarrow e' + \gamma'$).

c) Synchrotron Radiation by extreme relativistic $e^-(e^+)$ spiralling along $\vec{B}$. The emitted photons have a characteristic energy

$$\hbar\omega \sim \frac{\gamma_e^2 \, eB \, \hbar}{mc} \sin\theta \tag{3.3}$$

where θ is the spiral pitch angle.

d) The neutral pion decay $\pi^\circ \rightarrow \gamma + \gamma$. γ-rays of energy above 10^{15} eV are almost certainly not made by any of the above mechanisms or by any other radiation mechanism from the acceleration of a real charged particle. If such an energetic photon were made by an $e^-(e^+)$ the particle must have $\gamma_e > 10^{15}$ eV/mc^2 = 2×10^9. To achieve this, despite the radiation reaction associated with moving along curved $\vec{B}$ lines implicate in Eq. (3.2), needs a potential drop $e\gamma_e^4/s_c \sim 3\times10^{20}$ volts for $s_c \sim 10^{10}$ cm. Thus unless the local $\vec{B}$ were rather incredibly straight, the needed potential greatly exceeds (by four orders of magnitude for the Crab) the maximum potential drop which can be generated by the pulsar. If the source of radiation is an accelerated proton (or nucleus) the needed energy is achieved by a 10^{15}-10^{16} voltage drop which would give a Lorentz $\gamma_p \sim 10^6 - 10^7$. However, even in that case, neither Eq. (3.1) nor Eq. (3.2) would give photon energies approaching 10^{15} eV and mechanism b) with proton inverse Compton scattering has a cross section on X-rays only 10^{-7} that for electrons and is thus negligible. However the protons can make π° reasons by collisions with nucleons or from photopion production when 10^{15}-10^{16} eV protons pass through X-ray beams especially near the large, spin 3/2, isotropic spin 3/2 resonance at 300 MeV-center of mass energy where $\sigma(\gamma + p \rightarrow \pi^\circ + p) \sim 5\times10^{-28}$ cm^2 and $\sigma(\gamma + n \rightarrow \pi^\circ + n) \sim 4\times10^{28}$ cm^2. For π°-mesons produced in near resonance proton-X-ray collisions the meson laboratory energy $E(\pi^\circ)$, is about 1/3 that of the proton $E(p)$, and

$$E(p) \sim (3\text{-}6)\ E(\gamma) \tag{3.4}$$

In proton collisions the center of mass energy ~ 10^{12} eV so that multimeson production dominates. The energy of any γ-ray from π°-decay is then expected to be much less than that from Eq. (3.4).

We shall consider below in Sections 9, 10, and 12 the application to young pulsars of combinations of the above γ-ray creation and conversion mechanisms. First, however, we shall review the observations of energetic radiation from certain of these pulsars.

4. OBSERVATIONS OF ENERGETIC RADIATION FROM THE CRAB AND OTHER YOUNG PULSARS

About 10^{-3} of the total spindown energy loss rate of the Crab pulsar is emitted as X-rays and γ-rays. The spectrum between optical and GeV energies is shown in Fig. (12). In this entire range the emission is into two cuspshaped subpulses separated by 140° as shown in Fig. (1a). There is also a coincident cusped double pulse structure for part of the radioemission, the so-called microwave "mainpulse" and "interpulse". A γ-ray mainpulse has also been detected at 10^{12} eV; an interpulse contribution may also exist at this energy as shown in Fig. 1b. The power at this γ-ray energy is about 10^{-2} that radiated into lower energy X-rays and γ-rays. Weaker evidence exists for 10^{15} eV γ-ray radiation from the direction of the Crab. More extensive air showers at around this energy were observed at the ground by the Lodz group (Wdowczyk 1984) when the Crab was overhead during 1968-72 and 1975-82. The Utah "Fly's Eye" experiment (Boone et al., 1985) which observed Cerenkov light from γ-$e^{\pm}$ showers saw a similar peak on 9 December 1980, but not during early February of 1981. A compilation of relevant results by Wdowczyk (1984) is shown in Fig. 2. (There is no evidence that these 10^{15} eV γ-rays are coincident with the lower energy pulsed γ-rays from the Crab pulsar or even that they are pulsed at all.)

A double pulse structure (~70° separation) is also present in the observed optical and X-ray pulse structure from LMC 0540-693, a young pulsar in the Large Magellanic Cloud similar in age to the Crab. The Vela pulsar has a double pulsed γ-ray spectrum similar to structure to that from the Crab pulsar (~ 140° supbpulse separation), no, so far observed, X-ray intensity, and only a weak double pulse optical emission, whose separation (~ 90°) differs from that of its γ-rays (Fig. 3). The measured spectrum is shown in Figure 12.

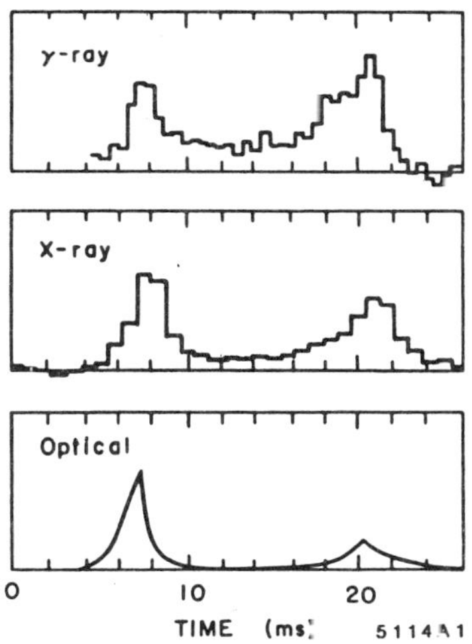

Figure 1a: Average optical, X-ray and γ-ray pulse profiles of the Crab pulsar (after Manchester and Taylor 1977).

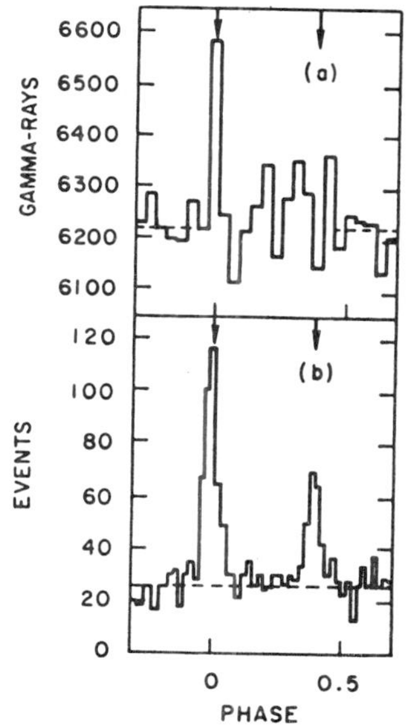

Figure 1b: γ-ray pulse profile of the Crab pulsar at 10^{12} eV (Dowthwaite et al. 1984) and 10^{8} eV (Wills et al. 1982).

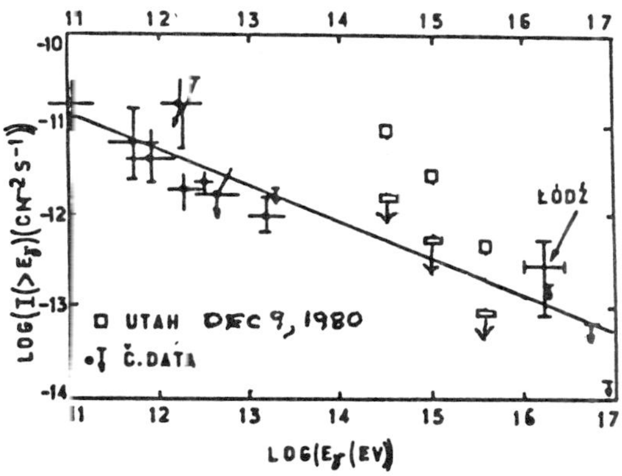

Figure 2

Energy spectrum of superenergetic γ-rays from the direction of Crab Nebula. (after Wdowczyk 1984) the solid line corresponds to the spectral index 0.4. Below the Utah, Dec. 9, 1980 data are the Feb. 1, 2, 6 1981 data.

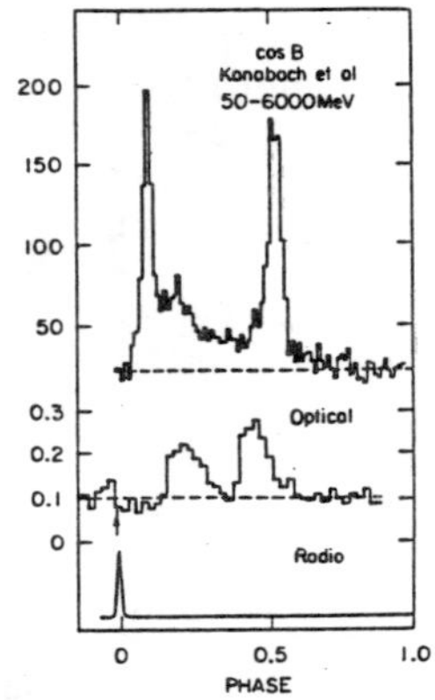

Figure 3
Average γ-ray, optical and radio profiles of the Vela pulsar (after Manchester and Taylor 1977 and Kanbach _et al._ 1980).

5. SOME IMPLICATIONS OF γ-RAY OBSERVATIONS

The γ-rays needed to create the $e^{\pm}$ wind of Section 2 and those that are observed directly in the double pulses indicate that associated with young high magnetic field rapidly spinning neutron stars are strong and efficient charged particle accelerators. There is no consensus on their structure (or even their location), or on whether they include static accelerators ($\vec{E}$ partly parallel to $\vec{B}$ and static in a corotating frame) or only dynamic ones (e.g. acceleration by electromagnetic waves). For accelerators positioned well inside the light cylinder one would expect some form of static acceleration based upon an electric field component along $\vec{B}$ whose source is the electric charge distribution (or lack of it) in the star and/or its magnetosphere. Then injected positive and negative charges would be oppositely accelerated. Accelerators positioned well outside the light cyclinder suggest a dynamic model with $\vec{E} \times \vec{B}$ of the star's magnetic dipole radiation responsible for an equal acceleration for injected e^- and e^+.

From observations and theory we take the following constraints:

(1) The $e^{\pm}$ production mechanisms and, presumably, the accelerator of the particles which produce the needed γ-rays are not effective at distances from the star (r) much greater than $c\,\Omega^{-1}$. This follows because all of the the Section 3 $e^{\pm}$ production mechanisms (a,b, or c) need either large $\vec{B}$ or large fluxes of IR photons, optical photons, or X-rays which may exist only nearer to the neutron star.

(2) The 10^{12} eV Crab (and Vela) γ-rays imply an emission distance $r \gtrsim c/3\,\Omega$; otherwise pair conversion by mechanism (1) of Section 3 in the local $\vec{B}$ will absorb the emitted γ-rays before they can propagate away from the star.

Therefore if $e^{\pm}$ production and the observed γ-ray emission are from the same or closely contiguous regions, the needed particle accelerator is in the outer "magnetosphere" (i.e. $r \lesssim c\,\Omega^{-1} \equiv$ light cylinder radius) or the trans-light-cylinder region. (For other proposed sites for the origin of some γ-rays cf. Tademaru 1973, Hardee 1977, Harding et al. 1978, Daugherty and Harding 1982).

(3) All of the observed energetic radiation from the Crab pulsar - from optical frequencies to 10^3 GeV - probably come from the same outer magnetosphere region since no differences in their pulse arrival times have been detected.

(4) The double beams of radiation are rotating "fan" beams (i.e., latitude coverage >> observed subpulse width) rather than filled cones. If they were filled cones, the probability for having detected both the Crab and Vela pulsars in searches for pulsars in young supernova remnants would be only about $(1/5)\times(1/5) = 1/25$.

Unfortunately there is not yet a complete and self consistent global theory of current flow, pair creation, and particle acceleration around a rotating magnetized neutron star (cf. Michel 1982). We shall therefore begin with a greatly over simplified approximation to a pulsar magnetosphere.

6. "STATIC" MAGNETOSPHERE MODELS

A "static" solution for the interior and exterior of a rotating magnetized neutron star is one in which there is no particle flow or time dependence when observed from a reference frame rigidly attached to the rotating star. Such a solution with realistic boundary conditions far from the star does not exist (Pilipp 1974, Michel 1982). In this section we shall discuss static solutions with a rather nonphysical boundary far from the star. In Section 7 we shall try to guess what happens when the unphysical boundary is removed.

We first consider why, mainly because of $e^{\pm}$ pair creation, the region around a young spinning magnetized neutron star (or, probably, any radiopulsar) is not expected to have any large region without plasma in t is near environment.

In the laboratory a conductor rotating uniformly with angular speed $\vec{\Omega}$ has an interior velocity

$$\vec{v} = \vec{\Omega} \times \vec{r} , \tag{6.1}$$

an interior electric field

$$\vec{E} = \frac{\vec{v}}{c} \times \vec{B} , \tag{6.2}$$

and a charge density

$$\rho = \rho_0 \equiv \frac{\vec{\nabla} \cdot \vec{E}}{4\pi} = \frac{\vec{\Omega} \cdot \vec{B}}{2\pi c} + \text{terms of order } \vec{\nabla} \times \vec{B} \qquad (6.3)$$

which depend upon the current distribution and vanish where there is none. From Eq. (6.2) it follows that

$$\vec{E} \cdot \vec{B} = 0 \qquad (6.4)$$

within the conductor, but not at its surface, both in the laboratory and co-rotating frames. In the non-relativistic corotatating frame the potential (V) satisfies

$$\nabla^2 V = 4\pi(\rho - \rho_0) \qquad (6.5)$$

If everywhere outside of the spinning solid, $\rho = 0$, $\vec{E} \cdot \vec{B} \neq 0$, and, between distant points near the star, the potential difference $\Delta V \sim \Omega^2 Bc^{-1}$. Maximum rotation speeds for a 10^2 cm 10^4 G rotator give E•B outside the rotator ~10 Volt cm^{-1}, and $\Delta V \sim 10^3$ Volts, quite insufficient to pull electrons or ions from the rotator surface. However, for the Crab pulsar, with surface $B \sim 3\times10^{12}$ G, the exterior $\vec{E} \cdot B$ would be 3×10^{12} Volt cm^{-1} if the exterior region were to remain empty. This is quite impossible in the presence of such a strong $\vec{E}$ along $\vec{B}$. Charge will be pulled from the star until Eq. (6.4) is achieved wherever $\rho \neq 0$, including the surfaces which separate $\rho \neq 0$ regions from those in which $\rho = 0$ (Goldreich and Julian 1969).

One type of solution which achieves this for a dipolar $\vec{B}$ aligned along the $\vec{\Omega}$ of a (charged) neutron star has been given by Michel and co-workers (Krause-Polstoroff and Michel 1985a and 1986b and references therein) and is sketched in Figure 4. A finite region of (charge separated) plasma forms above the magnetic poles of the spinning star. In it Eqns. (6.1) - (6.5) hold and the polar plasma acts like a rigid extension of the star. An oppositely charged equatorial plasma bulge extends out from the star. In it Eqn. (6.4) still holds, as in the static polar bulge solution. However, the angular velocity of the plasma, though constant along each $\vec{B}$ field line which threads this plama bulge, is not constant between different field lines; thus the equatorial bulge does not corotate exactly with the star. Outside of these polar and equatorial plasma extensions of the

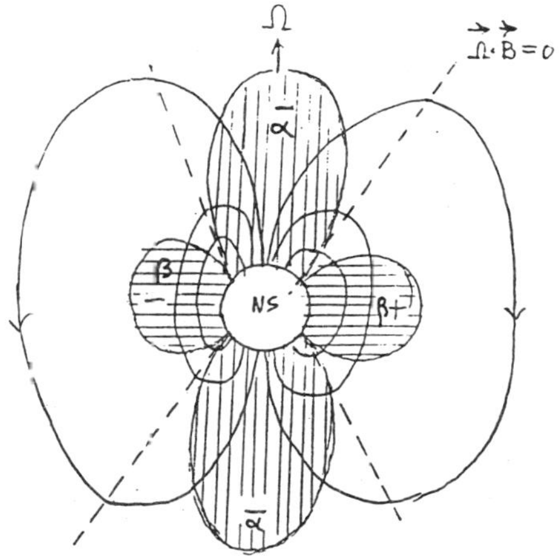

Figure 4

A possible magnetosphere of a rotating neutron star with an aligned dipolar magnetic field after Krause-Polstoroff and Michel (1985a and b). Region α corotates with the star and is, electrically, a rigid extension of it. Region β is oppositely charged and has a constant, but different, rotation along each field line through it. The charge density ρ = 0 in the exterior regions between and beyond α and β. The magnetic field lines are equipotentials within α, β, and the neutron star but not in the regions where ρ = 0.

star, $\rho \equiv 0$. It is only in this $\rho = 0$ region that $\vec{E} \cdot \hat{B} \neq 0$, but it must still vanish on the surfaces which separate the $\rho = 0$ region from that in which $\rho \neq 0$. [There is no static solution in which all of the extended exterior plasma has $\vec{E} \cdot \vec{B} = 0$ and exactly corotates with the star (Pillip 1974)]. It is not yet known whether a qualitatively similar solution exists for a non-aligned dipole or for much more complicated stellar magentic field distributions. However, even if aligned, a rapidly spinning neutron star with a strong magnetic dipole will begin to fill the $\rho \equiv 0$ void because of $e^{\pm}$ pair production there and subsequent separation of e^+ and e^- by $\vec{E} \cdot B$. Near the star the main production process is the combination 1-a of Section 3 which is expected to be a copious $e^{\pm}$ source wherever the potential drop along $\vec{B}$ between the polar and equatorial plasma extensions exceeds

$$\Delta V_{\pm} \sim 10^{12}\ B_{12}^{-1/7}\ s_6^{4/7}\ P^{-1/7}\ \text{Volts} \qquad (6.6)$$

(Sturrock 1971, Ruderman and Sutherland 1975), where s_6 is the $\vec{B}$ field line radius of curvatue (s_c) in units of 10^6 cm. In the stellar corotating frame the potential drop between the polar and equatorial plasmas through the $\rho = 0$ void between them of typical dimension the stellar radius R, can be of order

$$\Delta V \sim \Omega R^2 B_s / c \tag{6.7}$$

even along a $\vec{B}$ field line. This follows from Eqn. (6.5) with $\rho = 0$. For young pulsar parameters this ΔV exceeds 10^{18} volts and easily overwhelms the $\Delta V_\pm$ of Eq. (6.6) so that $e^\pm$ pair production in the plasma void can be very strong. Other $e^\pm$ creation processes which are effective in filling the $\rho = 0$ void far from the star will be discussed in Sections 9 and 10.

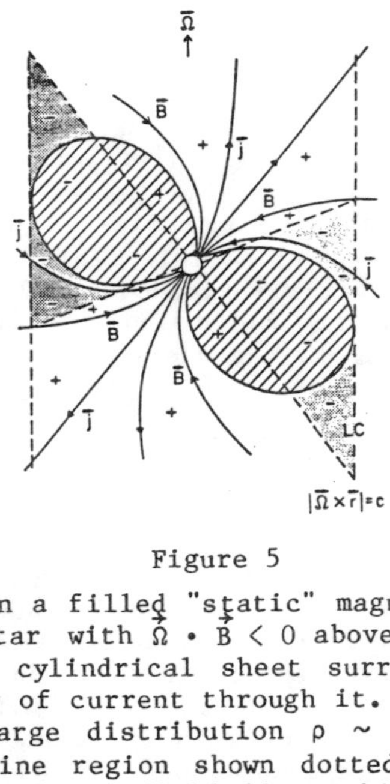

Figure 5

Charge distribution in a filled "static" magnetosphere of a spinning magnetized neutron star with $\vec{\Omega} \cdot \vec{B} < 0$ above the polar cap. Initially an insulating cylindrical sheet surrounds the magnetosphere and prevents the flow of current through it. (Its surface charge is not shown.) The charge distribution $\rho \sim \rho_0$ with the negatively charged open field line region shown dotted; $\rho_0 = 0$ on the "null surface" $\vec{\Omega} \cdot \vec{B}$. The closed field lines which do not penetrate the light cylinder ($|\vec{r} \times \vec{\Omega}| = c$) are in the shaded area. $\vec{B}$ field lines are represented by those of a static dipole, a bad approximation near the light cylinder and possibly near the star. Also shown is the assumed pattern for current flow when the insulating boundary sheet is removed. Current flow of positive charge away from the star (mainly via a non-charge-separated plasma) is assumed balanced by a return current which is mainly through the "null surface". The inward directed arrow indicates current flow direction only: negative electrons will be outflowing in the dotted region.

Since, because of very substantial pair creation, no strongly magnetized rapidly spinning pulsar magnetosphere solution exists with large plasma voids around the pulsar, we shall begin with an opposite initial magnetosphere approximation: no exterior plasma void at all around the star. Then a steady state solution which has $\vec{E} \cdot \vec{B} = 0$ wherever $\rho \neq 0$ is equivalent to having $\vec{E} \cdot \vec{B} = 0$ everywhere in the magnetosphere. However, such a plasma can have no natural boundary (i.e. one on which $\vec{E} \cdot \vec{B} = 0$ within the light cylinder and there is not yet a self consistent solution through the light cylinder. We shall, therefore, first consider the magnetosphere to be bounded by a thin insulating cylinder which does not permit current flow through itself (and whose radius is enough less than $c\,\Omega^{-1}$ that relativistic centrifual forces can be neglected). Current in the so bounded magnetosphere would flow only until $\vec{E} \cdot \vec{B} = 0$ is achieved everywhere. This in turn implies Eqs. (6.1) - (6.3); this static magnetosphere acts in every way as a rigid extension of the rotating star. It is sketched in Figure 5. However, this is not a unique solution. Plasma void ($\rho = 0$) regions can also exist in this static magnetosphere. On the void boundaries $\vec{E} \cdot \vec{B} = 0$ so that none of the boundary charge is pulled into these regions [cf. Holloway 1973, Ruderman and Sutherland 1975, Michel 1981, Cheng <u>et al.</u> 1976, Arons 1981]. A family of these is shown in Figure 6. The charge depleted slot shaped "gaps" are bounded by magnetic field lines along their length. Their ends do cross through field lines, but only near the magnetosphere's "null surface" where $\rho_0 = \vec{\Omega} \cdot \vec{B}/2\pi c = 0$ and ρ, in a filled static magnetosphere, changes sign. The exact position and shape of these ends cannot be fixed <u>a priori</u> but must be determined self consistently to give $\vec{E} \cdot \vec{B} = 0$ (Cheng <u>et al.</u> 1985a). Within the gap, away from its boundary $\vec{E} \cdot \vec{B} \neq 0$ and particle acceleration is possible. For a gap which begins and ends within the magnetosphere, the total potential drop (in a corotating frame) from end to end along $\vec{B}$ is zero. Such a gap will ultimately become filled with charge and destroyed, and is not a source for sustained particle acceleration through it. However a gap along "open" field lines which do not close within the light cylinder may accelerate particles out through the light cylinder and thus out of the magnetosphere. Because all parts of such gaps are a large fraction of the light cylinder radius away from the neutron star, the local magnetic fields are relatively very small, and the field line curvature radius is large, compared to those very near to the star. For a Crab parameter outgap $\Delta V_{\pm}$ of Eq. (6.6) exceeds 10^{16} volts, more than would be achieved along the entire length of a sufficiently thin outergap. Such a gap would not, therefore, be extinguished by the mechanism which fills the large near surface plasma voids. We argue below that such an "open" gap might be expected to form and remain in the non-static magnetosphere which would be formed if the insulating boundary cylinder were removed so that magnetospheric currents could flow through the light cylinder.

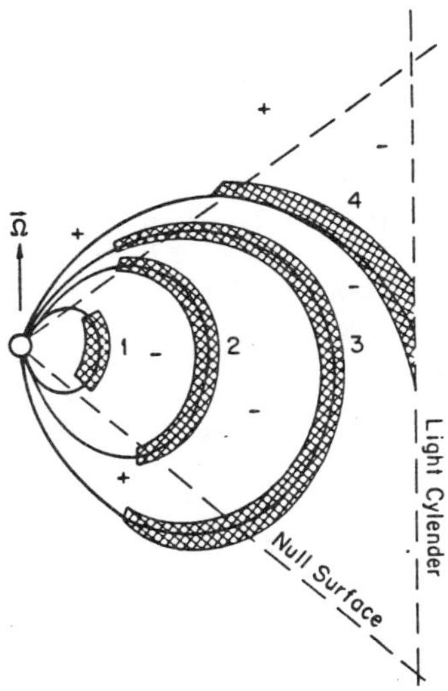

Figure 6

Four examples of thin magnetosphere gaps along $\vec{B}$ field lines in a rotating magnetosphere with axial symmetry. All charge has been removed from the interior of the gaps (shown cross hatched). In the corotating reference frame these gaps are imbedded in an equipotential conducting plasma. In gap 1 $\vec{E} \cdot B < 0$ at the gap ends; negative charge pulled into the gap through its end, would quickly replenish the negative charge intially removed and thus destroy the gap. (No charge flows across $\vec{B}$.) In gas 3, $\vec{E} \cdot B > 0$ at the gap ends (positive charge was initially removed there to form that part of the gap) and positive charge would flow into the gas to replenish what was initially removed from that gap region. Neither of these gaps is "static". Gap 2 has its end near where $\vec{E} \cdot B = 0$ and is "static". If $e^{\pm}$ production existed in it, the $\vec{E} \cdot B$ in the gap would accelerate e^{+} toward the gap ends but keep e^{-} bound thus replenishing the gap's charge deficit and annihilating the gap. Gaps of type 4 on open B field lines, but whose magnetospheric end is similar to that of gap 2, are the outergaps considered in Section 7.

7. "DYNAMIC GAPS" IN PULSAR OUTERMAGNETOSPHERES

We now remove the nonphysical thin cylindrical corotating insulating layer of Section 6, within which a static (corotating) magnetosphere could be established and maintained. Arguments for the subsequent behavior of the magnetosphere are of a handwaving (penwaving?) sort.

If the insulating sheet which, in the static magnetosphere model, separated the magnetosphere from the rest of the circumstellar region were removed, current would begin to flow along the open magnetic field lines which penetrated it. Strongly charge-depleted (i.e. $|\rho| << |\rho_0|$) or excess ($|\rho| >> |\rho_0|$) regions with $E \cdot B \neq 0$ may then form in several places in the magnetosphere for a variety of reasons: e.g. a near-surface positive ion depletion because of ion surface binding or a near surface positive ion excess from the space charge caused by ion iertia. The constraints on the location of the particle accelerator which powers the Crab pulsar's high energy radiation, as well as the large power it must supply, seem to rule out the significance for this purpose of any near surface accelerator (near surface accelerating gaps may, however, be the power source for "normal" radio emission from pulsars.) Much further from the star large charge depletion may result from charge separated flow out along upward curving open B-field lines (Arons 1981) or from current flow through the null surface. Although we shall discuss only the latter cause for charge depletion many of the results may well hold for either kind of outer magnetosphere gap origin.

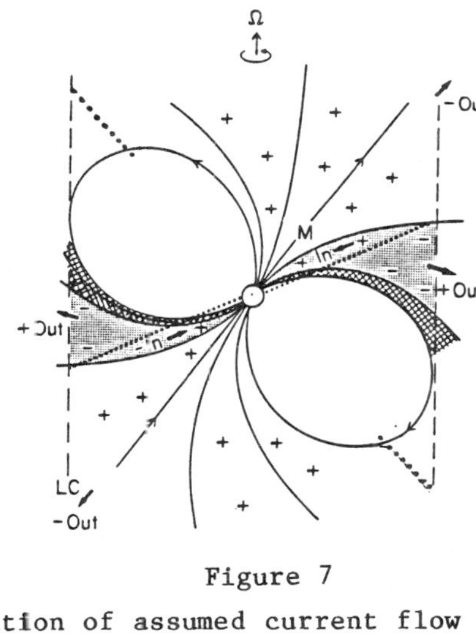

Figure 7

A schematic indication of assumed current flow patterns on open field lines. Current flow of positive charge away from the star (mainly via non-charge-separated plasma) is assumed balanced by a return current which is mainly through the "null surface". The directed arrows indicates paticle flow direction only: negative electrons will be outflowing in the dotted region. Negative charge is assumed to flow out leaving behind a partial void ($\rho \sim 0$) in the outermagnetosphere near the null surface. The coulomb field of $\rho - \rho_0 \sim -\rho_0$ pushes positive charge on the other side of the null surface toward the star. Pair production processes can fill all voids except the outgaps shown cross hatched. Two other possible outergaps which begin at the other null surface are not shown (cf. figure 10).

If the initial particle flow were only outward through the light cylinder in the general pattern of Figure 7, this flow would cause a charge depletion region to grow around the null surface as indicated in Figure 8. Negative (positive) charge moving out through the light cylinder leaves a void ($\rho = 0$) which, according to Eq. (6.5), has the electrostatic properties of positive (negative) charge in the depleted region. This pushes negative (positive) charge back toward the star from the starward side of the null surface. As the charge depleted regions (voids) grow so does $E \cdot B$ within them until pair production processes within a void, of the sort discussed in Section 3, limit further growth by supplying the charge carried away when current flows through the null surface. When γ-rays from one void region propagate through another one the latter void may be quenched by $e^{\pm}$ materialization there. However, all of the processes for making high energy γ-rays give rise mainly to γ-rays initially directed essentially tangent to the local B. As a result, a copious γ-ray production from voids in certain regions can quench voids in other regions but not vice-versa. The only void which cannot be quenched because, for geometrical reasons, points within it cannot be reached by high energy γ-rays created near its own upper (open) field line boundary or by those created in other regions are shown as A in

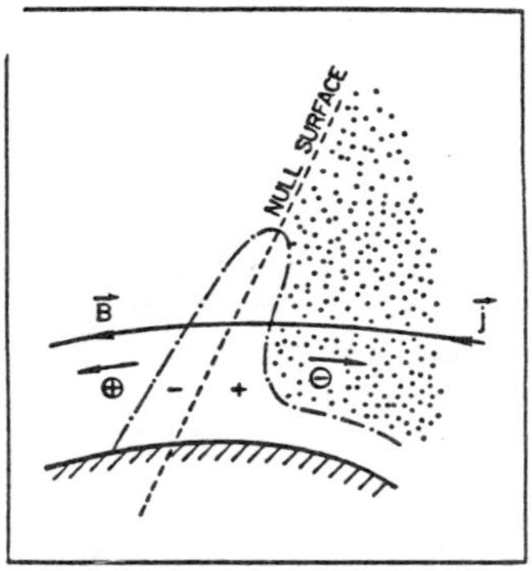

Figure 8

A growing deficient region ($\rho \sim 0$) in the outermagnetosphere near a null surface. The partial void near that surface from negative charge outflow is labelled + for the equivalent electrostatic effect of the absence of −. Its coulomb field pushes positive charge + on the other side of the null surface toward the star leaving behind a partial void which acts as a negative charge region (labelled −).

figure 9. It is the cross hatched region of figure 10. The right hand gap is bounded below by the last closed magnetic field line, bounded above by an open field line boundary, and ends within the magnetosphere, slightly inward of the null surface where the gap $E \cdot B = 0$. This is a special case of an open static gap of Section 6 and Figure 6. There are a pair of such gaps, one on each side of the star. (Another pair, not shown in Figures 7, is indicated in Figure 10 and discussed in Section 8.)

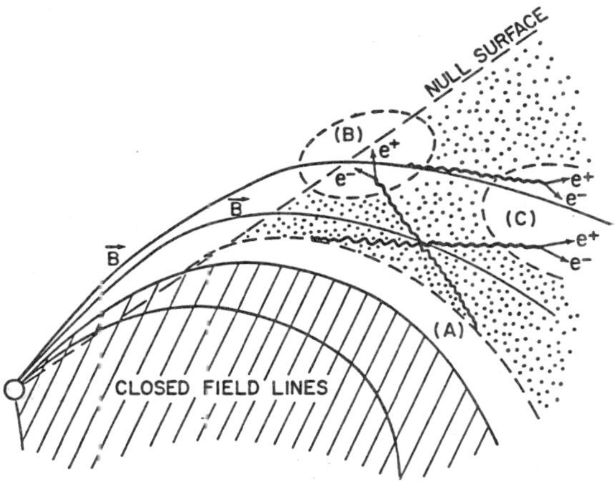

Figure 9

Three hypothetical outermagnetosphere charge depletion regions ("outergaps") in which γ-rays from other such regions make $e^{\pm}$ pairs. The lower boundary of region (A) is the last closed (= first open) B field line. Energetic γ-rays from (B) and (C) do not penetrate it. Gaps (A) and (C) fo not end within the magnetosphere light cylinder.

If the possibility of inward flow through the light cylinder is now included such a gap remains unquenched. It has a roughly constant electric field along B through its central region, and falls to zero only at its inner end. Negative (positive) charge sucked into this outergap through the light cylinder will be accelerated through the gap and flow down onto the star below. To quench an empty outergap such charges, and oppositely charged ones pulled into the gap from its inner end, would have to stop (or be strongly slowed) within the gap, a behavior quite opposite to their accelerated passage through it. Thus an outergap, as sketched in Figure 7, once formed, may be stable and continue to grow in thickness until the E • B through its interior (which, for thin gaps, is proportional to the square of the gap thickness) sustains enough $e^{\pm}$ production near the gap's open field line boundary to limit further growth. Details of the structrure of such dynamically limited gaps and the energetic radiation from them have been considered by Chen et al. (1985a,b). Their results are summarized below in Sections 8-12.

8. PULSE STRUCTURE OF ENERGETIC RADIATION SUSTAINED BY OUTERGAP PAIR PRODUCTION

Although details of outergap $e^{\pm}$ production can be complicated, and very sensitive to pulsar period, certain feature of the gap powered radiation emission are not. Symmetry of the magnetosphere structure always results in pairs of gaps on opposite sides of the magnetosphere as shown in Figures 7 and 10. The strongest part of the radiation is found to come from synchrotron and inverse Compton emission by pairs created just above the gaps' open field line boundary. A quite general feature of such radiation is that it is mainly emitted in both directions almost tangent to the local $\vec{B}$, reflecting the approximate symmetry between the flow of $e^-(e^+)$ outward and $e^+(e^-)$ inward for all $e^{\pm}$ created within the gap. [This fails markedly, of course, near the gap ends since the total $e^-(e^+)$ current flow there is the sum of all $e^-(e^+)$ creation within the gap and there has not yet been space enough a for large amount of $e^+(e^-)$ production.] The directions of emitted beams are shown in Figure 10.

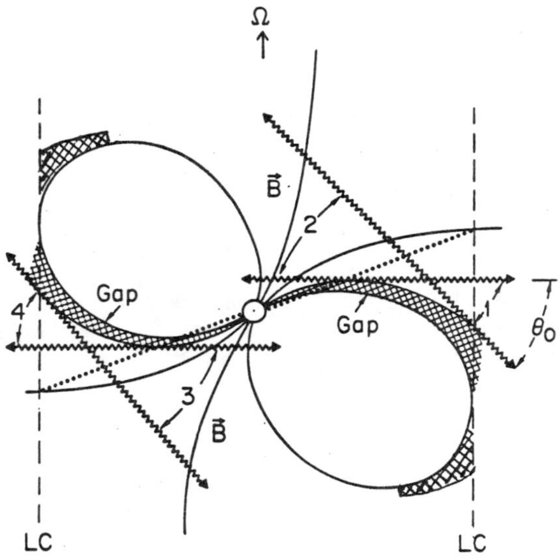

Figure 10

Model outermagnetosphere with indicated radiation beaming of secondary radiation from four emission regions near the outergap boundary. Any radiation from the outer pair of outergaps would not be observable unless an observer's line of sight were almost perpendicular to $\vec{\Omega}$.

These are fan beams whose longitudinal extent is, for thin gaps, almost 90°. An observer would see two of them whose arrival time separation comes from difference in time of flight of the radiation across the magnetosphere and from aberration effects since the azimuthal velocity of the local radiating region ($\Omega r_{\perp}$) is typically 0.6 c. The exact pulse time separation depends upon the inclination angle between the stellar spin and magnetic moment and upon the angle between the spin and the observer. Abberation also changes the shape of the emitted beam so that an observer should see a different (cusped) beam shape from that emitted at the varying r through the beam, in the corotating frame (Figure 11). The resulting double beam titled cusp resembles that of the well measured Crab optical profile of Figure 1a. For Vela (but not the Crab) the weak optical radiation originates closer to the star, where aberration is less, than does the γ-radiation. This results in closer optical subpulse spacing, but the midpoint between subpulses (corresponding to that of radiation from r = 0) should be the same for all radiation from the Vela

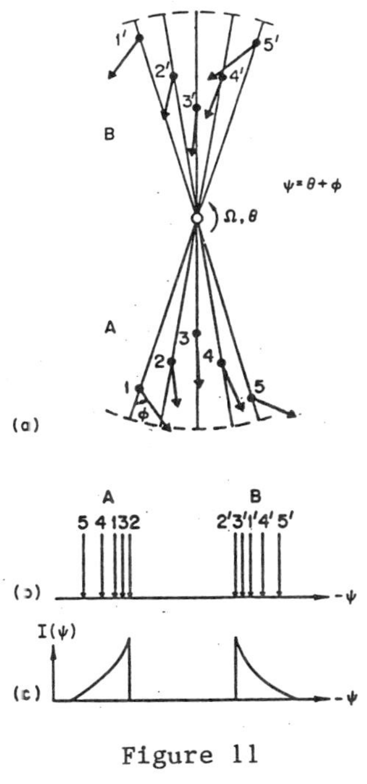

Figure 11

Effect of aberration in changing the order in which emission from neighboring field lines reaches an observer. A stationary phase exists for the beam from point 2 of beam A. The cusp shape from a similar argument applied to the second pulse (Figure 10), whose beam traverses the magnetosphere , is shown as B.

pulsar, as observed (cf. Figure 3). Possible radiation beams from the other possible outergap pair of Figure 10 will be ignored for two reasons. Because that pair is smaller and closer to the light cylinder both aberration and angular extent limit emission to directions nearly perpendicular to Ω so that the probability for observation is greatly reduced. Secondly, those gaps, may be quenced by radiation from the more powerful pair.

9. OUTERGAP RADIATION: THE CRAB PULSAR

The mechanism which is found to limit the growth of the Crab pulsar outergap and to supply enough $e^{\pm}$ pairs to prevent charge depletion elsewhere in the trans-null-surface magnetosphere is the following (Cheng et al. 1985). An $e^{+}(e^{-})$ accelerated through the outergap reaches an energy (limited by radiation reaction) at which it radiates around 10^5 multi-GeV curvature γ-rays per gap passage. These convert to $e^{\pm}$ pairs from collisions with soft X-rays (Section 3, mechanism 2 • a). When such pairs are created within the gap the e^{+}/e^{-} are separated, accelerated, and repeat the process. Beyond the gap they are unaccelerated and lose their energy by synchrotron radiation (MeV to optical) which supplies the needed X-ray flux through the gap. Inverse Compton scattering of the $e^{\pm}$ created outside the gap on these same X-ray photons boosts some of them into the MeV-GeV regime. The Crab outergap is limited by this combination of processes when the particle flow through it $\sim 10^{33}$ sec^{-1} and the (related) total potential drop in it along $\vec{B} \sim 10^{15}$ Volts. The total double fan beam power radiated from the gap transboundary region, which is the source of most of the gap powered observed radiation, $\sim 10^{36}$ erg s^{-1}. The total number of created $e^{\pm}$ pairs $\gtrsim$ the total number of multi-GeV γ-rays coming out of the gap (almost all completely converted to $e^{\pm}$ pairs) $\sim 10^5 \times 10^{33}$ sec^{-1} = 10^{38} sec^{-1}. This radiated power and $e^{\pm}$ flux are consistent with those "observed" from the Crab pulsar. The computed emission spectrum, arbitrarily adjusted to fit the integrated intensity, is given in Figure 12. The above mechanisms will not give 10^{12} eV γ-rays. These will be generated in this model by a process which begins with inverse Compton scattering of gap e^{-}/e^{+}. The resulting 10^{13} eV γ-rays will convert the gap trans-boundary region into $e^{\pm}$ pairs whose synchrotron radiation is typically in the 10^{12} eV regime. The beam which must cross the magnetosphere to be observed (beam 3 of Figure 10) may be absorbed by pair production on the stronger $\vec{B}$ through which it must pass. In this model the extent of inverse Compton scattering on optical photons deep within the Crab outergap depends upon some unknown details of the shape of the heavily plasma loaded magnetic field gap boundary lines near the light cylinder. If such inverse Compton scattering were to be sufficiently large the Crab pulsar outergap might be controlled in part or even mainly by the Vela pulsar mechanism of Section 10.

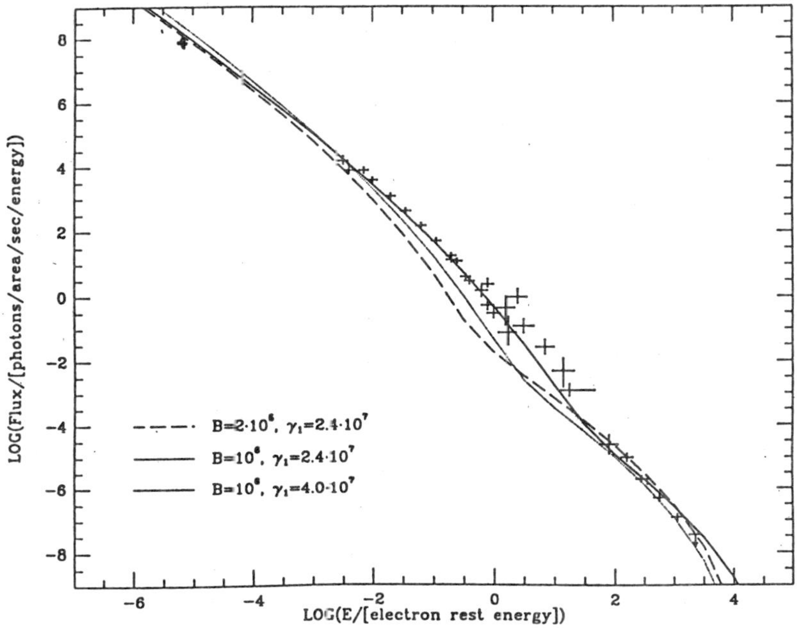

Figure 12

Calculated and observed spectra of the pulsed electromagnetic radiation from the Crab pulsar, from Cheng et al. (1985b). All of the calculated spectra are normalized to fit the integrated observed density. The theoretical curves differ mainly in the assumed local magnetic field at the emission point, 2×10^5 G for the dashed curve 10^6 G for the solid curve which fits the data best. Synchrotron radiation dominates the emission below about 15 MeV and inverse Compton scattered X-rays at higher energies.

10. OUTERGAP RADIATION: THE VELA PULSAR

The mechanism (2 • a) which limits the model outergap for Cab parameters is not effective in limiting a similar model outergap for Vela. This is mainly because Vela's outergap synchrotron radiation time scale ($\propto B^{-2} \propto P^6$) is of order 10^3 that of the Crab. There is then insufficient time for Vela $e^\pm$ to radiate enough synchrotron photons to sustain the 2 • a mechanism of Section 3 and 9 before the radition pairs flow out of the magnetosphere. Vela, instead, limits gap in the following, different, way. In Vela's outergap, each accelerated e^-/e^+ makes of order a hundred 10^{13} eV γ-rays by inverse Compton scattering on IR (10^{-1} eV) photons. These γ-rays then make $e^\pm$ pairs on this same IR flux. Some of these pairs are formed in the gap and thus supply e^-/e^+ there to repeat the process. Most convert beyond the gap boundary to give pairs (secondaries) whose crossed beams of synchrotron (secondary) radiation are the observed Vela beams. Additional (tertiary) pairs (10^{36} sec^{-1}) are made by pair production from the passage of low enegy γ-rays of one secondary beam through high energy X-rays of the other and vice-versa. These tertiary pairs give the IR synchrotron radiation which flows through the gap and supplies the IR flux whose inverse Compton scattering by gap e^-/e^+ initiates the connected sequence of outergap controlling processes. Calculations by Cheng et al. (1985b) give a self-limited outer gap for Vela parameters when the secondaries' radiated crossed fan beam power of of order 10^{35} - 10^{36} ergs s^{-1} and the total

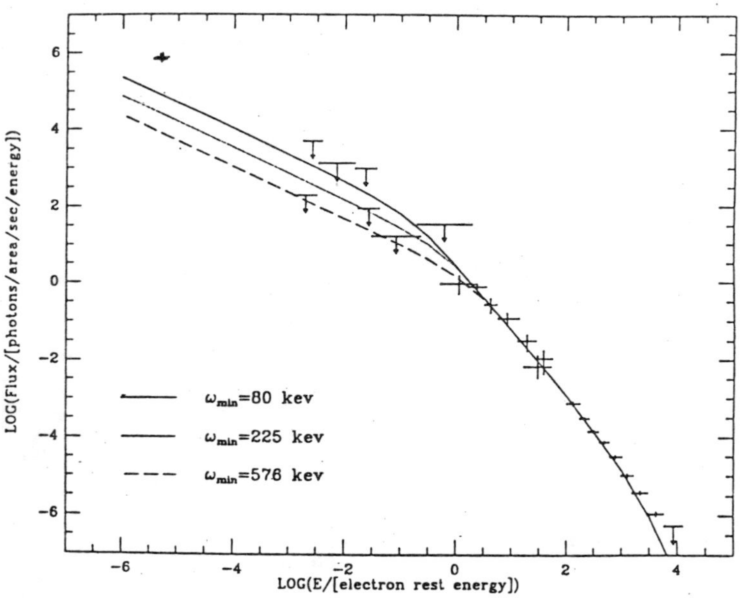

Figure 13

Calculated and observed spectra of the pulsed electromagnetic radiation from the Vela pulsar, from Cheng et al. (1985b). All of the calculated spectra are normalized to fit the integrated observed intensity. The curves with spectral breaks at various energies, discussed in Section 10, reflect mainly an uncertainty in outergap magnetic field. The tertiary optical emission (which agrees with that observed) is not included in the theoretical curve. The theoretical curve for modulated X-rays around 1 keV does not include an expected contribution between pulses from thermal polar cap emission. Its existence would raise the observationally derived upper limit by diminishing the amount of observed modulation through the pulse.

(tertiary) pair production rate $\sim 10^{36}$ sec^{-1}. The comptued emission spectrum, adjusted to fit the total integrated intensity, is given in Figure 13 and compared there to observations. The spectral break at around 10^2 keV occurs because $e^{\pm}$ pairs leave the magnetosphere before their characteristic synchrotron radiation energy falls much below this value. This model spectrum falls below the very weak observed optical intensity if the low energy (optical and IR) radiation from the tertiaries is not included. The latter gives double fan beams (total power $\sim 10^{30}$ ergs s^{-1}) emitted from regions enough closer to the neutron star that the pulse interpulse separation is only about 90° (cf. Section 8). In the Vela outergap model the secondaries lose almost all of their energy to synchrotron radiation, but inverse Compton scattering on the tertiary IR flux does take about 10^{-2} of it. This appears as 10^{12} - 10^{13} eV γ-rays with total emitted power about 10^{-2} that in the 10^2 - 10^3 MeV band.

The "millisecond pulsar", PSR 1937+214, and perhaps PSR 1953-29 have about the same $B_s P^{-2}$ as does the Vela pulsar. Therefore its outermagnetosphere current and voltage drop should be about the same as Vela's. It may, therefore, be the source of similar high energy γ-ray emission.

11. RADIOEMISSION

For Vela, radioemission (RF) is presumed to take place much nearer the star, possibly from polar caps at the stellar surface, through mechanisms entirely unrelated to those which support outergap emission. If the surface field is irregular, so is the distribution, the shape and the magnetic field directions of of polar caps. Then the observed radiopulse arrival time is unrelated to those of the higher energy outergap emitted pulse and interpulse. But if a somewhat irregular outergap secondary $e^{\pm}$ flow is illuminated by an RF beam which originated elsewhere, the beam can be coherently inverse Compton boosted by the motion along $\vec{B}$ of the moving "mirror" formed by the dense plasma of secondaries, even through the relatively strong local magnetic field strongly suppresses the incoherent scattering of RF photons by single $e^-(e^+)$. The resulting RF beam would have a higher frequency than that of the incident beam and be beamed in two directions coincident with those of the higher energy incoherent secondary synchrotron beams since the latter come from the same $e^{\pm}$ as those which coherently upscattered the incident RF. This suggests that the Crab outergap does not create the observed double pulsed RF emission of the Crab pulsar, but merely diverts and boosts a precursor like RF beam partially incident upon it.

12. 10^{15} eV NUCLEON BEAMS FROM THE CRAB PULSAR?

If the relative directions of stellar magnetic moment and spin are such as to give the charge distribution of Figures 5 and 7, a flow of gap accelerated e^- down onto the neutron star surface would pull protons (or positive ions) out of the stellar surface to balance the negative charge of e^- column. When these protons reach the outergap they will be accelerated through it with negligible energy loss from inverse Compton scattering or curvature radiation. They will, therefore, achieve the full 10^{15} eV gap potential drop. If emitted from the Vela pulsar such a beam would give no easily observable effect. If emitted from the Crab pulsar, such a beam might be a strong source of 10^{15} eV γ-rays since 300 eV soft X-rays of the X-ray flux through which the proton passes are near the 300 MeV center of mass peak for the $\gamma + p \rightarrow \pi^0 + p$ neutral photomeson production cross section. The calculated (superhigh energy) γ-ray emission rate from the $\pi^0 \rightarrow 2\gamma$ decay ($\sim 10^{33}$ sec^{-1}) gives a flux at the earth of 3×10^{-12} cm^{-2} s^{-1}, near that which has been reported from the direction of the Crab as discussed in Section 4. (Since 10^{15} eV γ-rays would not escape from within the Crab light cylinder the π^0 production cannot too close to the star.)

13. AGING OF PULSAR PARAMETERS AND ITS EFFECT ON ENERGETIC EMISSIONS

Older pulsars have very different energetic emission from that of very young ones (cf. Table 1). This is attributed to the evolution of the pulsar's period, magnetic field and, perhaps, also the angle between $\vec{\Omega}$ and the neutron star's magentic moment.

A pulsar's surface dipole magnetic field (B_s) is generally inferred in a somewhat model dependent way from the measured spindown rate of its angular frequency ($\dot{\Omega} \equiv 2\pi\dot{P}P^{-2}$) and the calculated neutron star moment of inertia (I):

$$- I\Omega\dot{\Omega} = - \dot{E} \sim R^6 B_s^2 \Omega^4/c^3 \sim \left(\frac{B_{12}}{P^2}\right)^2 10^{32} \text{ ergs s}^{-1} \quad (13.1)$$

where $B_s \equiv 10^{12} B_{12}$G. The observed distribution for $B_s(\propto (P\dot{P})^{1/2})$ as a function of pulsar period is shown in Figure 14. Because of selection effects and uncertain beaming corrections it is not yet possible to read with confidence evolutionary tendencies on pulsar birthrates from these data. (Many, perhaps most, may be born spinning much more slowly than the Crab pulsar.)

Initially the Crab and Vela would be expected to age mainly because of increasing P. The models of Sections 9 and 10 predict the observed great reduction of pulsed X-rays relative to γ-rays as P(Crab) $\rightarrow$ P(Vela). For the Vela outergap model calculation of Section 10 the outer-gap ultimately fills the entire part of the outermagnetosphere region where $\vec{B}$ field lines thread the null surface when the spin down energy loss rate of a pulsar drops to about 2×10^{34} ergs s^{-1}. At still longer pulsar periods the outermagnetosphere is no longer effective in maintaining copious $e^{\pm}$ production there. This is consistent wth the failure to observe X-ray synchrotron halos about such pulsars mentioned in Section 2.

Table 1

Observed Emissions from Aging Single Neutron Stars

pulsar	age (yrs)	number in Galaxy	strong radiation beams
Crab	930	$\lesssim 10$	RF, optical, X-ray, γ-ray
Vela	10^4	$\lesssim 10^2$	RF, γ-ray
typical pulsar	10^6–10^7	10^5	RF
"dead" pulsar	10^{10}	10^8	

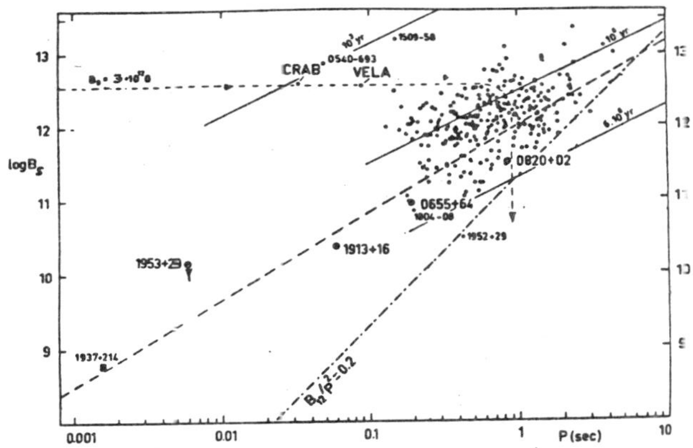

Figure 14

The surface dipole magnetic fields of pulsars (B_s) from Eq. (13.1) versus their periods (P) adapted from van den Heuvel (1984) and Rhadhakrishnan and Srinivasan (1981). An equivalent representation is in Alpar et al. (1982). The four binary pulsars are indicated by circled dots; the isolated millisecond pulsar is shown with a square. The binary 1953+29 has only an upper hound for $\dot{P}$ and B (Boriakcff 1984). Spin down ages $P\dot{P}^{-1}$ in (B_sP) space are solid diagonals labelled 10^3, 10^6, and 6×10^6 years. The dashed trajectory with $B_C = 3\times 10^{12}$G indicates the model evolution for P and $\dot{P}$ related by Eq. (13.1) when the neutron star magnetic field decays exponentially on a scale 3×10^6 years. The dashed-dot line is the theoretical radio-pulsar turnoff when all $e^{\pm}$ production is calculated to cease. The dashed diagonal is the theoretical birth curve of Eq. (15.3) for pulsars spun up by Eddington limit accretion from a companion.

The maximum effective potential drop which a single neutron star might sustain indefinitely on its open field lines is achieved when $\rho \sim 0$ is maintained there. Then along and across the open field lines

$$\Delta V \sim 10^{13} B_{12} P^{-2} \text{ volts .} \tag{13.2}$$

When this falls below the $\Delta V_{\pm}$ of Eq. (6.6) all $e^{\pm}$ pair production would be expected to cease on open field lines even near the neutron star surface. This occurs when $B_{12}\, P^{-2} \lesssim 0.2$ or, from Eq. (13.1) $|\dot{E}| \lesssim 4\times 10^{30}$ ergs^{-1}. It is plotted as the dash-dot line of Figure 14. For most theoretical models this would also imply that the spinning neutron stars would no longer be radiopulsars, and is consistent with the absence of radiopulsars below it. A common interpretation (Lyne 1985) of the distribution of B_s vs. P is that the surface dipole components of the magnetic fields of typical neutron stars decay to a much smaller value on a time scale of several million years. (This would more quickly carry them across the threshold of Eq. (6.6), than would be the case if only P increased.) Corroborative support comes from the observation that most radiopulsars are moving rapidly ($\gtrsim 100$ km s^{-1}) away from the Galactic plane where they were born, but are still sufficiently close to the plane that their true age since birth must be less than 10^7 years. The evolutionary track ($\dot{P}$ vs. P) has been measured for only two pulsars, the very young Crab and PSR 1509-58. They give $P\dot{P} \propto P^{\alpha}$ with $\alpha = 0.5$ and 0.2 respectively (Groth 1975; Manchester et al. 1985), and are consistent with a $B_s \sim 3\times 10^{12}$ and 10^{13} G each of which grows slightly with time. Most radiopulsars are much older and have a smaller $P\dot{P}$, interpreted as implying $B_s \sim 10^{12}$ G; many of these have very much smaller B_s. For the above two very young pulsars, $P\dot{P}$ and thus (perhaps) B_s will ultimately drop if they are to evolve into typical older pulsars. This is a popular interpretation because of the number of ways for B_s to diminish strongly after several million years as will be discussed in Section 14. An evolutionary path for a Crab-like pulsar with a decaying magnetic dipole is shown in Figure 14. Crucial questions about modeling possible very energetic radiation from some old neutron stars in accreting binaries may depend crucially upon the validity of such a suggested evolution. Therefore we turn next to theoretical support for it.

14. DECAY AND STRUCTURE OF OLD NEUTRON STAR MAGNETIC FIELDS: THEORY

Electrically, an older star is just a rigid metallic crust about a km thick, containing a cold fluid superconductor (protons and possibly pions) imbedded in a very strongly conducting, relativistically degenerate, electron sea. All of the usual processes which might generate magnetic fields have long since ceased, and the key question is how and for how long a magnetic field, once established, can continue to maintain itself.

A magnetized perfectly conducting classical fluid, if allowed to move freely will try to minimize its magnetic field energy, subject to the restriction that, within the fluid, moving elements carry with them their entrained magnetic flux. This constraint usually does not allow the fluid greatly to diminish the magnitude of its internal $|\vec{B}|$ but it may, and usually does, achieve internal rearrangements which greatly reduce its external magnetic field energy. It is easy to arrange for a perfectly conducting solid to be a permanent magnet but not so for a fluid conductor. Because of the reduction in the range of external $\vec{B}$ which goes with increasing multipolarities the allowed magnetohydrodynamical relaxations in a magnetized fluid conductor move with greatest vigor to diminish the lower longer range multipole components of the external $\vec{B}$ since this leads to the greatest reduction in external B field energy. Thus an initially rather highly ordered magnetic field in a liquid conductor can, after magnetohydrodynamic relaxation, give rise to a greatly altered external magnetic field: the typical surface field magnitude may be comparable to what it was initially, as is that of the internal magnetic field to which it is connected, but because its lower multipoles have been most effectively diminished the more complicated external surface field now has a much shorter range. A strong external dipole component may, therefore, be expected to be especially vulnerable to magnetohydrodynamic relaxation unless special internal configurations, e.g., an appropriate toroidal field, stabilize it.

The fluid interior of the neutron star is not a perfectly conducting classical fluid but rather a superconducting quantum fluid. One difference is that any magnetic field through the superconductor will be organized into separated tubes each of which has a quantized flux $\phi = \pi\hbar c e^{-1}$ (for a proton superfluid) and carries a magnetic field B_c approaching 10^{16} G. The average magnetic field density in the neutron star interior determines only the number of quantitized flux tubes per unit area. Flux tubes are separated by much more than their radii (10^{-12} cm) and do not significantly interact directly among themselves as long as the average B is very much less than the B_c inside of a flux tube. Flux tubes are constrained from moving relative to the degenerate electron sea in which they are imbedded because of strong momentum transfer to the sea from electron scattering by the flux tube magnetic field. Motion of flux tubes (and surrounding electrons) is driven internally by the very large (10^7 dyne) tension is each flux tube which tries to move the neutron star core's conducting fluid to reduce the total flux tube energy within it (proportional to the total flux tube length). Effectively the neutron star's magentized conducting fluid component will be subject to the same hydromagnetic relaxation motions as a purely classical one, except that the forces which drive it derive from an energy density $B^2/8\pi$ outside the star and a much greater energy density $BB_c/8\pi$ inside. (For details, references and other kinds of flux tube relaxation see, for example, Harvey <u>et al.</u> 1985).

Two circumstances which can greatly delay an internal magnetohydrodynamic relaxation of the external field of a cold neutron star are (Flowers and Ruderman 1977)

1) Gyroscopic forces from rapid stellar rotation.
2) Rigidity of the metallic crust surrounding the fluid interior so that whatever magnetic field passes through the crust remains fixed in and beyond it for the crustal (ohmic) eddy current dissipation time, estimated to be $10^6 - 10^7$ years.

Another suggested contributor to temporary external field stabilization in a warm neutron star is a convectively stable internal temperature distribution (Chanmugam 1984).

There are a variety of suggested scenarios for the generation of large neutron star magnetic fields. These include the compression of fossil fields by the implosion which forms the star, or a strong magnetohydrodynamic field generator active during the implosion. Initially spin stabilized, rapid crustal crystallization could freeze the external field for $10^6 - 10^7$ years even if the stellar spin rate falls by two or three order of magnitude. Blandford et al. (1983) showed that strong thermal gradients in the crust of young neutron stars may drive the growth of a seed field there until an external B $\sim 10^{12}$ G is sustained entirely by crustal currents. In either type of scenario when crustal currents die we expect to be left solely with the external field from the magnetohydrodynamically relaxed core fluid. (A similar endpoint is reached if the decrease in core fluid thermal gradients allows final relaxation to proceed). Thus there are plausible reasons to expect large reductions in the external magnetic field of a neutron star after several million years. However, as discussed above, the reduction is expected to be very much greater in the longer range (more highly spatially ordered) components. Thus the surface dipole component B_s may become very greatly diminished even though the (complicated) surface field might even retain some of the characteristic 10^{12} G inferred for the dipole component alone of younger pulsars. This may be the magnetic field configuration around most (all?) very old neutron stars except, possibly, for those in binaries whose surfaces are heated by strong accretion. In those cases large surface magnetic fields might be regenerated by crustal currents generated by outward thermal fluxes in some regions (Blandford et al. 1983). Such fields could not penetrate into the superconducting (and therefore "perfectly" diamagnetic) stellar cores and, may not add coherently to contribute much at large distances above the stellar surfaces. (The magnetic field energy is much less when adjacent surface regions have oppositely directed field directions.) The reason for emphasizing these possible but uncertain features of the magnetic field of older neutron stars is that accreting neutron stars with weak magnetic dipoles seem to be the most reasonable candidates for sources of superhigh energy γ-rays (section 16). We first turn, however to inferences from observations on the strength of the dipole moments of old neutron stars.

15. DECAY AND STRUCTURE OF OLD NEUTRON STAR MAGNETIC FIELDS: INFERENCES FROM OBSERVATIONS

Three kinds of observations have been used for inferring the magnetic fields of neutron stars:

1) Stellar spin up (down) rates.
2) Limiting spin up periods of accreting neutron stars.
3) Spectral features of X-ray emission from accreting neutron stars.

These will be applied below to various families of neutron stars. The first two imply weak magnetic dipole for many old neutron stars and probably strong dipoles for some others which may, however, be much younger. The third often suggests a high surface field even though the dipole may be small.

a) Isolated radiopulsars.

The measured spindown torque $I\dot{\Omega}$ was used to estimate the surface dipole field B_s of radiopulsars in figure 14. Those data seemed suggestive of a large drop in B_s after several million years (Lyne 1985) but other interpretations are not excluded.

b) Binary radio-pulsars; the millisecond pular.

A conducting Keplerian accretion disk around a neutron star (fed by massloss from a companion) will have an inner edge a distance r_m from the star which interacts with the outer edge field of the largely contained stellar magnetosphere (Lamb 1985; Ghosh and Lamb 1979). The former has a Keplerian angular speed near this interface

$$\Omega_k = (GMr_m^{-3})^{1/2} \qquad (15.1)$$

and

$$r_m \sim 3\times 10^8 \, \mu_{30}^{4/7} \, \hat{m}^{1/7} \, R_6^{-2/7} \, L_{37}^{-2/7} \text{ cm} , \qquad (15.2)$$

where μ_{30} is the stellar magnetic moment $(4\pi B_s R^3/3)$ in units of 10^{30} Gauss-cm^2 ($\mu_{30} = 1 \leftrightarrow B_s \sim 3\times 10^{11}$ G); L_{37} is the X-ray luminosity in units of 10^{37} ergs s^{-1} s(10^{-1} the Eddington limit), and 10^{-1} of the accreted rest mass energy is assumed to be radiated away as the accreted mass reaches the stellar surface; m is the neutron star mass in solar mass units. The neutron star will be spun up (down) by the accretion disk as long as $\Omega_k \gtrsim (\lesssim) \, \Omega$. The asymptotic steady state period

$$P_{\lrcorner} \sim 4 \, \mu_{30}^{6/7} \, R_6^{-3/7} \, \hat{m}^{-2/7} \, L_{37}^{-3/7} \text{ sec} \qquad (15.3)$$

The period P_L vs. B_s from Eq. (15.3) with an Eddington limit accretion rate ($L_{37} \sim 10$, for $\dot{m} \sim 1$) is the diagonal dashed line of Figure 14. The several binary pulsars are thought to have been spun up by accretion from their companions, to a period near that given by Eq. (15.3). The accretion has long ago ceased and they have since been spinning down as ordinary (i.e. isolated) radiopulsars. The millisecond pulsar (PSR1937+214) has probably had a similar history but has lost its companion. All of these are certainly old neutron stars (for PSR1937+214 the spinup phase alone took much more than 10^7 yrs) and would be dead pulsars if they had not been resurrected by accretion spin up. All of these old neutron stars have very much weaker magnetic dipole moments than do the younger isolated radiopulsars: the millisecond pulsar dipole field is estimated to be only about 5×10^8 G.

c) Quasiperiodic Objects; Low Mass X-ray Binary Galactic Bulge Souces

Newly observed and interpreted quasiperiodic objects (QPO's) appear to be accretion spun up neutron stars in binaries which are still accreting. The stellar spin (P) is not directly observable. Variations in the accretion rate (observed and interpreted as variations in L_{37}) are reflected as differences between the stellar P and that of the inner edge of its surrounding accretion disk (P_L). This frequency difference ($P^{-1} - P_L^{-1}$) is seen and interpreted as quasiperiodic X-ray emission from the interface between the magnetosphere and the accretion disk (Alpar and Shaham 1985; Lamb 1985). The stellar dipole field inferred from an analysis of GX5-1 ($\sim5\times10^9$ G) and other QPO's is in the same regime as those of the low dipole field accretion spun up radiopulsars. The QPO's are members of a wider class of accreting power X-ray emitting neutron stars fed by putative low mass Roche lobe filling companions. (The lack of directly observed neutron spin modulation of the X-ray emission has been taken as an indication of weak B_s. Successful modeling for X-ray bursts (Joos and Rappaport 1983) from stellar surface nuclear fusion also has been based upon the assumption of "negligible" B_s.) All seem to be extremely long lived systems ($\gtrsim 10^9$ yrs) so that the small B_s of QPO's and, presumably, of other Galactic bulge sources is to be associated with very old neutron stars.

d) Binary X-ray pulsars

This group of neutron stars in binaries observed while still strongly accreting from the relatively heavy companions have directly observable P. The active phase of the companion during which there is strong X-ray emission while the underlying pulsar period remains observable is typically estimated as 10^5 yrs. During that interval the torque on the neutron star could not spin it up to a period much below a second. Although the observable spinup interval may be small the age of the neutron star may be much longer. However, if formed from White Dwarf accretion over a critical mass, it need not be (van den Heuvel 1985; Helfand <u>et al.</u> 1983). Because $\dot{P}$ is also known in many cases the torque on the neutron star which tends to drive it toward the P_L of Eq. (15.1), can often be determined. The measured $\dot{P}$ can be related to the same neutron star parameters as those that determine P_L (Ghosh and Lamb 1979; Henrichs 1983). For an aligned dipole and an equatorial accretion disk

$$-\dot{P} = (6\times 10^{-5}\ \mathrm{s\ yr}^{-1})\ \mu_{30}^{2/7}\ \hat{m}^{-3/7}\ R_6^{6/7}\ I_{45}^{-1}\ L_{37}^{6/7} n(\omega_s) P^2 \quad (15.4)$$

where I is the stellar moment of inertia in units of 10^{45} g cm^2. $n(\omega_s)$ is a monatonically decreasing function of the "fastness parameter" ω_s, the ratio of stellar (and magnetosphere) angular frequency, Ω, to that of a Keplerian inner edge of the accretion disk, $\Omega_k(r_m)$. Thus n is a function of the combination $PL^{3/7}\ \mu_{30}^{-6/7}$. For $\Omega_k(r_m) >> \Omega$ $n \sim 1.4$ and passes through zero (i.e. the torque reverses) at $\Omega_k(r_m) \sim 0.4\,\Omega$. Figure 15 shows $\dot{P}$ from Eq. (15.4) as a function of $PL^{3/7}$ for various values of μ_{30}, together with observed long term average $\dot{P}$ for various binary X-ray pulsars. For each measured combination of

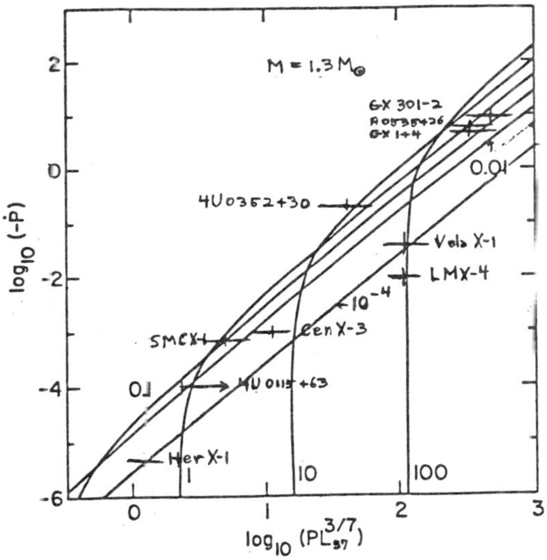

Figure 15

Typical spinup rates ($-\dot{P}$ in s yr^{-1}) as a function of $PL_{37}^{3/7}$ according to Eq. (15.4) of Lamb and Ghosh. The selection of data except for LMX-4 and 4U0115+ 63 and the theoretical curves labelled by the value of μ_{30}, the neutron star magnetic moment in units of 10^{30} G-cm^3, are those of Ghosh and Lamb (1979). The line $\mu_{30} = 3\times 10^{-4}$ passes through the Her X-1 and Vela X-1 (4U0900-40) data points in addition to curves of very much larger μ_{30}. The LMX-4 $\dot{P}$ is that reported by Bignami (1985).

P and $\dot{P}$ there are always two possible values for μ_{30}. The canonical choice [e.g. Lamb and Ghosh] has been that the appropriate μ_{30} is that which corresponds to $B_s \sim 10^{11} - 3\times 10^{13}$ G i.e. $0.5 \lesssim \mu_{30} \lesssim 100$. Then most X-ray pulsars would agree reasonably well with the kinds of magnetic field found in younger radiopulsars except for Her X-1, LMX-4, and 4U0900 (Vela X-1). The first would have $B_s \sim 2\times 10^{11}$ G and the latter two 3×10^{13} G. These three X-ray pulsars will be of special interest in Section 16 where it will be argued that their reported superhigh energy γ-ray emission would seem to argue for $B_s \lesssim 10^9$ G. We shall consider, therefore, the problems with the other possible interpretation of Figure 15, that LMX-4, Her X-1, and Vela X-1 could be interpreted as having $\mu_{30} \lesssim 10^{-3}$, and thus $B_s < 10^9$ G. Quantitative conclusions based upon Eq. 15.4 and Figure 15 may be quite fragile since different geometries (and models) for magnetosphere-disk transition layers may give rise to significant changes (e.g. Anzer and Börner 1983). A toy model, with all B_s so small that the accretion disk always reaches the stellar surface and transfers all of its Keplerian momentum there to the disk, gives

$$- \dot{P} = \frac{\dot{M}\ (GMR)^{1/2}\ P^2}{I} \tag{15.5a}$$

$$\sim 10^{-5}\ L_{37}\ P^2\ \mathrm{s\ yr}^{-1}\ . \tag{15.5b}$$

As emphasized by Anzer and Börner, it does not fit the observed data much worse than do other models.

e) Her X-1 and Vela X-1.

An argument for choosing the "small" dipole solution ($\mu_{30} \sim 3\times 10^{-4}$) for these X-ray pulsars is given in Section 16. We shall consider here the kind of evidence that supports the "large" dipole solution. For Her X-1 it consists of two main points: i) a sharp spectral dip in the X-ray emission spectrum near 40 keV has been interpreted as cyclotron absorbtion in $B \sim 4\times 10^{12}$ G (Trümpler 1985). ii) episodes in which $\dot{P}$ may fluctuate strongly and may even change sign are interpreted as external torque fluctuations which could relatively easily change the sign of $\dot{P}$ only if P is near the P_L of Eq. (15.3) (the "large" $\mu_{30} \sim 0.5$ solution) since the magnitude and sign of $\dot{P}$ from Eq. (15.4) are then very sensitive to the rate of accretion in $n(\omega_s)$. (Internal "glitch" type transitions do not appear to be large enough to be relevant.) Alternative interpretations are given below:

i) If the $\mu_{30} \sim 3\times 10^{-4}$ solution is adopted, $B_s \sim 10^8$ G and the stellar magnetosphere is terminated by the accretion disk at $r_m \sim 3\times 10^6$ (Eq. 15.2). This B_s is still strong enough to guide the accreting glow onto polar caps (i.e. $r_m > R$). Below much of the accreting polar caps heat flow is inward and will not generate magnetic field by the Blandford _et al._ mechanism of Section 14.

However there can be strong outward thermal flux inside and outside of the polar cap annulus in which accretion is expected to be strongest. This may generate a local surface $4\sim10^{12}$ G magnetic field through which 40 keV X-rays pass without necessarily giving a large long range dipole moment field (B_s). Alternatively the argument of Section 14 was that a large local surface B of this order may be expected to remain even after the decay of the well ordered dipole component.

ii) Short time scale (days-weeks) observations of $|\dot{P}|$, mainly for Vela X-1, indicate that the small value of $|\dot{P}|$ used in Figure 15 is qualitatively misleading: $|\dot{P}|$ can reach 1 s yr^{-1} for several days (and change sign, probably in response to reversal of the direction of the accretion rotation around it). Such a large $|\dot{P}|$ would put Vela X-1 far above the weak dipole line in Figure 15 and strongly support the canonical $B_s \sim 10^{12}$ G for it and perhaps for the other X-ray pulsars which otherwise allowed the weak dipole interpretation (F. Lamb, private communication). Only if, on the relevant time scales (extending probably to many months), the crust of Vela X-1's neutron star was not well coupled to its core could the short term large spin up (down) rates be interpreted in terms of a small spun up (down) crust moment of inertia rather than as resulting from a larger torque than that which could be achieved with the small dipole. If a weak magnetic field, $B \sim 10^8$, also coupled the crust to the interior conducting electrons and protons, the time to propagate crustal speed change to the core could be of order $R(4\pi\rho)^{1/2}/B \sim 10^6$ sec, but Ekman pumping of the electron-superconducting proton-superfluid neutron core boundary layer below the crust is expected to be orders of magnitude faster. The spinup time for the core neutron superfluid by spun up electrons is $\tau \sim 4\times10^2$ P (Alpar 1985). For Vela with P = 283 sec, $\tau \sim 1$ day, small enough that it is difficult to sustain the hypothesis that the large $|\dot{P}|$ is associated with a reduced effective stellar moment of inertia because of poor crust-core coupling. Thus the evidence from the measured $\dot{P}$ seem to suggest that despite their positions in Figure 15, Vela X-1 has a $B_s \sim 10^{12}$ G, Her X-1 probably has, and LMX-4 very possibly does also. If so these X-ray pulsars would presumably be relatively "young" ($\lesssim 10^7$ yrs) neutron stars.

f) γ-ray burst sources

We present below arguments that γ-ray burst sources are old neutron stars with weak magnetic fields. [For a recent review and references see Liang and Petrosian (1985)]. Cosmic bursts of soft γ-rays (typically 10^{-5} ergs cm^{-2} of 10 keV-10 MeV X-rays over several seconds) are detected at the earth at a rate of about 1 every 64 hours. They are isotropically distributed and have a number versus intensity distribution that appears consistent with a homogeneous distribution. Thus the sources may be so close that the disk structure of our Galaxy is irrelevant, or part of a halo population, or extragalctic. The short time scale for intensity variations - typically 10^2 ms (but less that 0.3 ms in one atypical case) - argues that the spatial size of the radiating region is far less than canonical stellar radii. Recent observations of substantial 10-40 MeV components in some bursts support the local source hypothesis. An extra-galactic source would need an intrinsic X-ray luminosity over 10^4 times larger than a local one to give the same observed bursts. However the X-ray photon density near a compact extra-galactic source would then be so large that the short mean free path for $e^- - e^+$ production by 10-40 MeV γ-rays passing through it might not permit those γ-rays to escape from the source. Therefore modelers are now most comfortable with the existence of 10-40 MeV components to γ-ray bursts from compact sources only when the characteristic distance to them is not greater than several hundred parsecs. The needed compactness of the source size and the relatively large (compared to ordinary stellar events) burst energy ($\sim 10^{36}$ ergs) are additional evidence pointing to part of the local neutron star population as the source of γ-ray bursts. Although isolated old neutron stars may occasionally have structural "glitches" these should not be sufficiently frequent, large, or release energy rapidly enough to be the γ-ray burst sources. Much speculation and modeling has been directed toward neutron stars in binaries, accreting so slowly from underluminous companions that neither are yet detectable as steady sources. Burst energy would come either from accretion disk instabilities or from subsurface nuclear fusion explosions. A further clue to the source population comes from observational bounds on the repetition rate for bursts from any one source. Observations suggest that the repetition rate does not exceed about one every several years but that some, perhaps all, burst source do repeat. [At least one source (the anomalous "March 5 event") and possibly a second (the "40 longitude source") have been observed to repeat in γ-rays. Other evidence for repetition comes from historical records of optical transients presumed coincident with γ-ray bursts (e.g. Schaefer, 1981).] The maximum average repetition rate gives a needed population of γ-ray burst sources of at least a few hundred within 500 light years. Since this population is comparable to (and perhaps greater than!) the total number of neutron stars in binaries expected within that volume, old neutron stars (age $\gtrsim 10^9$ yrs) would have to dominate the γ-burst sources. Indeed even if the burst sources are isolated neutron stars the number of "young" ones, i.e. age $\lesssim 10^7$ years, is too small for them to be a significant contributor.

The release of 10^{37} ergs of γ-rays would require an accreted mass of 10^{17} g if all the energy came from gravitational binding (accretion disk instability), 10^{18} g if hydrogen fusion was the source, and 10^{19} g for explosive helium burning. For a nominal 10 year burst repetition interval the explosive fusion models would then imply thermal X-ray sources during the accretion in the range 10^{30} - 10^{31} ergs s^{-1}, low enough so that they might escape detection as steady X-ray sources. If the accretion is through the intermediary of a Keplerian accretion disk around the neutron star, the neutron star's spin will ultimately approach the period P_L given by Eq. (15.3). Since the accretion rate $(10^{17} - 19^{19})g(10\ yr^{-1})^{-1} \sim (10^{-16}$

$- 10^{-14})$ $M_{\odot}$ yr^{-1}, $P_L \sim 3^{\pm 1} B_8^{6/7}$ sec (with B_8 the surface dipole field B_s in units of 10^8 G. Thus these old neutron stars should, perhaps, be rotating with a period that depends upon their dipole magnetic fields in a manner similar to that for the accretion spinup resurrected pulsars. There are some indications of periodicity in some γ-ray burst observations (Hurley 1985). The 1979 "March 5 event" certainty has an 8 sec periodicity (e.g. Barat et al., 1979) but it is a very atypical repeating burstor. Wood et al. (1981) discuss a 4.2 sec period for the 1977 October 29 burst. Barat et al. (1984) consider the plausibility of a 5.7 sec period in the 1979 January 13 burst data. Evans et al. (1980) discuss others in the several second range. If these periods are those of an accretion disk spin stabilized neutron star they would indicate $B_s \sim 10^9$ G. This is similar to that of those old neutron stars which, with enormously stronger accretion rates, seem to have been spun up to be millisecond pulsars. Moreover such a weak magnetic field would easily allow 20-40 MeV γ-rays to escape even from near the neutron star surface without pair creation, much more difficult to accomplish if B_s had the 10^{12} G of younger neutron stars (at least those which are radiopulsars). However Woosley and others (1984) (Petrosian and Liang, 1985) have martialed arguments that the qualitative features of γ-ray bursts from thermonuclear explosions are difficult to interpret unless there exists some surface $B \sim 10^{12}$ G. Most significantly perhaps, a notch in the burst photon energy spectrum has been interpreted as evidence for cyclotron absorption with B ~ a few times 10^{12} G (Liang 1985). Thus there are also indications that the proposed γ-burst family of old neutron stars, whether in binaries or isolated has at least some strong magnetic field regions.

16. SUPERHIGH ENERGY RADIATION (10^{15} eV) FROM OLD NEUTRON STARS IN ACCRETING BINARIES

Now, after the long diversion of Sections 14 and 15 to support the hypothesis that neutron stars older than about 10^7 years have small magnetic dipole moments, we turn to the problem of very high energy radiation reported from some neutron stars which are accreting X-ray pulsars. The low counting rates (and poor angular resolution) for superenergetic γ-ray sources ($E_\gamma > 10^{12}$ eV) relative to background so far restricts the identification of such neutron star sources to those with an already measured spin or orbital period. Claimed observations of sources of superenergetic γ-rays are summarized in Table 2. Five are known strongly accreting X-ray pulsars in binaries. Among these details about Cen X-3 have not been circulated (Aug. 1985). In Cyg X-3 the nature of the collapsed object is not yet known. It is the only source whose 10^{15} eV radiation has been observed by more than one group and the statistical evidence for it is much stronger than for the others. For them the reported fluxes are all about three standard deviations above background despite the large variation among the inferred absolute luminosities. This cause for skepticism may be balanced by the observation that the five X-ray pulsars which have been reported as 10^{15} eV γ-ray sources are shown in Figure 15 as pulsars whose long time scale average spin up torque per unit mass accreted ($\propto \dot{P}P^{-2}L_x^{-1}$) is far below that of the other 5 X-ray pulsars for which no such γ-radiation has been reported. This may be associated with an accretion torque which fluctuates in time or, at least in some cases, with a small B_s, but it is difficult to see why statistical fluctuations in γ-ray backgrounds would single out that group.

Table 2

Superhigh Energy Radiation from X-ray Pulsars in Binaries

Pulsar	P	L(X-ray)	L(10^{12}eV)	L($\sim 10^{15}$eV)	$\langle E_\gamma \rangle$
Her X-1	1.2 s	10^{37} ergs s^{-1}	10^{35} e)	10^{37} a)	5×10^{14} eV
Vela X-1	283	2×10^{36}	--	2×10^{34} b)	3×10^{15}
LMC X-4	13.5	4×10^{38}	--	10^{38} c)	10^{16}
Cyg X-3	--	10^{37}	10^{36}	$(3-6)\times10^{36}$ d)	2×10^{15}
4U0115+53	3.6	$< 3\times10^{37}$	6×10^{35} e)	--	
Cen X-3	4.8	5×10^{37}		?	10^{15} ?

a) Baltrusaitis et al. 1985
b) Protheroe et al. 1984
c) Protheroe and Clay 1985
d) Samorski and Stramm 1983, Lloyd-Evans et al. 1983
e) Douthuaite et al. 1984a, 1985

Radiopulsars with $\Omega^2 B_s$ much smaller than those of the Crab Vela pulsar would no longer be expected to sustain the huge outergap potential drops of Sections 7, 9, 10, and 15. It is not surprising that none have been reported as candidates for 10^{12} - 10^{15} eV γ-ray sources. The old neutron stars in binaries which are, spin relatively slowly; Vela X-1 is hardly spinning at all! Even with large 10^{12} G fields it appears very difficult to understand how pulsar spin motion alone could generate the needed $\Delta V \gtrsim 10^{15}$ Volts. (Even without any magnetosphere at all and a B_s of 10^{12} G, Vela X-1 could still only give rise to a maximum $e\Delta V \sim eB_s R^2 \Omega c^{-1} \sim 2\times10^{14}$ volts which does not exceed the 3×10^{15} eV γ-ray energies reported from it.)

A conventional way to generate the huge ΔV needed to give nucleon beams (Section 3d) enough energy to produce the needed π -

mesons, is for ionized matter to move at near relativistic speeds across a sufficiently strong magnetic field. The fastest velocity in the binary system is that of the inner edge of the accretion disk; the strongest magnetic field is that of the neutron star. The largest potential drop might be expected near the magnetosphere accretion disk interface at $r = r_m$ where (Brecher and Chanmugam 1985)

$$\Delta V \sim |\vec{B}(r_m) \times \vec{\Delta} v| r_m f . \qquad (16.1)$$

The numerical factor f depends upon details of the magnetosphere-accretion disk transition and the disk structure. The main velocity difference in the transition region

$$|\Delta \vec{v}| \sim \left| \left(\frac{GM}{r_m}\right)^{1/2} - \Omega \right| \qquad (16.2)$$

For $\Omega << 2\pi P_L^{-1}$ of Eq. (15.3), the large velocity mismatch will cause pulling and reconnection of $\vec{B}$ at the outer edge of the magnetosphere by the much more rapidly rotating inner edge of the accretion disk, and thus large electric fields. In this limit Eq. (16.2) becomes

$$\Delta V \sim B_z(r_m)[GMr_m]^{1/2}\ c^{-1}\ f \qquad (16.3a)$$

$$\sim 10^{16}\ (10^{-8}\ B_s)^{-3/7}\ (L_{38})^{5/7}\ \text{volts}\ , \qquad (16.3b)$$

with $f \sim 1$, r_m from Eq. 15.2 ($r_m > R$), and $B_z(r_m) \sim B_s(R/r_m)^3$. A maximum $\Delta V \sim 10^{16}$ volts for an Eddington limit accretion rate ($L_{38} \sim 1$) could be achieved with $B_s \sim 10^8$ G (which also gives $r_m \sim R$). A stronger surface dipole field B_s would only give a smaller ΔV as emphasized by Brecher and Chanmugan. If $B_s \sim 10^{12}$ G and $L_{38} \sim 10^{-1}$, typical for the large dipole field ($B_s \sim 10^{12}$ G) intepretation for an accreting X-ray pulsar, the maximum $\Delta V \sim 10^{13}$ volts, insufficient even for the generation of Her X-1's 10^{15} eV γ-rays (cf Eq. 3.4). On the other hand, the small dipole model for the neutron stars of Table 2 suggests a maximum ΔV near what is needed to account for the observations:

a) For Her X-1 and Vela X-1, arguments for and against $B_s \sim 10^8$ G were given in Section 15 d and e. From Eq. 16.3 b, $\Delta V \sim 2\times 10^{15}$ volts for Her X-1 and 5×10^{14} volts for Vela X-1 when $B_s \sim 10^8$ G.

b) For LMC X-4 the reported γ-ray energies $\sim 10^{16}$ eV, Eq. 3.4 and Eq. 15.2 (with $L_{38} \sim 4$) then also give $B_s \sim 10^8$ G.

c) The X-ray pulsar 4U0115+53 is a transient with no clear average L(X-ray). For the $\dot{P} \sim 10^{-4}$ s yr^{-1} of Figure 16 and

$L_{37} \sim 1$ it would join Her X-1 and Vela X-1 on the $\mu_{30} \sim 3\times10^{-4}$ line of Figure 15.

d) Finally, for Cyg X-3, since neither the collapsed object nor the period (if it is a neutron star) are, at present, known there are no compelling arguments against assuming it to be in this same category.

A differently powered accelerator for 10^{15} - 10^{16} eV protons might be the accretion shock generated by the infall of matter onto the neutron star. Ellison (1985) and Eichler and Vestrand (1985) have investigated the conditions under which this can occur (Cesarsky 1985). In order to keep electron synchrotron radiation from removing energy too rapidly the ambient $B < 10^8$. But the accretion shock is expected to be well within a stellar radius of the neutron star surface so that only weak field neutrons stars would be candidates for such a proton acceleration mechanism.

A general argument that the $L_\gamma(10^{15} - 10^{16}$ eV) comes from near the neutron surface is based upon its large intensity, reaching 10^{-1} L_x or greater. Such a large fraction of the accretion energy ($L_x \sim GM\dot{M}R^{-1}$) would not be available beyond 10 R. For this reason, and because of the apaprent difficulty in achieving the needed particle energy when $B_s \sim 10^{12}$ G, it seems that the γ-ray emission accelerator implied by the data of Table 2 puts the accelerator near the surface of a small B_s neutron star. This would appear to conflict with those arguments of Section 15e which supported large B_s for Vela X-1, and probably for Her X-1.

In the above view of the location of proton beam acceleration at $r \sim r_m$ the superenergetic protons would move out through a KeV X-ray flux of number density n_x such that

$$\int_r^\infty n_x dr \sim 10^{29}\, r_6^{-1}\, L_{38}\ \text{cm}^{-2} \qquad (16.4)$$

Since the resonant neutral photopion production peak (at CM energy 300 MeV) $\sim 5\times10^{-28}$ cm^2 the reaction X-ray + p → π° + p may contribute in a major way to the π° production. (The $e^\pm$ pair production absorbtion cross section of 10^{16} eV γ-rays on keV X-rays - at a laboratory energy 10^7 times threshold - has so low a cross section that those energetic γ-rays will generally not be absorbed by these same X-rays.) In that case the observed maximum E(γ) may be only a factor of several below the maximum $e\Delta V$. If collisions with limb matter of the companion star are the dominant π° production mechanism a considerably greater $e\Delta V$ would be called for.

An important clue to the location of the origin of the 10^{15} - 10^{16} eV γ-rays is the fact that $B \sim 10^3$ G is the maximum magnetic field which they can cross without being converted to $e^\pm$ pairs. This is very much weaker than $B(r_m)$, and suggests that even if the (proton) particle accelerator is mainly near the inner edge of the accretion disk, the π° production occurs much further away.

Because π^+ production will accompany π° production with comparable intensities, the rapid decay $\pi^+ \to \mu^+ + \nu$ will result in μ-mesonic neutrino beams whose directions and fluxes are similar to those of the 10^{15} eV γ-rays.

17. ENERGETIC γ-RAYS (10^2 - 10^3 MeV) FROM ACCRETING NEUTRON STARS OR MAGNETIC WHITE DWARFS?

The interpretation in Section 16 of the 10^{15} - 10^{16} γ-rays from certain strongly accreting binary X-ray pulsars implied the existence (at least transiently) of proton accelerating potentials $\Delta V \sim 10^{16}$ volts near the neutron star while it is being spun up (or down). Protons, moving through that ΔV, could acquire a kinetic energy equal to this full potential drop. Electrons injected into the opposite end of the accelerator would be accelerated through this same potential drop. If the neutron star's magnetic dipole moment ($\vec{\mu}$) is reversed while the spin directions of the accretion disk around it ($\hat{\Omega}_K$) and the neutron star spin $\vec{\Omega}$, if relevant, remains unchanged, all disk plasma motions are unchanged. However, ΔV reverses sign if $\vec{\Omega}_A \times \vec{\mu}$ does, so that wherever protons were previously injected and accelerated electrons would now be. In either case, electrons, unlike protons cannot achieve the full ΔV because of the strong radiation reaction forces which retard their motion as they become superenergetic. Most effective should be inverse Compton scattering on the large flux of accretion disk generated X-rays. These would take away the energy $e\Delta V$ in the form of γ-rays whose energy would greatly exceed the threshold for $e^{\pm}$ production on these same X-rays (~ 1 GeV). These pairs will in turn produce more γ-rays through the mechanisms of Section 3. Whatever the details of the ΔV accelerator and the processes it powers, electrons which enter it will have their energy ultimately converted to γ-radiation below 1 GeV since only then can these γ-rays escape from the disk region around the neutron star without further $e^{\pm}$ production on the local X-ray flux.

There are three reasons why such radiation from certain X-ray pulsars in binaries might be considered candidates for the 20 unidentified Cos B Galactic γ-ray sources mentioned in Section 1.

1) Intensity. These X-ray binaries might be expected to give a 10^2 - 10^3 MeV γ-ray luminosity comparable to the 10^{15} eV γ-ray luminosity supported by their accelerating proton relatives. From Table 2 this is in the range of 2×10^{34} - 10^{37} ergs s^{-1}. The Cos B sources are in the range 4×10^{35} to 5×10^{36} ergs s^{-1} with the lower limit set by detection threshold.
2) Galactic Distribution and Number. The observed distribution of the unidentified 10^2 - 10^3 MeV Cos B γ-ray sources is shown in Figure 16. Figure 17 shows the observed distribution of Galactic low companion-mass X-ray binaries (LMXB) and of the more massive companion X-ray binaries (MXRB) which include most X-ray pulsars. Here the latter group included Vela X-1 and Cyg X-3 (circled) and Her X-1 (although in many similar compilations Cyg X-3 and Her X-1 are included in the LMXB). When

account is taken of the region of the sky search by Cos B (which excluded Her X-1) the distribution of observed MXRB neutron stars closely resembles that of the unidentified Cos B sources in its very small spread in Galactic longitudes and also in total number of sources. The longer lived LMXB sources have a much broader distribution in Galactic longitude. There are too few isolated radiopulsar neutron stars younger than the Crab for either family to be as attractive candidates for the Cos B sources as are a subfamily of X-ray pulsars. However the Cos B γ-ray sources do not coincide with any subset of the catalogued X-ray pulsars or the MXRB of Figure 17b. This may be because of the much smaller L_x necessary to allow $10^2 - 10^3$ meV γ-rays to escape. A 10^2 MeV photon has a cross section $\sigma \sim 2\times 10^{-25}$ cm^2 for making an $e^{\pm}$ pair on a 10 keV X-ray. With $\sigma \int_{\mathrm{km}}^{\infty} n_x$ (10 keV) dr $\sim 10^{27}$ L_{38} [cf. Eq. 16.4] we need $L_x \lesssim 10^{35}$ ergs s^{-1}. Thus the X-ray pulsar Cos B candidates are expected to be less luminous in X-rays than all but one of the identified binary X-ray pulsars (4U0352+30) and the data points of Figure 18b. Such L_x would be expected before and/or after the very strong accretion phase during which most X-ray pulsars are observed. For $L_x \sim 10^{35}$ ergs s^{-1} and $B_s \sim 10^8$ G Eq. 16.3b gives $\Delta V \sim 10^{14}$ volts, just large enough to sustain the γ-ray production processes of Section 3.

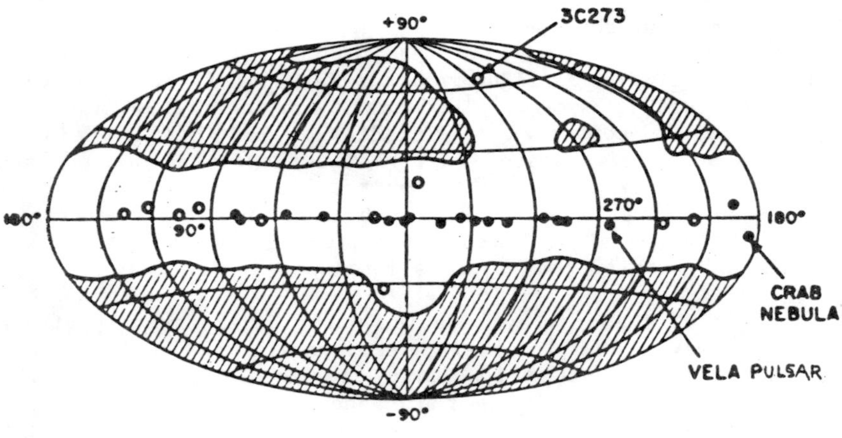

Figure 16

The Region of the sky, in Galactic latitude and latitude, searched for gamma-ray sources by Cos B (unshaded) and sources detected above 100 MeV. The closed circles denote sources with intensities > 1.3×10^{-6} photons cm^{-2} sec^{-1}; open circles denote sources below this level.

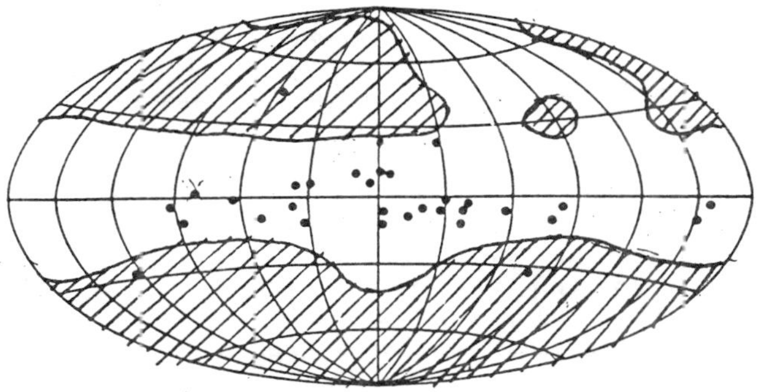

(a)

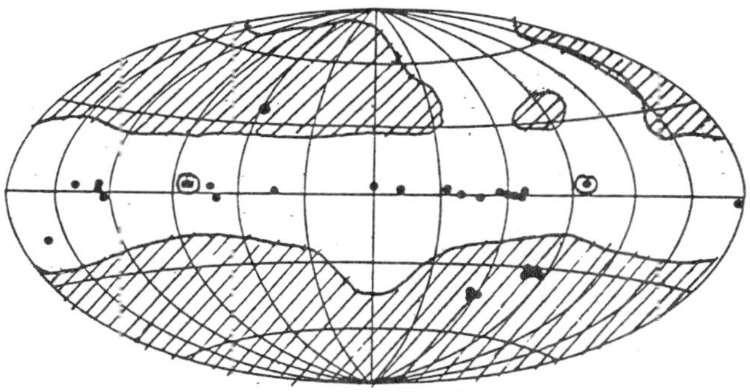

(b)

Figure 17

Distribution in Galactic longitude and latitude of observed X-ray accreting neutron stars in binaries. a) low (companion) mass (LMXB) X-ray binaries: b) more massive companions form the (MXRB) family which include Cyg X-3 and Vela X-1 (both circled) and Her X-1 (in the upper left unsearched region).

Spectrum. For an accelerator with $|\Delta V| \sim 10^{14} - 10^{15}$ volts and a characteristic B around it, several radii above the neutron star surface, of 10^7 G, both the magnetic field and ΔV are comparable to those of the outermagnetosphere accelerating gap of the Vela and Crab pulsars (Sections 9 and 10). These were calculated to give a $10^2 - 10^3$ MeV γ-ray spectrum $dN/dE \sim E^{-2}$. In the Crab outergap the primary e^-/e^+ also initially lose their energy to inverse Compton scattering on soft X-rays, as expected for the accretion disk accelerator. However, even the rather different detailed mechanisms in the Vela outergap (with smaller B) gave a similar γ-ray spectrum in this region. The unidentified Cos B sources have a very similar spectrum, $dN/dE \sim E^{-2}$, about the same as that from the pulsar outergaps with similar parameters.

Another possible candidate for the seat of an "outergap" accelerator which sustains strong $10^2 - 10^3$ MeV γ-ray emission is a rapidly rotating strongly magnetized White Dwarf. Magnetic White Dwarfs constitute about 2% of all White Dwarfs, and are thus about as numerous as neutron stars. Because they have a 10^3 times larger radius the eddy current decay time of their magnetic fields is very much larger than the crustal decay in neutron stars: magnetic White Dwarfs will keep their magnetic fields for longer than the age of the Universe. A rapidly rotating magnetic White Dwarf with P ~ 10 sec and $B_s \sim 2\ 10^6$ G (comparable in magnetic flux to 10^{12} G radiopulsar) has a corotating magnetosphere with a net charge density (ρ_0), maximum potential drop, and field line curvature which can be comparable to those in the outermagnetosphere of the Vela pulsar. If it develops a similarly structured outergap this would be a strong source of γ-rays with the required spectrum. The needed rapid spin may possibly be achieved in some cataclysmic binaries in which White Dwarfs may be spun up by strong accretion instead of being spin locked into the orbital rotation of their companions. (In such binaries about half the White Dwarfs are strongly magnetized.)

It is a pleasure to thank the Aspen Center for Physics, where these lecture were prepared, for their hospitality. I also wish to thank Professors J. Shaham, D. Eichler, and F. Lamb for stimulating conversations on some of the points discussed. This is Columbia Astrophysics Laboratory Contribution Number 304.

REFERENCES

Arons, J. 1981, IAU Symposium 74: Origin of Cosmic Rays 175.

Alpar, M., and Shaham, J. 1985, Nature, (in press).

Alpar, M. 1985, Proceedings of NATO Advanced Study Institute "High Energy Phenomena Around Collapsed Stars" F. Pacini, ed. (Reidel 1987).

Alpar, M., Cheng, A., Ruderman, M., and Shaham, J. 1982, Nature, **300**, 728.

Anzer, U. and Börner, G. 1983, Astron. and Ap..
Baltrusaitis, R., Cassiday, G., Cooper, R., Elbert, J., Gerhardy, P., Loh, E., Mizumoto, Y., Sokosky, P., and Steck, D. 1985, University of Utah preprint.
Barat, C., Hurley, K., Niel, M., Vedrenne, G., cline, T., Desai, U., Schaeffer, B., Teegarden, B., Evans, W., Fenimore, E., Klebasadel, R., Laros, J., Estulin, I., Zenchenke, V., Kuznetsov, A., Kurt, V., Ilovaisky, S., and Motch, C. 1984, Ap.J. (Lett.), (in press).
Barat, C. et al. 1979, Atron. Astrophys. (Lett.), **79**, L27.
Bignami, G. 1985, Proceedings of NATO Advanced Study Institute "High Energy Phenomena Around Collapsed Stars" F. Pacini, ed. (Reidel 1987).
Blandford, R., Applegate, J., and Hernquist, L. 1983, M.N.R.A.S., **204**, 1025.
Boone, J., Cady, R., Cassiday, G., Elbert, J., Loh, E., Sokosky, P., Steck, D., Waserbaech, S. 1983, Proceedings of the Cosmic Ray Workshop, University of Utah, p. 268-84.
Boriakoff, V. 1984, Proceedings of NRAO Millisecond Pulsar Workshop, D. Stinebring, ed.
Brecher, K., and Chanmugam, G. 1985, Nature, **313**, 767.
Cesarsky, C 1985, Proceedings of NATO Advanced Study Institute "High Energy Phenomena Around Collapsed Stars" F. Pacini, ed. (Reidel 1987).
Chanmugam, G. 1984, Proceedings of NRAO Millisecond Pulsar Workshop, D. Stinebring, ed.
Cheng, A., Ruderman, M., and Sutherland, P., 1976, Ap.J., **212**, 808.
Cheng, A., K.-S., Ho, C., and Ruderman, M. 1985a, Ap.J., (in press).
Cheng, A., K.-S., Ho, C., and Ruderman, M. 1985b, Ap.J., (in press).
Daughterty, J., and Harding, A. 1982, Ap.J., **252**, 337.
Dowthwaite, J., Harrison, A., Kirkman, I., Macrae, H., McComb, T., Oxford, K., Turner, K., and Walmsley, M. 1984a, Ap.J., **286**, L35.
Dowthwaite, J., Harrison, A., Kirkman, I., Macrae, H., McComb, T., Oxford, K., Turner, K., and Walmsley, M. 1984b, Nature, **309**, 691.
Dowthwaite, J., Harrison, A., Kirkman, I., Macrae, H., McComb, T., Oxford, K., Turner, K., and Walmsley, M. 1985, preprint.
Eichler, D., and Vestrand, T. 1985, University of Maryland preprint.
Ellison, D. 1985, University of Maryland preprint.
Evans, W. Klebesadel, R., Laros, J., and Terrel, J. 1980, Nature, **286**, 784.
Flowers, E., and Ruderman, M. 1977, Ap.J., **215**, 302.
Ghosh, P., and Lamb, F. 1979, Ap.J., **234**, 296.
Goldreich, P., and Julian, W. 1969, Ap.J., **157**, 869.
Groth, E. 1975, Ap. J. Suppl., **29**, 431.
Hardee, P. 1977, Ap.J., **216**, 873.
Harding, A., Tademaru, E., and Esposito, L. 1978, Ap.J., **225**, 226.
Harvey, J., Ruderman, M., and Shaham J. 1985, Phys. Rev. D, (in press).
Helfand, D., Ruderman, M., and Shaham, J. 1983, Nature, **304**, 423.
Helfand, D. 1982, Supernova Remnants and their X-ray Emission, IAU Symposium 101.

Henricks, H. 1983, Accretion-Driven Stellar X-ray Sources, W. Lewin and W. van den Heuvel, eds., (Cambridge: Cambridge Univesity Press).

Holloway, N. 1973, Nature, **246**, 6.

Hurley, K. 1985, Proceedings of NATO Advanced Study Institute "High Energy Phenomena Around Collapsed Stars" F. Pacini, ed. (Reidel 1987).

Joos, P., and Rappaport, S. 1983, Accretion-Driven Stellar X-ray Sources, W. Lewin and W. van den Heuvel, eds. (Cambridge: Cambridge University Press).

Kanbach, G., Bennet, K., Bignami, G., Buccheri, R., Caraveo, P., D'Amico, N., Hermson, W., Lichti, G., Masnou, J., Mayer-Hasselwander, H., Paul, J., Sacco, B., Swanenberg, B., and Wills, R. 1980, Astron. Astrophys., **90**, 163.

Kennel, C., and Coroniti, F. 1984, Ap.J., **283**, 710.

Krause-Polstorff, J., and Michel, F. C. 1985a, M.N.R.A.S., **213**, 43.

Krause-Polstorff, J., and Michel, F. C. 1985b, M.N.R.A.S., **144**, 72.

Lamb, F. 1985, Proceedings of NATO Advanced Study Institute "High Energy Phenomena Around Collapsed Stars" F. Pacini, ed. (Reidel 1987).

Liang, E., and Petrosian, V. 1985, γ-ray and Neutron Star Phenomenon, Proceedings of Stanford Work, July 31 - August 3, 1984, (New York: AIP), (to be published).

Liang, E. 1985, Proceedings of NATO Advanced Study Institute "High Energy Phenomena Around Collapsed Stars" F. Pacini, ed. (Reidel 1987).

Lloyd-Evans, J., Coy, R., Lambert, A., Lapikens, J. Patel, M., Reid, R., and Watson, A. 1983, Nature, **305**, 784.

Lyne, A. 1985, Proceedings of NATO Advanced Study Institute "High Energy Phenomena Around Collapsed Stars" F. Pacini, ed. (Reidel 1987).

Manchester, R., Durdin, J., and Newton, L. 1985, Nature, **313**, 374.

Manchester, R., and Taylor, J. 1977, Pulsars, (San Francisco: W.H. Freeman and Co.).

Michel, R.C. 1981, Pulsars, IAU Symposium 95, ed. W. Sieber and R. Wielebinski.

Michel, R. C. 1982, Rev. Mod. Phys., **54**, 1.

Phillip, W. 1974, Ap.J., **190**, 391.

Protheroe, R., and Clay, R., 1985, Nature, **315**, 205.

Protheroe, R., Clay, R., and Gerhardy, P. 1984, Ap.J. (Lett.), **280**, L47.

Radhakrishnan, V., and Srinivasan, G. 1981, Proceeding of the 2nd Asian-Pacific Regional IAU Meeting, ed. B. Hidayat.

Reynolds, S. 1984, Proceedings of NRAO Conference on Millisecond Pulsars, Green Bank, 6-8 June 1984.

Ruderman, M., and Sutherland, P. 1975, Ap.J., **199**, 51.

Samorski, M., and Stamm, W. 1983, Ap.J. (Lett.), **268**, L17.

Schaeffer, B. 1981, Nature, **294**, 722.

Sturrock, P. 1978, Ap.J., **164**, 529.

Tademaru, E. 1973, Ap.J., **183**, 625.

Trümper, J. 1985, Proceedings of NATO Advanced Study Institute "High Energy Phenomena Around Collapsed Stars" F. Pacini, ed. (Reidel 1987).
van den Heuvel, E. 1977, Ann. of New York Acad. Sci., **302**, 14.
van den Heuvel, E. 1984, preprint.
van den Heuvel, E. 1985, Proceedings of NATO Advanced Study Institute "High Energy Phenomena Around Collapsed Stars" F. Pacini, ed. (Reidel 1987).
Wills, R. et al. 1982, Nature, **296**, 723.
Wood, K., Byram, E., Chubb, T., Friedman, H., Meekins, J., Share, G., and Yentis, D. 1981, Ap.J., **247**, 263.
Woosley, S. 1984, Proceedings of AIP Santa Cruz Summer Workshop, "High Energy Transients", July 1983, (New York: AIP), p. 51.
Wdowczyk, J. 1984, Recontre de Moriond: High Energy Astrophysics, J. Tran Thanh Van, ed.

NEUTRON STARS AS COSMIC HADRON PHYSICS LABORATORIES

David Pines
Department of Physics
University of Illinois at Urbana-Champaign
1110 West Green Street
Urbana, Illinois 61801

ABSTRACT. Recent developments have radically changed our understanding of both the dynamics of neutron-star superfluids and the free precession of neutron stars. Astrophysical constraints on neutron matter are derived from observation and compared with microscopic theory.

1. INTRODUCTORY REMARKS

The identification of radio pulsars as rotating neutron stars in 1968, together with the observations of glitches (sudden spinups) followed by comparatively long ($\gtrsim$d) post-glitch relaxation processes in the Vela and Crab pulsars in 1969, led theorists to explore in some detail the states of matter inside neutron stars, in the hope and expectation that by combining theory with observation it might prove possible to obtain information concerning the internal structure and physical behavior of these highly condensed objects. As a result, by the summer of 1970 it was argued that glitches constitute a direct probe of the stellar interior, and that since the long relaxation times observed in the post-glitch behavior of both the Vela pulsar[1,2] and the Crab pulsar[3] were too slow to be explained in terms of the viscous processes of normal matter, these observations provided direct evidence for the existence of cosmic hadron superfluidity.[4]

In similar fashion, the discovery of pulsating compact x-ray sources in 1971, and their subsequent identification as accreting rotating magnetic neutron stars in close binary systems, led both observers and theorists to explore whether timing observations of these objects might likewise provide information on their internal structure. Of particular interest was the possibility that the 35 day periodicity of the high-low x-ray cycle observed in Her X-1[5] was caused by free precession of the neutron star, analogous to the Chandler wobble of the earth;[6] as we shall see, the wobble frequency of a neutron star depends on the thickness of its solid outer crust and hence is a sensitive barometer of the neutron matter equation of state for densities $\gtrsim \rho_0$, the density of nuclear matter.

During the past fifteen years, both the Vela and Crab pulsars have exhibited further substantial glitches in their rotational

F. Pacini (ed.), High Energy Phenomena Around Collapsed Stars, 193–208.

periods (six for the Vela pulsar, one for the Crab pulsar), as have four other pulsars (PSR 0525+21, 1385-43, 1641-45, 2224+65), while detailed analyses of the post-glitch behavior of the Vela pulsar have been carried by Downs,[7] and of the Crab pulsar by Lohsen.[8] Over this same period considerable theoretical work was carried out on the nature of the neutron superfluids in both the stellar crust and core, and on the coupling of these superfluids to the stellar crust. These developments have radically changed our understanding of the dynamics of neutron star superfluids. In the early 1970's it was believed that the glitches were caused by starquakes, and the post-glitch behavior reflected internal torques on the crust produced by the core neutron superfluid, which comprises the bulk of the moment of inertia of the star; today the picture is that (for a review, see Ref. 9)

- the core superfluid rotates rigidly with the crust
- because of the pinning of vortices in the crustal superfluid to crustal nuclei, the crustal superfluid, which contributes only $\sim 10^{-2}$ of the stellar inertial moment, can be out of equilibrium with the crust for timescales of weeks to years and is responsible both for pulsar glitches and post-glitch relaxation.
- vortex creep theory, which describes the motion of pinned vortex lines in the crustal superfluid, provides an excellent fit to the Vela pulsar timing observations of Downs,[7] which span the decade 1969-79 and include four giant glitches.
- the observed post-glitch relaxation times in the Vela and Crab pulsars, and PSR 0525+21 provide a means of estimating the interior temperature of these pulsars.
- interior temperatures of the Crab and Vela pulsars obtained in this way, while consistent with observations, are not consistent with a standard cooling scenario.
- the dominant source of heat in pulsars old enough to have radiated away their original heat content is vortex creep.

Over this same time period, explanations of both the origin of the 35^d periodicity in Her X-1, and its implications for stellar structure have fluctuated substantially. Thus at the end of 1972 large wobble amplitudes resulting from a wobble-operated accretion gate were suggested for Her X-1, while the rather soft equations of state for neutron matter available at that time led theorists to the conclusion that Her X-1 must have a solid inner core.[6] Subsequent observations of its optical and x-ray behavior tended to favor an interpretation based on an accretion disk which precessed with a 35 day period and obscured the primary x-rays.[10] Moreover, it was shown on the one hand that the hadron equation of state could be comparatively stiff, and hence a 1.4 $M_\odot$ neutron star could have a crust which was sufficiently thick to enable it to precess with a 35 day period,[11] but that on the other hand, any appreciable pinning of neutron superfluid vortices in the stellar crust would make its wobble period ~ hours.[12] Finally, it was shown that the 35 day clock of Her X-1 possesses a stability of only ~ a few percent, which is far less than might be expected for free precession.[13]

Today, as a result of extensive observations of Her X-1 with the EXOSAT satellite, a definitive picture has emerged[14] in which:

- the 35 day cycle results from large amplitude (~ 25°) free precession of the magnetic dipole axis of the neutron star
- the large amplitude wobble causes massive unpinning of vortex lines, so that the wobble period is determined primarily by the initial oblateness and crustal extent of the neutron star, but that
- unpinning and repinning events are responsible for the observed poor stability of the 35 day clock.

These results, taken together with limitations on the precession frequency imposed by the neutron matter equation of state, lead me to conclude that the hadron equation of state must be comparatively stiff.

In these lectures I summarize briefly the detailed microscopic calculations on neutron matter and the properties of the pinned crustal superfluid which provide a basis for comparing theory with observation on neutron stars. I then discuss the interplay between theory and observation which I have summarized in Table I, and, in so doing, hope to persuade you that this interplay has been sufficiently extensive and successful to justify the title of these lectures, "Neutron Stars as Cosmic Hadron Physics Laboratories."

2. HADRON MATTER EQUATION OF STATE AND NEUTRON STAR STRUCTURE

Neutron stars typically possess masses ~ 1.4 $M_\odot$. Extensive theoretical studies have shown that such stars have radii of the order of 10-16 km, with a solid outer crust, ~ 1-5 km thick, containing increasingly neutron-rich nuclei in a periodic array, free electrons, and free neutrons, beneath which is a quantum liquid interior which begins at densities somewhat less than the density of nuclear matter, $\rho_o = 2.8 \times 10^{14}$ g cm^{-3}, and contains largely superfluid neutrons, with an admixture of a few percent of superconducting protons and relativistic electrons. The constituents of neutron stars at densities well in excess of ρ_o are uncertain (see Ref. 15 for a review of the possibilities). The equation of state of neutron star matter in the crustal region is comparatively well understood; the remaining and key ingredient in determining the maximum mass and internal structure of neutron stars is the calculation of a reliable equation of state for the neutron liquid phase. It is a difficult calculation, in part because the basic interaction between neutrons is still not perfectly known, in part because, for a given interaction model, calculation of the ground state energy of a system at nuclear densities and beyond is a far-from-trivial many-body problem. Pandharipande et al.[11] have shown that one can construct a variety of models for the neutron interaction which are consistent with terrestrial constraints (the free nucleon scattering data and the experimentally known energy, equilibrium density, and symmetry energy of nuclear matter), and which span a range from the so-called "soft"

Table 1. Neutron Stars as Cosmic Hadron Matter Laboratories

Density range of Hadron Matter	Theoretical Predictions	Observational Consequences
$\frac{\rho_o}{500} \lesssim \rho \lesssim \rho_o$	• Superfluid crustal neutrons • Pinning of superfluid vortices to crustal nuclei • Vortex creep theory of post-glitch relaxation • Internal temperature of glitching pulsars • Vortex creep heating of old neutron stars • Wobble period fluctuations	• Origin of macroglitches in the Vela and other pulsars • Postglitch behavior and surface temperatures of Vela and Crab pulsars and PSR 0525+21 • Surface temperatures of old neutron stars • Changes in wobble period
$\rho \gtrsim \rho_o$	• Equation of state of neutron matter can be soft or stiff • Crustal extent, wobble period, M-R relation, M_{max}, and starquake frequency are sensitive e.o.s. barometers	• 35^d periodicity observed in Her X-1 • Initial spins of Her X-1 like stars • Starquakes in Crab and other very young pulsars
$\rho \gtrsim 2\rho_o$	• Possible exotic new forms of matter, such as pion condensates, quark liquids, which lead to novel cooling mechanisms	• Reduced surface temperatures of young neutron stars

equations of state (derived from models for which the average system interaction energy is attractive at nuclear densities) to "stiff" equations of state (derived from models for which the average system interaction energy becomes repulsive at sub-nuclear densities). Representative examples of the corresponding models are the phenomenological Reid (R) potential for neutron-neutron interaction, and the tensor interaction (TI) model which assumes that the attractive part of the neutron-neutron interaction comes from higher order pion exchange. Stars calculated with a stiff equation of state have lower central density, a larger radius, and a considerably larger crustal volume, than do stars of the same mass calculated with a soft equation of state. The cross-sections of R and TI model stars of 1.4 $M_\odot$ are shown in Fig. 1.

At present perhaps the principle uncertainty in the nuclear matter equation of state comes in the role which is assigned to three-body or four-body interactions. In calculations based on the best current two-body potentials, one finds too much binding and a too-high saturation density. As Friedman and Pandharipande (FP)[16] have shown, one corrects for this by introducing three-body (or four-body) forces which are both short range and repulsive and which become increasingly important at nuclear matter densities and beyond. While the early calculations of the neutron matter equation of state differed substantially, more recent calculations have tended to differ far less, and a theoretical consensus may be emerging that the equation of state is comparatively stiff, somewhat stiffer, say, than the FP equation of state. One way of understanding this trend is to note that as many-body techniques improve, energies calculated from the best two-body potentials become increasingly negative; hence one needs more in the way of short range repulsion arising from three or four-nuclear interactions (Pandharipande, private communication).

3. HADRON SUPERFLUIDS

At least two distinct pairing states of neutron superfluids are expected:

(i) In the inner part of the crust, where the interparticle spacing is greater than the range of the repulsive forces ($\lesssim$ 1 fm), ^1S_o pairing dominates, with an average energy gap, $\Delta \sim 1$ MeV.[17,18]

(ii) At the higher densities ($\gtrsim \rho_o$) characteristic of the quantum liquid interior, the combination of short range repulsion and an attractive tensor force leads to a $^3P_2$ paired superfluid.[17,19]

The microscopic calculation of the the energy gap as a function of density is far from trivial, as many-body corrections to both the mass of the neutrons and their effective interaction play a significant role. Thus quite recently, Chen et al.[20] have found that when a perturbation theoretical calculation based on correlated basis functions is carried out for the ^1S_o pairing state, the calculated average energy gap is $\lesssim 0.25$ MeV. Their results for the transition temperature are compared with earlier work in Fig. 2. As we shall see, the observational consequences of large vs. small energy gaps are substantial, so that just as is true for the neutron matter equation of state, it might prove possible to use observations on neutron stars as a guide to which theoretical calculation is more likely correct.

Because neutron stars rotate, neither neutron superfluid will be spatially uniform; the neutron superfluids contain an array of quantized vortex lines, with circulation $\kappa = h/2m$, parallel to the rotation axis, of sufficient number that on a macroscopic scale the neutrons will appear to be rotating as a rigid body and thus have their classical moment of inertia. In the stellar crust, the normal cores of the vortices, where the condensate wave function decreases to zero, either may be pinned to the lattice of nuclei present there or may thread the spaces between them. It is favorable for a vortex line to pin to a nucleus if the cost per particle to create its normal core is thereby reduced. Since the nuclei also contain superfluid neutrons, whether or not one has pinning is then a matter of comparative superfluid energetics; the pinning energy per nucleus is[21,22]

$$E_p = \frac{3}{8}\left[\frac{\rho_{out}\Delta^2_{out}}{(E_F)_{out}} - \frac{\rho_{in}\Delta^2_{in}}{(E_F)_{in}}\right]V \ . \tag{1}$$

Here "out" and "in" denote local values of Δ and the Fermi energy, E_F, for superfluid neutrons outside and inside the nuclei. V is the overlap volume of the vortex core and the nucleus of radius R_N; the core radius is the coherence length of the "outside" neutron superfluid, $\xi = 2E_F/\pi k_f\Delta$. The maximum available pinning force per unit length of vortex line is $f_p = (E_p/\xi b)$, where b, the spacing between successive pinning centers along the vortex line, depends on the physical circumstances under which pinning takes place.

Pinning will be maximal in strong pinning layers, where E_p is greater than a typical lattice energy, $E_L \sim Z^2e^2R_N/b_z^2$ (b_z is the lattice spacing), so that the pinning force is strong enough to dislodge nuclei from their equilibrium sites in the lattice, and b is comparable to b_z. When $E_p \lesssim E_L$, one has a transition to <u>weak pinning</u> layers, in which the vortex core pins only to those sites through which it passes; the spacing between the sites, b, is $\sim (E_L/E_p)b_z$, and the effective pinning energy is reduced to $\sim E_p(E_p/E_L)$.[22] In the higher density regions of the pinning layers, where the coherence length, $\xi \gtrsim b_z$, a vortex core can "pin" to more than one nucleus at a time. Moving a vortex line requires still less change in the pinning energy; one is in a region of <u>superweak pinning</u>, in which the effective spacing, b, between multiple pinning sites is far larger than $(E_L/E_p)b_z$.[23]

The nature and extent of the pinning layers depends sensitively on the variation of the gap through the inner stellar crust. Thus for the typical "large gap" microscopic calculations, one finds strong pinning for stellar densities $3 \lesssim \rho_{13} \lesssim 8$, where ρ_{13} is the stellar density in units of 10^{13} g cm^{-3}, weak pinning for $8 \lesssim \rho_{13} \lesssim 11$, and

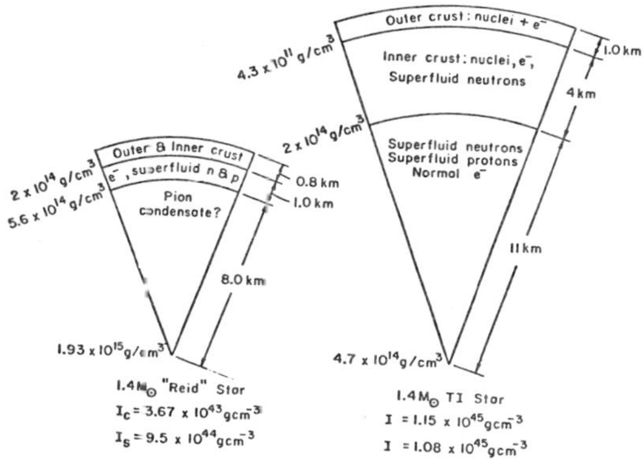

Fig. 1. Cross section of 1.4 $M_\odot$ Reid and TI stars.

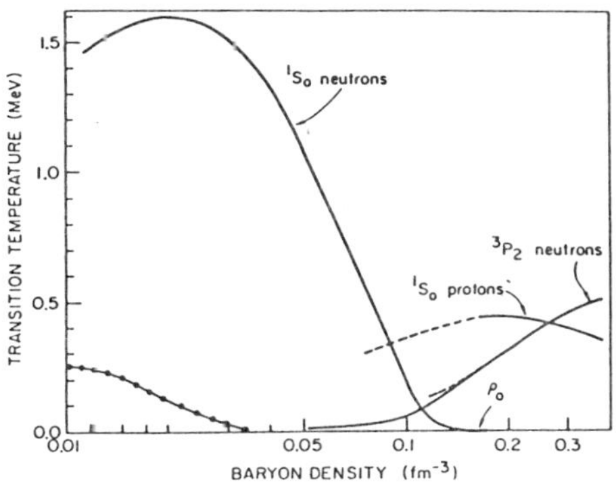

Fig. 2. Calculated superfluid transition temperatures as a function of density. The solid lines represent variational calculations; the dotted line depicts the results of Chen et al.[20]

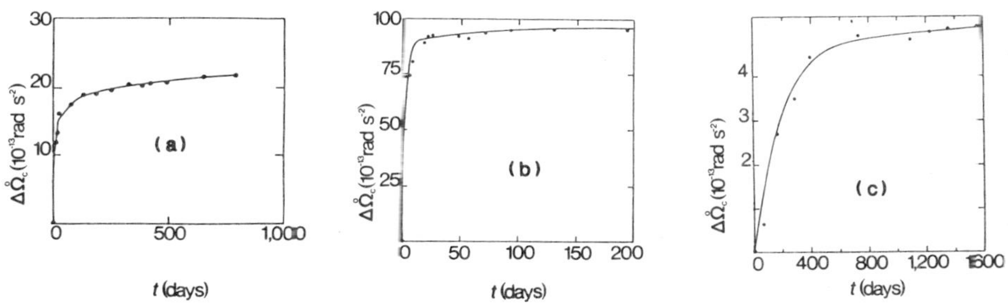

Fig. 3. Fits to the post-glitch $\dot{\Omega}_c(t)$ (from Ref. 9). (a) Representative post-glitch fit for the Vela pulsar: $\dot{\Omega}_c(t)$ following the glitch of July 1978. (b) Fit to $\dot{\Omega}(t)$ following 1975 Crab pulsar glitch. (c) Fit to $\dot{\Omega}(t)$ following the glitch observed in PSR0525+21.

superweak pinning at still higher densities.[24] On the other hand, with the "small gap" calculations of Chen et al.,[20] there are no regions of strong pinning, weak pinning exists at densities $0.2 \lesssim \rho_{13} \lesssim 3$, and superweak pinning at both higher and lower densities.[25]

4. HOW DO THE CORE NEUTRONS COMMUNICATE WITH THE OUTSIDE WORLD?

The principle energy source of a pulsar is the rotational energy of the neutrons. How do the superfluid core neutrons, which make up the bulk of the neutron star, follow changes in the crust rotation frequency, Ω_c? The electrons and protons in the interior must co-rotate by electromagnetic coupling, irrespective of whether or not the protons are superfluid, while the charged particles in the crust and those in the interior communicate readily through the viscosity and/or the magnetic field in the interior. Since the rotational state of the core neutron superfluid is determined by the distribution of its quantized vortices, the timescale for changes in this distribution and hence for crust-core coupling, must come from the coupling of interior electrons and protons to the normal cores of the vortex lines. It was initially believed that the crust-core timescale was comparatively long (~ days to years); however Alpar et al.[26] have shown that when one takes into account the change in the structure of a neutron vortex line that results from the interaction between the neutron and proton condensates in the quantum liquid interior, the dynamical coupling time is reduced to $\tau \cong 400\ P(s)$ where P is the rotational period. This work provides crucial support for invoking only the crust superfluid in a model of pulsar glitches and post-glitch relaxation.

5. VORTEX CREEP THEORY

The coupling between the pinned crustal superfluid and the crust is achieved by thermal creep of the vortex lines against pinning energy barriers, by analogy with flux creep in type II superconductors.[27] In the steady state the pinned superfluid is coupled to the rest of the star and shares the effect of external torques. A glitch perturbs vortex creep away from its steady-state rate. Part of the pinned superfluid is thereby decoupled, and its gradual recoupling leads to observable behavior.

The angular velocity of the neutron superfluid, $\Omega(r)$, is determined by the distribution of the pinned vortex lines. $\Omega(r)$ can change only if vortices move. Their motion is described by an internal torque, $N_{int} \equiv \int_p dI_p \dot{\Omega}(r,t)$ while at the rate at which thermally activated vortex lines are able to tunnel between adjacent pinning configurations is

$$v_r = v_o \exp\left[-\frac{E_p}{k_B T}\frac{(\omega_{cr} - \omega)}{\omega_{cr}}\right] \tag{2}$$

where T is the crustal temperature, $\omega \equiv \Omega - \Omega_c$ is the superfluid angular velocity relative to that of the crust, and

$$\omega_{cr} \equiv (\Omega - \Omega_c)_{max} = f_p/\rho\kappa r \equiv E_p(\rho)/\rho\kappa r\xi b \tag{3}$$

is the maximum superfluid lag which can be supported by the pinning force, f_p. In a given pinning layer, the approach to steady state is characterized by a relaxation time, τ_i, proportional to the temperature[23]

$$\tau_i = \left(\frac{\omega_{cr}}{E_p}\right)_i \frac{k_B T}{|\dot{\Omega}_c|} \equiv \frac{\alpha_i T}{|\dot{\Omega}_c|} \tag{4}$$

Glitches are catastropic unpinning events extraneous to the creep process which itself involves a continuous slow vortex motion. The internal torque produced by the neutron superfluid in response to a glitch, $(\Delta\Omega_c)_0$, takes the simple form

$$N_{int}(t) = \sum_i I_i \dot{\Omega}_c \frac{1}{1 + (e^{t_{oi}/\tau_i} - 1)e^{-t/\tau_i}} \tag{5}$$

where I_i is the inertial moment of the ith pinning layer and t_{oi}, is the associated post-glitch delay time. For pinning layers where there is no glitch-associated vortex motion, $t_{oi} = (\Delta\Omega_c)_o/|\dot{\Omega}_c|$, and in those regions where there is vortex motion, $t_{oi} = \delta\Omega_i/|\dot{\Omega}_c|$ where $\delta\Omega_i$ is the glitch-associated change in the superfluid rotation frequency.

The vortex creep theory of post-glitch response differs from the two-component model of Baym et al.:[28] (1) It involves a number of distinct components which, however, make up only a comparatively small fraction (~ a few percent) of the neutron superfluid, namely the pinned superfluid in the crust. This is borne out by the observed values of $(\Delta\dot{\Omega}_c/\dot{\Omega}_c)_0 \sim 10^{-3} - 10^{-2}$. (2) It is a fundamentally nonlinear theory of superfluid response; glitch-associated vortex creep depends exponentially on the size of the glitch (in regions in which no vortex motion occurs) and on the change in the superfluid velocity (in regions in which vortex motion occurs during the glitch). (3) It leads to relaxation times proportional to the internal temperature of the star, so that timing observations can provide a unique probe of the internal stellar temperature.

Thermally activated vortex creep leads to an internal dissipation of the stored rotational energy of the neutron star. The magnitude of the continuous energy dissipation resulting from creep follows directly from the superfluid equations of motion, and is

$$\dot{E}_{dis} \cong |\dot{\Omega}_c| \int_p dI_p \omega_\infty \cong |\dot{\Omega}_c| I_p \bar{\omega}_{cr} \tag{6}$$

where I_p is the moment of inertia of the entire set of pinning layers, $\bar{\omega}_{cr}$ is an average through the pinning layers and ω_∞ is the steady-state value of $\Omega - \Omega_c$. This dissipation is the principal source of heat for old pulsars which have radiated away their original heat content.

6. VELA PULSAR GLITCHES

Alpar et al[24] have shown that the observed complex post-glitch behavior of the Vela pulsar following the four glitches which took place in the decade 1969-79 finds a straightforward explanation in terms of vortex creep theory. They fit the observational data of Downs[7] with the following expression for $\dot{\Omega}_c(t)$,

$$I_c\dot{\Omega}_c(t) = N_{ext} - N_1 - N_2 - N_A \qquad (7)$$

Here N_1 and N_2 are torques which represent an average response of two distinct pinning layers through which no vortices move and are specified by Eqs. (5) and (6), with relaxation times τ_1 and τ_2. N_A is associated with the third region through which vortices move and varies linearly with time. A choice of τ_1 = 3 days and τ_2 = 60 days, for each of the four glitches, gives an excellent fit to the observational data (Fig. 3). Alpar et al.[24] conclude that the two distinct pinning layers correspond to weak pinning and superweak pinning respectively, characterized by $\langle\alpha^{sw}\rangle \sim 2\times10^{-17}$ days $K^{-1}s^{-2}$ and $\langle\alpha^{w}\rangle \sim 4\times10^{-16}$ days $K^{-1}s^{-2}$ (see Eq. (4)). Within each pinning region the actual pinning parameters may vary either way by factors of ~2. From the observed relaxation times, Alpar et al. estimate the internal temperature of the Vela pulsar to be $(1.5\pm1)\times10^7$ K. The quoted errors do not refer to a particular calculation, but rather reflect an estimate of the degree of uncertainty introduced by lack of knowledge of various pinning parameters.

7. OTHER GLITCHING PULSARS

Alpar, Nandkumar and Pines[29] find, surprisingly, that the post-glitch behavior of the Crab pulsar following its two large glitches in 1969 and 1975 could be fitted (see Fig. 3) with the extreme (Occam's razor) hypothesis that both the internal structure and the vortex pinning sites in the Crab pulsar are essentially the same as those hypothesized for the Vela pulsar, so that the only significant physical differences between the two pulsars reside in the spin down rate and internal temperature. In this fashion they deduce an internal temperature for the Crab pulsar of 3.8×10^8 K, with, again, an uncertainty either way of a factor of ~2. For a 1.4 $M_\odot$ neutron star with a reasonably stiff equation of state, they estimate the surface temperatures observed at infinity to be $T_{\infty,Vela} \simeq 3\times10^5$ K and $T_{\infty,Crab} \simeq 1.6\times10^6$ K. For both pulsars, these surface temperatures are well below the upper limits obtained by the Einstein observatory. On comparing these estimates with calculations on neutron star cooling and evolution, one finds that an internal temperature ratio of 25 is not consistent with the standard cooling calculations, which, for a given Crab temperature would predict a considerably hotter Vela pulsar. Therefore, these results may provide the first indication that non-standard, more rapid cooling scenarios (perhaps reflecting the presence of a pion condensate, if a pion condensate indeed leads to sufficiently rapid cooling) are in operation in neutron stars.

Consider now pulsars which are old enough to have radiated away their original heat content so that their dominant source of heat is due to internal dissipation by vortex creep. On equating the rate of dissipation E_{dis}, Eq. (6), to the stellar luminosity, one finds a limiting black-body surface temperature of

$$T_s \simeq 1.1\times10^5[I_{p,43}\bar{\omega}_{cr}\dot{\Omega}_{-14}/R_6^2]^{1/4} \quad (8)$$

From the upper limit to the surface luminosity of the Vela pulsar in the soft X-ray region and the Einstein detection of PSR1929+10, $I_{p,43}\bar{\omega}_{cr}$ is estimated to be $\lesssim 1$.[30] For such a pulsar, the thermal flux given by equation Eq. (8) is small enough for the internal temperature to be close to the surface temperature. If all old pulsars possess essentially the same pinning layers as the Crab and Vela pulsars, they would possess creep relaxation times that depend only on their spin-down rate,

$$\tau_{sw}^{old} \sim 220|\dot{\Omega}_{-14}|^{-3/4} \text{ days} \quad (9)$$

$$\tau_{w}^{old} \sim 4.4\times10^3|\dot{\Omega}_{-14}|^{-3/4} \text{ days} \quad (10)$$

for the superweak and weak pinning layers, respectively.

Among the old pulsars which have exhibited glitches, post-glitch timing observations are available only for PSR0525+21. Alpar et al.[29] find that vortex creep theory provides a good fit to the observed post-glitch behavior of this pulsar with a 'fast' relaxation time of ~150 days and is consistent with the additional presence of a much slower 3,000-day relaxation process (Fig. 3). The vortex creep theory prediction of τ for PSR0525+21 is, from Eq. (9), some 140 days, in excellent agreement with this fit. Thus vortex creep theory can explain the post-glitch behavior of both very young and very old pulsars. What is striking is that it does so without involving new parameters for each glitch or pulsar.

8. ASTROPHYSICAL CONSTRAINTS ON NEUTRON MATTER ENERGY GAPS

Since vortex creep theory provides an excellent fit to observation, it is natural to ask what constraints this fit places on microscopic models of neutron superfluidity, and whether the current microscopic energy gap calculations are compatible with these constraints. The astrophysical constraints provided by vortex creep theory include the following: $(I_p/I)_{weak} \gtrsim 2\times10^{-2}$ (from the fits to the Vela pulsar glitches using a weak coupling relaxation time ~60^d); $\langle I_{p,43}\,\omega_{cr}\rangle \lesssim 1$) (from the post-glitch behavior of PSR 0525+21); these imply a further constraint, that there is no appreciable layer with strong pinning (i.e. $E_p \gtrsim 1$ MeV, $\omega_{cr} \gtrsim 1$). No current gap calculation is compatible with these constraints. The "large" energy gap results lead to substantial strong pinning regions, while the "small" energy gap results predict too little weak pinning, no matter how stiff the neutron matter equation of state. However, work in progress with Ali Alpar and Radha Nandkumar suggests that with a "modest" energy gap, i.e. a $\Delta(\rho)$ relation which is intermediate between that given by the large and the small gap results, and a neutron matter equation of state which is somewhat stiffer than the MF model (see below), it is possible to satisfy the above observational constraints.

9. THE 35 DAY PERIODICITY OF HER X-1 AND THE NEUTRON MATTER EQUATION OF STATE

Trumper et al.,[14] on the basis of an analysis of their EXOSAT observations of Her X-1 have shown that the clock mechanism for the 35-day cycle is the free precession of the neutron star, and that the amplitude of this stellar wobble is substantial, being ~ 25°. A wobble origin for the 35^d clock was first suggested by Brecher;[31] it was examined in some detail by Pines et al.[6] and by Lamb et al.[32] Since Trumper will be discussing the observations and their interpretation in some detail later during this Institute,[14] I shall concentrate in these lectures on the significance of this observation for the neutron matter equation of state, and for formation scenarios for the neutron star in Her X-1.

Since the existence of a stable free precession mode requires a solid crust astrophysical object, it is intuitively obvious that the wobble frequency will depend on the extent of the solid outer crust of the star. According to Baym and Pines,[33] the energy of a rotating star with a solid crust may be written as

$$E = E_o + L^2/2I\,(1 - \varepsilon) + A\varepsilon^2 + B(\varepsilon_o - \varepsilon)^2 \tag{11}$$

where ε is the stellar oblateness, and ε_o is a reference oblateness which corresponds to that oblateness at which the crust will be strain free. For a given stellar mass, A and B, which respectively measure the gravitational and elastic energy content of the star, may be calculated microscopically; both quantities are sensitive to the neutron matter equation of state. The wobble frequency, Ω_W, may be simply expressed in terms of their ratio as[34]

$$\frac{\Omega_w}{\Omega} = \frac{3}{2}\,\frac{B}{A + B}\,\varepsilon_o \approx \frac{3}{2}\,\frac{B}{A}\,\varepsilon_o\ , \tag{12}$$

since $B \sim \mu V_{crust}$ is quite small compared to $A \sim 3\ GM^2/25\ R$, for all stellar models. As we have seen, the crustal extent varies significantly depending on whether the neutron matter equation is soft or stiff; to the extent then that we can use theoretical arguments to pin down ε_o, observation of stellar wobble in Her X-1 provides information on the neutron matter equation of state.

To use these results to learn more about the neutron matter equation of state, we follow the arguments given by Pandharipande et al.[11] concerning the current value of the reference oblateness, ε_o. At the time of formation of the star, ε_o is given by the initial

Table 2. Stellar parameters for a 1.33 $M_\odot$ Hercules X-1

Model	I_{45}	B_{48}	A_{52}	B/A	ε_o	P_o^{max}(ms)	ω_q^2
Reid	0.93	0.5	9.4	5.3×10^{-6}	5.2×10^{-2}	1.4	3.8×10^{13}
BJ	1.33	3.4	7.2	4.7×10^{-5}	5.8×10^{-3}	5.7	2.3×10^{12}
MF	1.97	6.7	5.4	1.2×10^{-4}	2.2×10^{-3}	12.8	4.4×10^{11}
TI	2.09	14.2	5.2	2.7×10^{-4}	1.0×10^{-3}	19.6	1.8×10^{11}

stellar oblateness, and is $\varepsilon_o^o = I\,\Omega_o^2/4A$, where Ω_o is the initial angular velocity of the star. There are then two possibilities: either the current oblateness is essentially identical to ε_o^o; or starquakes and/or plastic flow act to reduce ε_o^o to its current lower value. The importance of the latter physical processes depends on the critical strain angle, σ_c/μ. If $\sigma_c/\mu \gg \varepsilon_o^o$, no appreciable plastic flow or oblateness changing starquakes will have taken place since the crust solidified in the first days after the formation of the neutron star and from the 35^d periodicity one obtains the initial rotational periods of the star shown in Table 2. If, on the other hand, $\sigma_c/\mu \lesssim \varepsilon_o$, oblateness reducing crustquakes are possible as the star spins down, since the starquake condition is $\varepsilon_o - \varepsilon = \sigma_c/\mu$, where $\varepsilon = I\Omega^2/4A$ is the stellar oblateness at a given rotational frequency, Ω. Since the current value of $\varepsilon(\sim10^{-7})$ is negligible compared to σ_c/μ or ε_o^o, it follows that if $\sigma_c/\mu \lesssim \varepsilon_o^o$, the current reference oblateness is σ_c/μ, while the initial reference oblateness, ε_o^o must have been larger, and the initial spin period shorter, than the periods given in Table 2.

Let us consider next the question of the damping and excitation of stellar wobble in a pulsating x-ray source, and the extent to which the precessional frequency we have calculated above might be altered by the pinning of crustal vortices. As discussed by Pines and Shaham,[35] the most likely source of wobble damping is friction between the crust and the liquid core of the star; one finds a damping time

$$\tau_w = \tau(\Omega/\Omega_w) \tag{13}$$

where τ, the relaxation time for crust-core coupling, is ~ 400 P(s). Hence the quality factor, Q, of the stellar wobble is given by $Q = \tau_w\Omega_w \sim \tau\Omega \cong 2500$. Hence without an effective pumping mechanism, stellar wobble cannot be expected to persist for long periods. Fortunately in accreting neutron stars such a mechanism exists: a wobble-operated magnetic accretion gate makes possible the resonant pumping of large amplitude wobble.[6] As Lamb et al.[32] have discussed in some detail, there are a number of ways in which a comparatively small wobble-induced modulation of mass flow to the stellar surface could act to pump large amplitude stellar wobble.

The possibility of large amplitude wobble has interesting consequences as well for the results of Shaham[12] on the influence of pinned vortices in the crust on the precession frequency, Ω_w. Shaham showed that as long as stellar wobble did not act to unpin vortices, crustal vortex pinning would give rise to a dramatic increase in the wobble frequency from the value given by Eq. (12) to $(\Omega_w/\Omega) \sim (I_p/I)$. Since we have seen that values of $(I_p/I) \sim 10^{-2}$ are to be expected for pulsars, the existence of such a pinned superfluid inertial moment in Her X-1 would reduce the wobble period from ~ months to ~ hours and make untenable the attribution of the 35^d periodicity to stellar wobble. On the other hand, if the stellar wobble is of sufficient amplitude to unpin vortices, then pinning plays only a minor role in determining the wobble frequency. That situation will exist provided the amplitude α of the stellar wobble is such that $\alpha \gtrsim \omega_{cr}/\Omega \sim \omega_{cr}/5$. For the glitching neutron stars we have thus far considered, there are no regions of strong pinning; the weak pinning region possesses an $\omega_{cr} \lesssim 1/4$, and hence once the wobble amplitude exceeds ~ 3°, wobble will act to unpin vortices. As a result, the wobble frequency will be determined primarily by the reference stellar oblateness and crustal extent, according to Eq. (12). Since wobble must be of comparatively large magnitude to be observable we are thus led to the interesting conclusion that a wobbling neutron star will not glitch, and that a glitching neutron star will not possess an observable stellar wobble.

We are now ready to consider the consequences of the identification by Trumper et al.[14] of large amplitude wobble in Her X-1. First, we note that the amplitude of the wobble is sufficiently large ($\alpha \sim 1/2$) that only regions with $\omega_{cr} \gtrsim 2.5$ would remain pinned; since the mass of Her X-1 ($\sim 1.33 \pm 0.15\ M_\odot$) is comparable to that ($\sim 1.4\ M_\odot$) we have hypothesized for the glitching pulsars, we would not expect such strongly pinned regions to exist in the star, and hence would expect the wobble period to be given by (12).

Further consequences of the identification of the 35^d periodicity as stellar wobble come from the interrelated constraints which are provided by Eq. ( ). It is straightforward to show that:

- $\sigma_c/\mu \gtrsim 10^{-3}$, since a lower bound to σ_c/μ is the current reference oblateness, and a lower bound to $\varepsilon_o \sim 10^{-3}$ is obtained from the TI model which characterizes the stiffest neutron matter equation of state compatible with laboratory nuclear physics constraints.
- the initial rotational period, P_o of Her X-1, was at most ~ 20 msec, since an upper bound for P_o is likewise obtained from the TI model. Hence Her X-1 must have functioned as a pulsar, rather than an accreting x-ray star, for some considerable time before slowing down to a period ($\gtrsim$ .1 sec) which permits accretion of appreciable amounts of matter.
- the Reid equation of state is too soft to characterize neutron matter since, as may be seen in Table 2, it implies an initial rotational period ($P_o \lesssim 1.4$ msec) too rapid to be accomodated by the astrophysical evidence concerning the formation of pulsars.
- to the extent that one can conclude from the astrophysical evidence that $P_o \gtrsim 15$ msec, the extrapolated initial period of the Crab pulsar, the neutron matter equation of state must, in fact, possess a stiffness which lies between that of the MF model

and the TI model. To see this, note that since $\varepsilon_o = (2/3)(A/B)(P_R/P_W) \lesssim \varepsilon_o^o \lesssim I(\Omega_o^{max})^2/4A$, the parameters A and B for Her X-1 must be such that

$$\omega_q^2 = 2A^2/BI) \lesssim 3.2 \times 10^{11}[.015/P_o^{min}/sec]^2. \quad (14)$$

From the tabulated values of ω_q^2 in Table 2, the above conclusion follows.

- to the extent that microscopic calculations yield a given equation of state of neutron matter, the initial spin of Her X-1 is determined. For example, should the MF model be shown to provide a correct account of neutron matter, it follows from Table 2 that the initial rotational period of Her X-1 was 12.8 ms.

To conclude this discussion, I note that while the astrophysical evidence concerning the nature of the pinning layers and the observed amplitude of the wobble are such that pinning would seem to play no role in determining Ω_w, its influence on Ω_w is so large that the temporary existence of a minute fraction of strongly pinned vorticity ($I_p/I \sim 10^{-9}$) could produce an observable shift in the wobble period. Shaham[36] has suggested that the observed fluctuations in the wobble period are due to just such an effect.

10. CONCLUSIONS AND ACKNOWLEDGEMENTS

Since both the post-glitch fits and the identification of the 35^d periodicity in Her X-1 as stellar wobble require a quite stiff neutron matter equation of state, the astrophysical evidence for such an e.o.s. would seem quite strong, while the evidence for an "intermediate" $\Delta(\rho)$ curve would seem equally strong. It is encouraging that the recent microscopic results tend to point in the same direction, and we would seem well on the road toward using neutron stars as cosmic hadron physics laboratories.

These notes represent, in part, a condensation of the review written earlier this year with Ali Alpar, whom I thank for many helpful discussions on the topics contained herein; I should also like to thank John Clark and Vijay Pandharipande for stimulating discussions on microscopic theories. This work has been supported in part by NSF Grant PHY80-25605, by NASA Grant NSG-7653, and by NATO Grant RG 18681.

11. REFERENCES

1. Radhakrishnan, V. and Manchester, R. N. Nature **222**, 228 (1969).
2. Reichley, P. E. and Downs, G. S. Nature **222**, 229 (1969).

3. Boynton, P. E., Groth, E. J., Hutchinson, D. P., Nanos, G. P., Partridge, R. B. and Wilkinson, D. T. Ap. J. **175**, 217 (1972).
4. Pines, D. In Proc. of the 12th Intl. Conf. on Low Temperature Physics, ed. E. Kanda, p. 5, (1971).
5. Tananbaum, H., Gursky, H., Kellog, E., Levinson, R., Schreier, E. and Giacconi, R. Ap. J. **174** L143 (1972).
6. Pines, D., Lamb, F. K. and Pethick, C. J. In Proc. Sixth Texas Symp. on Relativistic Astrophysics, Annals NY Acad. Sci. **224**, 237 (1973).
7. Downs, G. S. Ap. J. **249**, 687 (1981).
8. Lohsen, E. Astron. Astrophys. Suppl **44**, 1 (1981).
9. Pines, D. and Alpar, M. A. Nature **316**, 27 (1985).
10. Gerend, D. and Boynton, P. E. Ap. J. **209**, 562 (1976).
11. Pandharipande, V. R., Pines, D. and Smith, R. A. Ap. J. **208**, 550 (1976).
12. Shaham, J. Ap. J. **214**, 251 (1977).
13. Boynton, P. E., Crosa, L. M. and Deeter, J. E. Ap. J. **237**, 169 (1980).
14. Trumper, J., Kahabka, P., Ögelman, H., Pietsch, W. and Voges, W, (preprint); Trumper, J. (this volume).
15. Baym, G. and Pethick, C. J. Ann. Rev. Astr. and Ap. **17**, 415 (1979).
16. Friedman, B. and Pandharipande, V. R. Nucl. Phys. A **361** 502 (1981).
17. Hoffberg, M., Glassgold, A. E., Richardson, R. W. and Ruderman, M. Phys. Rev. Lett. **24**, 775 (1970).
18. Yang, C.-H. and Clark, J. W. Nucl. Phys. A **174** 49 (1971).
19. Takatsuka, T. Prog. Theor. Phys. **48**, 1517 (1972).
20. Chen, J., Clark, J., Krotscheck, E. and Smith, R. A. (Preprint.)
21. Alpar, M. A. Ap. J. **213**, 527 (1977).
22. Anderson, P. W., Alpar, M. A., Pines, D. and Shaham, J. Phil. Mag. A. **45**, 227 (1982).
23. Alpar, M. A., Anderson, P. W., Pines, D. and Shaham, J. Ap. J. **276**, 325 (1984).
24. Alpar, M. A., Anderson, P. W., Pines, D. and Shaham, J. Ap. J. **278**, 791 (1984).
25. Alpar, M. A., Nandkumar, R. and Pines, D. (In preparation.)
26. Alpar, M. A., Langer, S. and Sauls, J. A. Ap. J. **282**, 533 (1984).
27. Anderson, P. W. Phys. Rev. Lett. **9**, 309 (1963); Anderson, P. W. and Kim, Y. B. Rev. Mod. Phys. **36**, 39 (1964).
28. Baym, G., Pethick, C., Pines, D. and Ruderman, M. Nature **224**, 872 (1969).
29. Alpar, M. A., Nandkumar, R. and Pines, D. Ap. J. **288**, 191 (1985).
30. Pines, D. and Alpar, M. A. In Proc. NRAO Workshop No. 8, 'Birth and Evolution of Neutron Stars,' ed. S. P. Reynolds and D. R. Stinebring p. 161 (1984).
31. Brecher, K. Nature **239**, 325 (1972).
32. Lamb, D. Q., Lamb, F. K., Pines, D. and Shaham, J. Ap. J. **198**, L21 (1975).
33. Baym, G. and Pines, D. Ann Phys. **66**, 816 (1971).
34. Pines, D. and Shaham, J. Phys. Earth Planet. Interiors **6**, 103 (1972).
35. Pines, D. and Shaham, J. Nature **248**, 483 (1974).
36. Shaham, J., in preparation.

RECENT DEVELOPMENTS ON THE CRAB NEBULA

L. Woltjer
European Southern Observatory
Karl-Schwarzschild-Strasse 2
D-8046 Garching b. München

ABSTRACT. The spectrum of the Crab Nebula is strongly affected by interstellar absorption, especially in the ultraviolet. For a B-V color excess of $0\overset{m}{.}4$ a smooth spectrum is obtained from the radio to the X-ray frequencies, except for a "bump" in the infrared, presumably caused by dust emission. The composition of the filamentary shell depends on the validity of the photoionization models which still have much uncertainty. Apart from a large and variable helium abundance, no clear anomalies have been firmly established. Because of this and because of the uncertainty in the distance, the nature of SN 1054 remains in doubt. There are important asymmetries in the kinematics and composition of the shell and in the energy input into the Nebula.

INTRODUCTION

Early analysis of the fragmentary historical data on SN 1054 led to the conclusion that it was a supernova of type I. Minkowski (1971) noted that this conclusion was not justified. In the subsequent literature this has sometimes erroneously been taken as indicating that it was of type II. All that may be said, however, is that the type is unknown and will remain so, unless new informative records are discovered. The magnitude at maximum has generally been estimated as V = -5 (Clark and Stephenson 1977), but the uncertainty should be at least a magnitude.

The distance of the Crab Nebula remains a greater source of uncertainty in its parameters than is generally appreciated. If the Nebula were a spherical shell, its distance could be determined from a comparison between the transverse angular motions at the edge and the radial velocities at the center. However, the projected image is already far from circular, and the true three dimensional shape of the Nebula is unknown. A distance of 2 kpc (Trimble 1968) has usually been adopted, but as pointed out by Trimble and Woltjer (1971), no value in the range of 1 - 3 kpc can be excluded on the basis of the kinematical data.

F. Pacini (ed.), High Energy Phenomena Around Collapsed Stars, 209–221.

Recently, the radial velocities of numerous filaments have been measured by Clark et al. (1983). From these the three dimensional structure of the shell may be determined if a distance is adopted. In figures 1 and 2, cross sections through the major and minor axes are shown for an assumed distance of 1.5 kpc. If a different distance were chosen, the horizontal scale in these figures would, of course, change in proportion, while the vertical scale obtained by multiplying the observed radial velocities with the expansion age of the Nebula would remain constant. It should be noted that the proper motions indicate that the expansion age of the Crab Nebula is 85 years less than its real age (Trimble 1968) corresponding to an accelerated motion of the shell.

From figures 1 and 2 the thickness of the shell is much in evidence. In addition, there is a clear asymmetry with the stronger filaments and larger velocities occurring towards the NW along the major axis. Later it will be seen that similar asymmetries are found

Figure 1: Cross section through a strip ±30" wide around the major axis. NW is to the right and positive radial velocities (converted to the same linear scale) are up. Filled circles represent stronger filaments.

Figure 2: Same as in figure 1 along the minor axis. NE is to the right.

in the synchrotron radiation and probably also in the composition of the filaments. From the figures it is again clear that for any distance between 1 and 3 kpc the shape of the Nebula does not look particularly implausible; a much enlarged data set is unlikely to change this conclusion. The only way to properly determine the distance would be a parallax measurement of the pulsar, but this is still outside the range of present day technology.

One of the interesting features recently discovered in the Crab is the "jet", a tubular feature somewhat over an arc minute in length. It is not unlikely that it represents a place where relativistic particles may leak out of the filamentary shell. For a further description see Véron et al. (1985).

THE CONTINUOUS SPECTRUM OF THE NEBULA

The polarization of the radiation from the Nebula at radio, optical and X-ray (Weisskopf et al. 1978) wavelengths has confirmed that the dominant emission mechanism is synchrotron radiation. More recently data on the ultraviolet (Wu 1981) and infrared (Marsden et al. 1984) have filled in some of the remaining "holes" in the spectrum.

At radio wavelengths between 1 and 35 GHz the spectral index α ($F_\nu \propto \nu^{-\alpha}$) is equal to 0.30±0.01 (Baars et al. 1977). Data down to 10 MHz seem compatible with a continuation of this spectrum. At the high energy end of the spectrum the situation appears to be equally clear. Between 2 and 60 kev the total flux of the Crab has a much steeper spectrum with $\alpha = 1.10 \pm 0.03$ (Toor and Seward 1974) or after subtraction of the contribution of the pulsar $\alpha = 1.15$ for the nebula. Beyond 80 kev the spectrum steepens further to $\alpha = 1.4$ (Strickman et al. 1979, Hasinger 1984).

Between the radio and the X-ray domains complications arise from the effects of interstellar dust both in absorption and in emission. An attempt to determine the absorption was made by Miller (1973) who measured the ratio of the intensities of the lines of [SII] around 4070 and 10,300 Å. These lines originate from the same upper levels and therefore their intensity ratios are determined by atomic parameters independent of density and temperature. For two filaments Miller obtained colour excesses E_{B-V} of 0.58 and 0.46. The method is sensitive to uncertainties in the atomic parameters and in the calibration of the line intensities. In fact, more recent calculations summarized by Draine and Bahcall (1981) would reduce these values by 0.06. Wu (1981) determined the colour excess from the strength of the absorption feature at 2200 Å in the Crab Nebula and obtained $E_{B-V} = 0\overset{m}{.}50$. However, the ratio of the absorption at 2200 Å and the colour excess is not necessarily constant from star to star (Meyer and Savage 1981). Woltjer (1958), O'Dell (1962) and Brodskaya (1963) estimated the absorption from data for stars in the neighbourhood of the Crab. While their results were compatible with $E_{B-V} \simeq 0\overset{m}{.}5$ no very high accuracy can be obtained this way.

Adopting $E_{B-V} = 0\overset{m}{.}5$, as has been done by most authors in recent times, leads to a peculiar spectrum for the nebula: the spectrum is steep in the X-ray range, flattens in the uv - with an even steeper portion in between, then steepens towards the IR to flatten towards the radio wavelengths. While such a spectrum cannot be a priori excluded, it is not very plausible as most loss mechanisms for synchrotron electrons increase with energy and thereby lead to a gradual steepening of the spectrum towards higher frequencies. It is therefore of interest to see if a reasonable change in the adopted interstellar absorption could lead to a smoother spectrum.

In table 1 are shown the values of the average spectral index in various spectral ranges for different values of E_{B-V}. These are based on uv measurements with ANS by Wu (1981), IR measurements at 12 µ with IRAS by Marsden et al. (1984), and other data mentioned before. The normalization of the uv data will be discussed later.

Table 1

Average spectral index values for some spectral ranges as a function of the (interstellar) colour excess E_{B-V}.

E_{B-V}	$0^m.6$	$0^m.5$	$0^m.4$	$0^m.3$
Radio wavelengths	0.30	0.30	0.30	0.30
3300 Å - 12 μ	0.59	0.72	0.85	0.97
1555 Å - 3300 Å	0.02	0.44	0.86	1.28
2 kev - 1550 Å	1.35	1.21	1.07	0.94
60 kev - 2 kev	1.15	1.15	1.15	1.15

In table 1 the strong dependence of the ultraviolet spectrum on the precise value of the colour excess is much in evidence. When the absorption is high, the uv spectrum is flatter than that in adjacent spectral regions, while if the absorption is low, the inverse is the case. For a colour excess $E_{B-V} = 0^m.40$, however, the spectrum is smooth. In the following, we therefore adopt this value, which within the errors does not seem to be incompatible with the results from other methods.

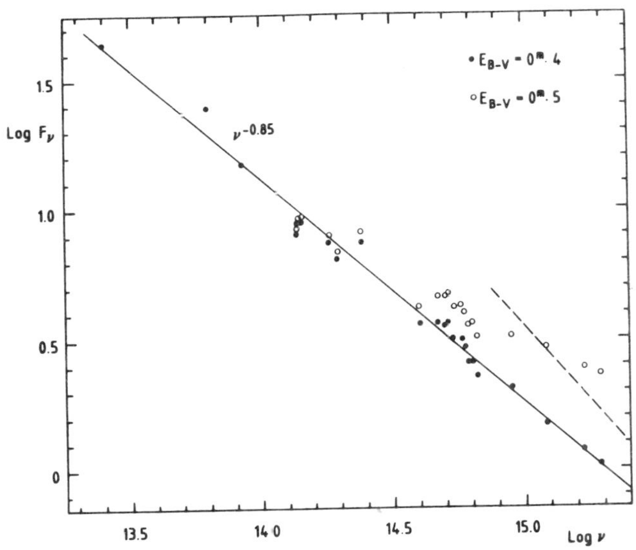

Figure 3: The uv - IR spectral data for two different values of the color excess. The dashed line is the extrapolated X-ray spectrum.

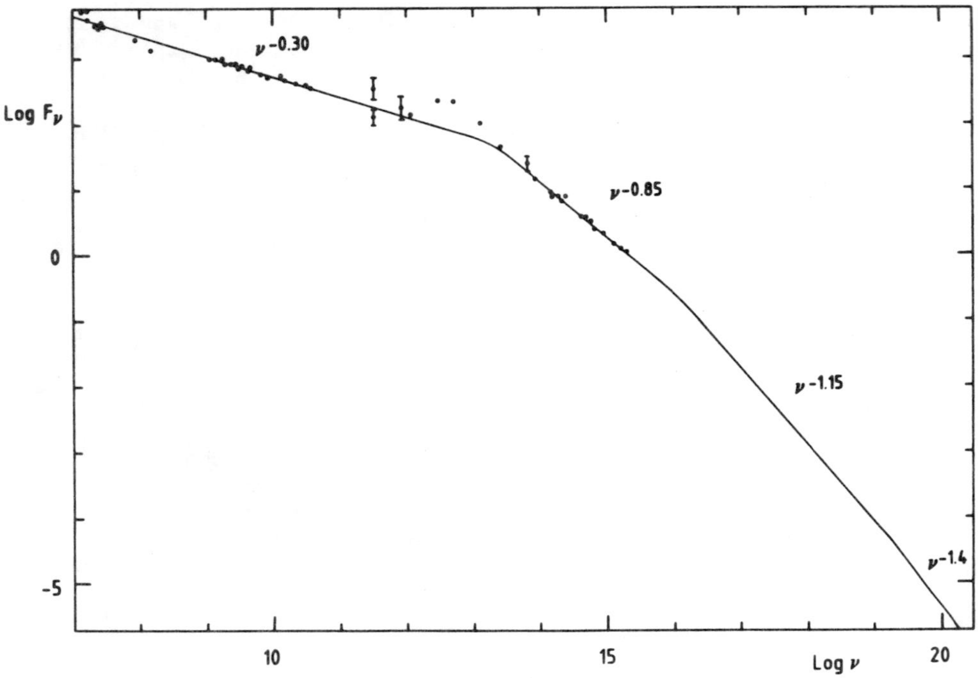

Figure 4: The overall spectrum of the Crab Nebula for $E_{B-V} = 0^m.4$. The flux values are expressed in Jansky (1 Jy = 10^{-23}erg cm^{-2} sec^{-1}).

In figure 3 we show the near IR, optical and uv parts of the spectrum. The optical data have been taken from Kirshner (1974). While his instrumentation appears to have been superior to that used in earlier measurements, the systematic difference with the results of O'Dell (1982) and the small scale "wiggles" in the spectrum remain disturbing. The data of Wu have been renormalized to fit to Kirshner's flux at 3390 Å (12 % lower than in Wu 1981). It is to be noted that there are rather large differences between the results of Wu and the earlier OAO data of Johnson (1972). The near IR data have been taken from Becklin and Kleinman (1968), Grasdalen (1979) and Ney and Stein (1968). Because the measuring beams used by the former two were much smaller than the Nebula, the fluxes had been corrected on the basis of Woltjer's (1957) optical map on the assumption that the optical and IR distributions are the same. Some qualitative support for this assumption may be found in the IR map of Grasdalen. We have increased the fluxes of Ney and Stein by 10 % to take into account the light lost with their 4 arcminute beam. Of course, in all these data and also in the uv data of Wu possible spectral variations over the Nebula (Scargle 1969) remain a source of uncertainty. It is seen from figure 3 that a constant spectral index of 0.85 gives an acceptable fit to all IR-uv data for $E_{B-V} = 0^m\!.4$; for $E_{B-V} = 0^m\!.5$ no fit with constant α is possible, nor can a smooth fit between the uv and X-ray spectra be made. If instead of Kirshner's (1974) optical data we had used those derived by Wu (1981) from Scargle's (1969) colour measurements in the central part of the Nebula, these results would have been very nearly the same. Obviously a redetermination with modern detectors of the optical and near IR fluxes and their variation over the nebula is in order.

The overall spectrum of the nebula is shown in figure 4. Almost everywhere the spectrum is remarkably smooth with gradually increasing spectral index. Of course, the precise frequencies at which the steepening takes place may still change a bit if corrected fluxes become available. The only exception is in the range of 25-100 μ where the IRAS data indicate excess emission. Marsden et al. (1984) ascribe this excess to emission by dust in the Crab Nebula heated by X-ray photons as suggested by Dwek and Werner (1981). A mass of several times 10^{-3} $M_{\odot}$ of dust with a temperature around 70°K would probably suffice; however, the dust emission depends sensitively on the sizes and surface emissivities of the dust particles and larger values for the mass cannot be entirely excluded. The main parts of the spectrum of the Nebula are summarized in table 2.

Table 2

The spectrum of the Crab Nebula.

$10^7<\nu<10^{12}$ Hz	$F\nu = 1040\ (\nu/10^9)^{-0.30}$ Jy
$2\times10^{13}<\nu<3\times10^{15}$	$F\nu = 1.82\ (\nu/10^{15})^{-0.85}$ Jy
$10^{16}<\nu<10^{19}$	$F\nu = 1.25\times10^{-3}\ (\nu/10^{18})^{-1.15}$ Jy
	$= 9.7\ E_{kev}^{-1.15}$ kev/(kev cm² sec)

Thus far, we have considered the integrated spectrum of the Nebula. It is clear, however, that there are important variations over its surface, the overall spectra being steeper near the edges. In fact, from figure 5 it is seen that the Nebula becomes progressively more centrally condensed towards higher frequencies, while at the same time the distribution becomes more asymmetric, with the center of gravity shifting towards the NW. At X-ray energies, this is presumably related to the short lifetimes (few years) of the radiating electrons, combined with an asymmetric energy input; this asymmetry is also seen in the optical "wisps", features which move from the neighborhood of the pulsar towards the NW with speeds of the order of 0.1 c and which presumably represent hydromagnetic waves. Part of the asymmetry, however, seems also to be due to the structure of the magnetic field, since it is already seen in the radio emission which is generated by electrons with long lifetimes. Accurate multifrequency maps are needed, especially near the breaks in the spectrum to disentangle the effects of particle acceleration and propagation and magnetic field structure.

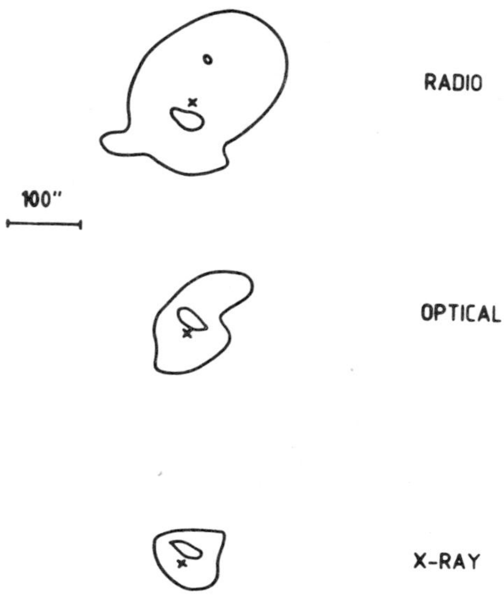

Figure 5: Isophotes in the continuous spectrum of the Crab Nebula at different frequencies. The cross marks the pulsar and the outer isophote is for half the surface brightness of the inner isophote.

COMPOSITION OF THE FILAMENTARY SHELL

Several spectroscopic studies of selected filaments have been made during the last few years; these have yielded intensities of emission lines of H, He, C, N, O, Ne, S, A, Fe and Ni (Davidson 1978, 1979, Fesen and Kirshner 1982, Davidson et al. 1982 a, Dennefeld and Péquignot 1983, Henry et al. 1984). The accurate measurement of the line intensities is not always easy because of the superposition of filaments with different radial velocities; their interpretation is particularly difficult because of large optical depth effects.

Recent models are based on the assumption that photoionization is responsible both for the ionization and the energy input into the filaments. This assumption is questionable, since the sweeping up of interstellar or circumstellar matter could be an important source of heating. In fact, if a filament moves at 1000 km/sec through a medium with 1 hydrogen atom per cm^3, the energy available for thermaliza tion is about 1 erg cm^{-2} sec^{-1}, nearly the same as the total radiative energy impinging on the filament.

In the pure photoionization models it is assumed that a slab of matter is exposed on one side to a flux of radiation. A spectrum has to be adopted for this flux, and in most recent calculations this has been taken to be of the wavy shape that results from the adoption of a high color excess as explained in the previous section. The changes in the flux as one moves deeper into the filament are then calculated taking into account the ionization (essentially of H, He and He^+) and recombination processes. The principal uncertainty results from the assumptions which have to be made about the pressure distribution. An isobaric assumption has frequently been adopted, but since the filaments are probably confined by magnetic forces, this may well be entirely wrong. In addition, many filaments may be subject to the ionizing flux not only from one direction, while also the atomic physics parameters - especially those relating to dielectronic recombination processes - still may have much uncertainty.

The most complete models to-date have been made by Henry and MacAlpine (1982) and by Péquignot and Dennefeld (1983). The former conclude that elements like O are substantially underabundant (per nucleon), the latter that they are normal or slightly overabundant. Part of the difference may be related to the values adopted for the atomic parameters, which were adjusted by Péquignot and Dennefeld in such a way that at least for a well studied planetary nebula the "right" results were obtained. There is full agreement, however, by all investigators that helium is substantially overabundant with respect to hydrogen (by a factor 3-10).

In several supernova remnants (Cas A, Kepler, Pup A) there is evidence of an enhanced nitrogen abundance - particularly in slow moving filaments. Presumably, this results from the CNO cycle operating in material that has subsequently been removed from the presupernova by a stellar wind. The nitrogen rich material currently being expelled by η Car (Davidson et al. 1982 b) may be a good example of this process. In the Crab Nebula, however, there is no evidence for nitrogen enrichment; in fact, Péquignot and Dennefeld find an underabundance of a factor of two with respect to oxygen. Apparently, the high helium abundance is not the result of H burning in the CNO cycle.

On the basis of 65 hours of ultraviolet spectroscopy with the IUE satellite, Davidson et al. (1982 a) conclude that there is also no important overabundance of carbon. Dennefeld and Péquignot (1983) seem to come to the same conclusion on the basis of a measurement of the [CI] lines in the near infrared, but everything there depends on very uncertain atomic parameters. From the point of view of stellar models it is difficult to produce much helium while leaving C, N and O unaffected. Nomoto (1981) has shown that stellar models with a mass of around 9 $M_\odot$ have some favorable characteristics in this respect, but difficulties remain.

The abundance of the iron group elements are of particular interest in connection with the possible powering of supernova light curves by radioactive decay. Available data suggest that the iron abundance (per nucleon) is not above solar. However, Dennefeld and Péquignot (1983) and Henry et al. (1984) found that the [Ni II] λ 7380 line is surprisingly strong, indicating a Ni/Fe ratio a factor of 10-100 above solar. However, the former authors already noted that older supernova remnants like IC 443 in which most of the material is swept up interstellar matter show similar apparent Ni/Fe ratios, while Henry (1984) noted a significantly high value in the Orion Nebula. It therefore seems probable that there is some problem in the atomic physics of nickel - or possibly in its chemistry - and the question of its abundance remains open.

Large variations in the relative line intensities are observed in different filaments, but more complete models are needed to see to what extent real abundance differences are involved. For helium, however, the situation is clearer. From the photoionization models of Henry and MacAlpine (1982) it is seen that the intensity ratio of He I λ 5876 to Hβ is nearly equal to the He/H ratio up to values of 0.4. In figure 6 the available data for this intensity ratio are shown, and it seems apparent that there is a systematic difference between the N and S parts of the Nebula. The fact that there is no correlation with the emission like flux further supports this interpretation as an abundance rather than an excitation effect.

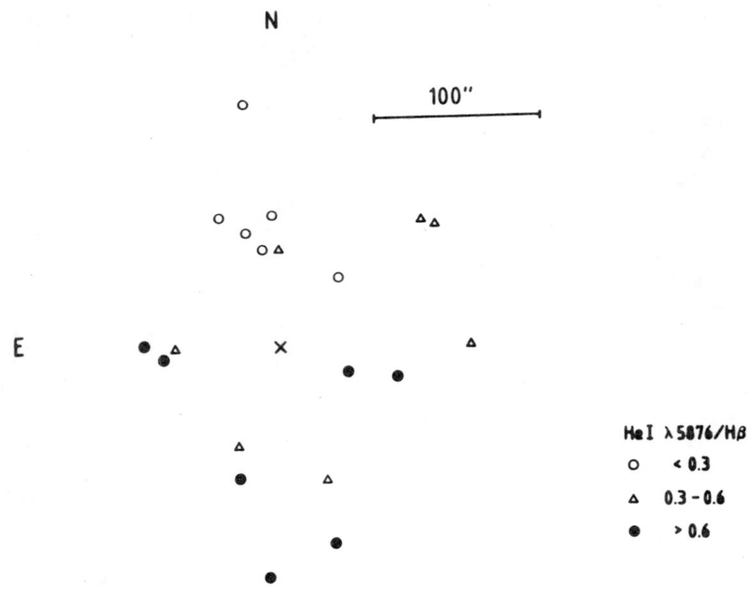

Figure 6: Variation of the ratio of He I λ 5876 / Hβ over the face of the Crab Nebula.

It is clear from the foregoing discussion that more detailed studies of the emission lines over the face of the Crab Nebula are needed, while spatially resolved observations of filaments could give valuable checks on the models. Also the possible role of dust in the abundance determinations remains to be evaluated.

THE NATURE OF THE CRAB NEBULA

The filamentary shell consists of parts where hydrogen is predominantly ionized and others where it is mainly neutral. From the Hβ flux measured by Kirshner (1974) and assuming $E_{B-V} = 0\overset{m}{.}4$, $D = 2$ kpc, $N_e = 1000\ cm^{-3}$ (as an uncertain average from [O II] λ 3727 doublet observations) and $T_e = 15{,}000°$ (from the [O III] lines), we obtain for the mass of ionized hydrogen 0.3 $m_\odot$. The regions where hydrogen is neutral emit [S II] lines, from which Henry and MacAlpine (1982) infer $N_e = 1200\ cm^{-3}$, $T_e = 8000°$ and a neutral hydrogen mass which corresponds to 0.2 $m_\odot$. From the total hydrogen mass of 0.5 $m_\odot$ and a He/H ratio by a number of 0.4, the total mass of the shell then becomes 1.3 $m_\odot$; the uncertainty in this figure is large, since the appropriately averaged value of N_e over bright and faint filaments is not yet well determined.

In table 3 we have assembled various data on the Crab Nebula for different assumptions about the distance. The magnitude of the supernova at maximum M_V(SN) may be compared with (for H = 50 km/sec/Mpc) M_V = 19.6 ± 0.4 for SN I and M_V = −18.2 ± 1.4 for SN II (Tammann 1982, where we have derived the SN II value as the average for those objects for which the estimated visual absorption is less than 1^m). While the SN 1054 data seem to fit best to a type II supernova, the distance to the galactic plane z seems to be rather large for a massive progenitor. It is interesting to note that 3C 58 which rather closely resembles the Crab Nebula also has a substantial distance to the galactic plane (150-400 pc, depending on the distance) and a low M_V (−13.5 to −16), at least if the identification with SN 1181 is correct.

Table 3

Parameters of the Crab Nebula for different distances

D(kpc)	1	2	3
M_V(SN)	-16.2	-17.7	-18.6
z(pc)	100	200	300
M(Nebula in $m_\odot$)	0.2	1.0	2.2
$<V^2>^{1/2}$ (km/sec)	1000	1400	1900
$1/2M<V>^2(10^{49}$erg)	0.2	2	8
$L(10^{38}$erg/sec)	0.2	0.7	1.6
$d(1/2M\ V^2)/dt(10^{38}$erg/sec)	0.2	2	9

Of course, there is also the possibility to associate the Crab Nebula and 3C 58 with subluminous type I supernovae if such objects do, in fact, exist.

If the distance to the Crab Nebula is low, it is very well possible to have the total mass of the Nebula and the pulsar below the Chandrasekhar limit for a single white dwarf. At larger distances, it still seems difficult to bring the total mass up to the 9 $m_{\odot}$ value proposed by Nomoto unless stellar winds have swept away most of the mass to distances in excess of a few pc. The r.m.s. velocity of the filaments $\langle V^2 \rangle^{1/2}$ is low, especially if it is taken into account that the motion has been accelerated after the supernova outburst. As a consequence, the kinetic energy of the shell comes out rather low by supernova standards.

For plausible neutron star parameters the loss of rotational energy from the pulsar amounts to about 5 x 10^{38}erg/sec. Most of this energy is converted into relativistic particles and magnetic fields. The former lose part of their energy in radiation, while both suffer adiabatic losses in accelerating the shell. The total luminosity of the Nebula L and the increase in the kinetic energy of the shell on the assumption of a uniform acceleration are given in the last two lines of table 3. Of course, the assumption of a uniform acceleration of the shell is an oversimplification, and more realistic models have been discussed (Woltjer 1972). For a distance of 1 kpc we conclude that most of the pulsar energy input must have been lost to the Nebula - perhaps by the escape of relativistic particles. For a three times larger distance, the powering of the Nebula begins to be a problem.

Of particular importance are the conclusions about the asymmetry of the Crab Nebula event. The main energy input, the largest velocities, the strongest filaments and the lower helium abundances seem to be found towards the NW of the pulsar. Further detailed studies of these asymmetries could be worthwhile.

REFERENCES

Baars, J.W., Genzel, R., Pauliny-Toth, I.I. and Witzel, A., 1977 Astron. Astrophys. **61**, 99

Becklin, E.E. and Kleinmann, D.E. 1968 Astrophys. J. **152**, L25

Brodskaya, E.S. 1963 Izv Crimean Astrophys. Obs. **30**, 126

Clark, D.H. and Stephenson, F.R. The Historical Supernovae (Pergamon 1977)

Clark, D.H., Murdin, P., Wood, R., Gilmozzi, R., Danziger, J. and Furr, A.W. 1983 Month. Not. Roy. Astron. Soc. **204**, 415

Davidson, K. 1978 Astrophys. J. **220**, 177

Davidson, K. 1979 Astrophys. J. **228**, 179

Davidson, K., Gull, T.R., Maran, S.P., Stecher, T.P., Fesen, R.A, Parise, R.A., Havel, C.A., Kafatas, M. and Trimble, V.L. 1982 a Astrophys. J. **253**, 696

Davidson, K., Walborn, N.R. and Gull, T.R. 1982 b Astrophys. J. **254**, L47

Dennefeld, M. and Péquignot, D. 1983 Astron. Astrophys. **127**, 42

Draine, B.T. and Bahcall, J.N. 1981 Astrophys. J. **250**, 579

Dwek, E. and Werner, M.W. 1981 Astrophys. J. **248**, 138
Fesen, R.A. and Kirshner, R.P. 1982 Astrophys. J. **258**, 1
Grasdalen, G.L. 1979 Publ. Astron. Soc. Pacific **91**, 436
Hasinger, G. 1984 Report 186 Max-Planck-Institut f. Extraterrestrische Physik Garching
Henry, R.B. and MacAlpine, G.M. 1982 Astrophys. J. **258**, 11
Henry, R.B., MacAlpine, G.M. and Kirshner, R.P. 1984 Astrophys. J. **278**, 619
Henry, R.B. 1984 Astrophys. J. **281**, 644
Johnson, H.M. 1972 The Scientific Results from the Orbiting Astronomical Observatory (OAO-2), NASA SP-310, 207
Kirshner, R.P. 1974 Astrophys. J. **294**, 323
Marsden, P.L., Gillet, F.C., Jennings, R.E., Emerson, J.P., de Jong, T. and Olnon, F.M. 1984 Astrophys. J. **278**, L29
Meyer, D.M. and Savage, B.D. 1981 Astrophys. J. **248**, 545
Miller, J.S. 1973 Astrophys. J. **180**, L83
Minkowski, R. 1971 IAU Symposium 46, 241
Ney, E.P. and Stein, W.A. 1968 Astrophys. J. **152**, L21
O'Dell, C.R. 1962 Astrophys. J. **136**, 809
Péquignot, D. and Dennefeld, M. 1983 Astron. Astrophys. **120**, 249
Scargle, J.D. 1969 Astrophys. J. **156**, 401
Strickman, M.S., Johnson, W.N. and Kurfess, J.D. 1979 Astrophys. J. **230**, L15
Tammann, G.A. 1982 Supernovae: A Survey of Current Research (ed. M.J. Rees and R.J. Stoneham, Reidel Publ.), 371
Toor, A. and Seward, F.D. 1974 Astron. J. **79**, 995
Trimble, V.L. 1968 Astron. J. **73**, 535
Trimble, V.L. and Woltjer, L. 1971 Astrophys. J. **163**, L97
Véron-Cetty, M.P., Véron, P. and Woltjer, L. 1985 Astron. Astrophys. **151**, 101
Weisskopf, M.C., Silver, E.H., Kestenbaum, H.L., Long, K.S. and Novick, R. 1978 Astrophys. J. **220**, L117
Woltjer, L. 1972 Pulsars and High Energy Activity in Supernova Remnants, Proc. Accademia Nazionale del Lincei, Quad. 162, 131
Woltjer, L. 1957 Bull. Astron. Inst. Netherlands **13**, 301
Woltjer, L. 1958 Bull. Astron. Inst. Netherlands **14**, 39
Wu, C.C. 1981 Astrophys. J. **245**, 581

RADIO SUPERNOVAE

K. W. Weiler
E. O. Hulburt Center for Space Research
Naval Research Laboratory, Code 4131
Washington, D.C. 20375-5000
U.S.A.

ABSTRACT. We present extensive new results on radio supernovae (RSN) and collect, for completeness, most other available results. With this extensive and unique data set, we investigate the class properties of the RSN and derive descriptive parameters for the gross features of their radio "light curves." These are then applied to the several models available for the phenomenon to investigate their relative merits and their relations to the various types and subtypes of supernovae.

The results, based on an albeit very limited statistical sample, are that all RSN turn on very sharply in the radio with the turn-on occurring first at shorter wavelengths and slowly progressing to longer wavelengths with time. This part of the phenomenon appears best explained, in all cases, by a change of optical depth of external thermal free-free absorption. The time delay until this radio turn-on is obviously affected by the amount of thermal material in the vicinity of the supernova and the velocity of the supernova shock wave and has been observed to differ at 6 cm wavelength from 11 days before (SN1983n, Type I_{SL} supernova) to 1 month (SN1980k, Type II_L) to 1 year (SN1979c, Type II_L) after maximum optical light. Through models, this optically thick part of the light curve also provides us with information on mass loss rates from the stellar systems by winds in the last stages of stellar evolution.

After this initially sharp turn-on, which probably has the same physical origin for all RSN, a differentiation occurs. The Type I_{SL} supernova, SN1983n, shows a very steep power law decline of radio flux density ($t^{-1.6}$) at 6 cm with time after maximum and a relatively steep spectral index ($\alpha = -1.0$; $S \propto \nu^{+\alpha}$) while the two Type II_L supernovae, SN1979c and SN1980k, show much slower rates of decline ($t^{-0.7}$) and flatter radio spectral indices ($\alpha \sim -0.6$). When available models are applied to this optically thin part of the radio light curves, either external shock-driven "mini-shell" or centrally-driven "mini-plerion" models can provide reasonable fits to the gross properties.

1. INTRODUCTION

F. Pacini (ed.), High Energy Phenomena Around Collapsed Stars, 223–234.

Several hundred extragalactic supernovae have been identified by their optical light and catalogued since the first modern detection of a supernova (S Andromedae) in the nearby Andromeda Nebula (M31) by E. Hartwig at the Dorpat Observatory on 1885 August 20. However, numerous searches at radio frequencies for many tens of these objects for years revealed no detections to the limits of the largest and most sensitive radio telescopes available (see e.g., de Bruyn, 1973; Brown and Marscher, 1978; Ulmer *et al.*, 1980; Weiler *et al.*, 1981; Cowan and Branch, 1982) with the single exception of the supernova SN1970g in M101 (Gottesman *et al.*, 1972). This last, showing itself as a weak increase in the integrated flux density from a region containing the position of the supernova and a nearby confusing HII region, nevertheless allowed the determination of a very rough radio "light curve" (Allen *et al.*, 1976). In the following decade, no more information on the radio emission from supernovae was obtained until SN1979c in M100 (NGC4321) was found to be a powerful emitter of 6 cm radio emission in 1980 April (Sramek, van der Hulst, and Weiler, 1980; Weiler *et al.*, 1981). Stimulated by this clear evidence that supernovae are significant sources of radio emission and that the study of these "radio supernovae" (RSN) provides important information on the properties of the progenitor stellar systems and their immediate circumstellar environments, the detection and study of additional RSN has progressed rapidly.

2. RESULTS

The general properties of all historical radio supernovae (i.e., those with which a modern historical optical supernova can be associated) known at present, along with those of their parent galaxies, are listed in Table 1. Two of these, SN1979c and SN1983n, we will discuss in particular detail here. For a more extensive discussions see Weiler *et al.* (1986).

2.1. SN1979c

SN1979c was discovered near maximum light on 1979 April 19 ∿ 100" SE of the nucleus of the galaxy M100 (NGC4321) by G. E. Johnson (1979) of Swanton, Maryland. It quickly became an object of intensive study in many wavelength ranges and has been firmly established as a Type II_L supernova of the "L" or "linear" subclass. Its properties in many wavelength ranges have been summarized by Panagia *et al.* (1980) and in the radio range by Weiler *et al.* (1981, 1982, 1983). Additional references are given in Table 1 and the radio data are plotted in Figure 1.

2.2. SN1983n

SN1983n was reported in the spiral galaxy NGC5236 (M83) by Evans (1983) and reached maximum optical light on 1983 July 17 (see Panagia *et al.*, 1986; Sramek, Panagia, and Weiler, 1984). Subsequent optical observations determined it to be a peculiar, apparently subluminous Type

TABLE 1: Radio Supernovae

SN Name	Optical SN Type	Observed Radio Max. at 6 cm: Age (yrs)	Flux Density (mJy)	Spectral Luminosity (erg s^{-1} Hz^{-1})	SN Optical Max.: Date (YY/MM/DD)	Brightness: m_B[39]	A_B	M_B	Parent Galaxy: Name	Class	Red Shift[1] (km s^{-1})	Dist[1] (Mpc)	References
A. Detections													
SN1950b	II?	30	0.5[6]	$\sim3*10^{25}$	50/03	≲14.5pg	>0.4	<-15.2	NGC5236 (M83)	ScI-II	515	∼7	6
SN1957d	II?	23	1.9[6]	$\sim1*10^{26}$	57/12	≲15.0pg	>0.4	<-15.7	NGC5236 (M83)	ScI-II	515	∼7	6,7,8,9
SN1970g	II	1.4	∼2.5	$\sim1*10^{26}$	70/08/01	∼11.7	0.6[45]	-18.2	NGC5457 (M101)	Sc	402	7[2]	10,11,12,13
SN1979c	II_L	1.2	8.3	$\sim2*10^{27}$	79/04/19	11.6	0.4	-20.0	NGC4321 (M100)	ScI	1546	17[2]	13,14,15,16,17, 18,19,20,21,22
SN1980k	II_L	0.4	2.6	$\sim1*10^{26}$	80/10/30	11.6	1.3[46]	-18.9	NGC6946	ScI	43	7[2]	20,21,22,23,24
SN1981k	II?	0.5	∼2[27]	$\sim1*10^{26}$	81/08/15	<16	>0.8	<-14	NGC4258	Sc	470	6.6[27]	25,26,27
SN1983n	I_{SL}	0.08	18.5	$\sim1*10^{27}$	83/07/17	11.8[3]	1.1	-18.5	NGC5236 (M83)	ScI-II	515	∼7	28
SN1984ℓ	I_{SL}	0.14	0.7	$\sim4*10^{26}$	84/08/30	13.9[30]	0.5	-18.5	NGC991	ScII	1607	∼24	29,30,31,32
B. New Upper Limits													
SN1980n	I	0.2	<0.6	$<7*10^{26}$	80/12/11	12.5	0.6[46]	-20.7	NGC1316	D3-D4	1635	33	33,34,35,36
SN1981b	I	0.05	<0.3	$<3*10^{26}$	81/03/12	12.0	1.0[4]	-21.6	NGC4536	Sc	1809	33	20,21,37
		1.6	<0.2										
SN1983g	I	0.2	<0.2	$<1*10^{26}$	83/04/04	13.0	1.5[5]	-20.2	NGC4753	SO per	1288	22	38,39
SN1984a	I	0.2	<0.3	$<1*10^{26}$	84/01/10	<13.1[41]	2.1[5]	<-20.7	NGC4419	SBab	-229	22	40,41,42,43,44

1 Sandage and Tammann, 1975
2 Benvenuti, Dopita, D'Odorico, 1984
3 Panagia et al., 1985
4 Barbon, Ciatti, Rosino, 1982
5 Panagia, 1985
6 Cowan and Branch, 1984, 1985
7 Cowan and Branch, 1982
8 Pennington and Dufour, 1983
9 Pennington, Talbot, Dufour, 1982
10 Gottesman et al., 1972
11 Allen et al., 1976
12 Detre, 1970
13 Sandage, 1961
14 Weiler et al., 1981
15 Johnson, 1979
16 Rosino et al., 1979
17 de Vaucouleurs, 1979
18 Panagia et al., 1980
19 Weiler and Sramek, 1980
20 Weiler et al., 1983
21 Weiler et al., 1982
22 Sramek, Weiler, van der Hulst, 1982
23 Wild, 1980
24 Sramek, van der Hulst, Weiler, 1980
25 Wild, 1983
26 Hummel et al., 1983
27 van der Hulst et al., 1983
28 Sramek, Panagia, Weiler, 1984
29 Panagia, Sramek, Weiler, 1985
30 Evans, 1984
31 Ishida, 1984
32 Argyle, 1984b
33 Fomalont and Geldzahler, 1984
34 Wischnjewsky, 1980
35 Schweizer, 1980
36 Landolt and Connolly, 1980
37 Wood, 1981
38 Kosai and Kozai, 1983
39 Barbon et al., 1984
40 Aksenov, 1984
41 Rosino, 1984
42 Rafanelli and Iijima, 1984
43 Argyle, 1984a
44 Meikle, Graham, Bode, 1984
45 Kirshner and Kwan, 1975
46 Panagia, 1982

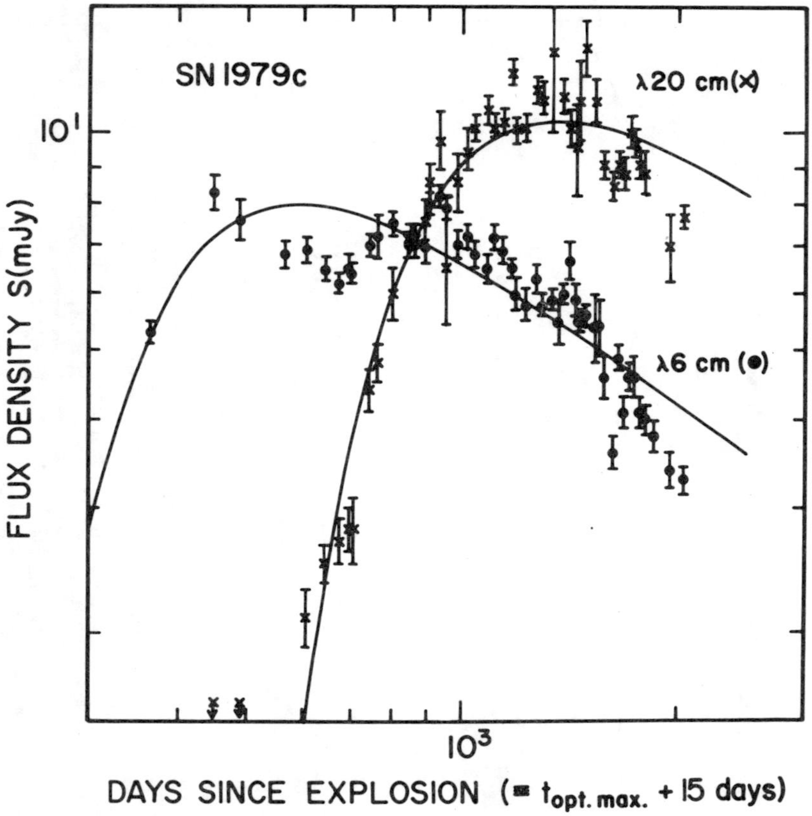

Figure 1: Radio "light curves" for the Type II_L supernova SN1979c in NGC4321 (M100). The two wavelengths 20 cm (x) and 6 cm (•) are shown together. The age of the supernova is measured in days from the estimated date of explosion on 1979 April 04, 15 days before the date of maximum optical light on 1979 April 19. The solid lines represent the best fit light curves.

I (therefore designated Type I_{SL}) supernova (Panagia et al., 1986). Radio observations were started quickly and already on 1983 July 6, 11 days before maximum optical light, a detection at 6 cm with the VLA was obtained with a flux density of 2.0 ± 0.5 mJy. Subsequent monitoring at 6 cm, and after 1983 December at 20 cm, at a rate of approximately once per month, established a rapidly declining radio light curve. [See Table 1, Sramek, Panagia, and Weiler (1984), and Chevalier (1984b)]. These data are plotted in Figure 2.

3. DISCUSSION

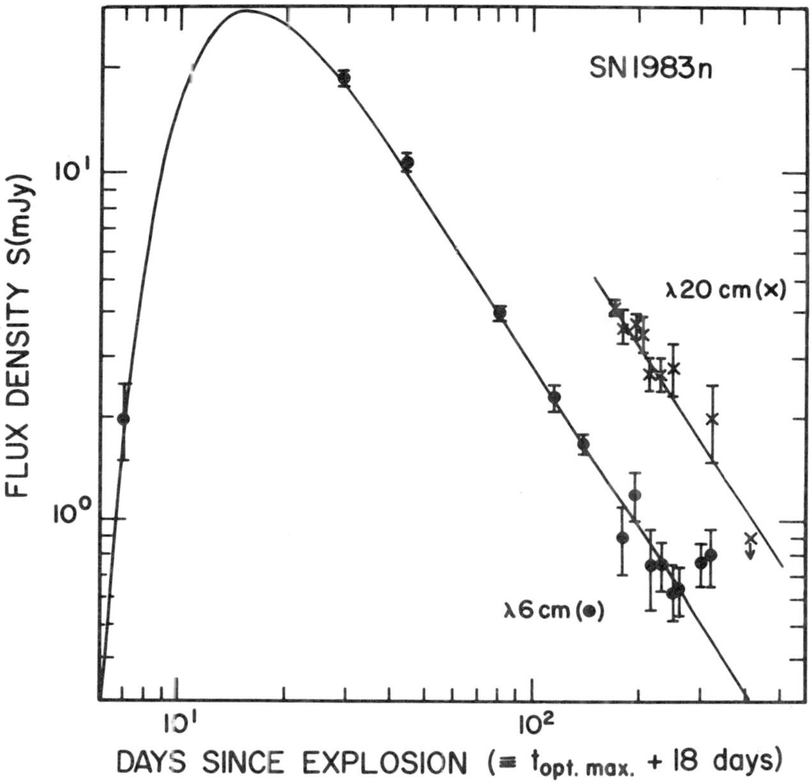

Figure 2: Radio "light curves" for the Type I_{SL} supernova SN1983n in NGC5236 (M83). The two wavelengths 20 cm (x) and 6 cm (●) are shown together. The age of the supernova is measured in days from the estimated date of explosion on 1983 June 29, 18 days before the date of maximum optical light on 1983 July 17. The solid lines represent the best fit light curves.

3.1. Light Curves

The RSN have all had their radio light curves fitted by the relation

$$S(\mathrm{mJy}) = K_1 \left(\frac{\nu}{5\ \mathrm{GHz}}\right)^{\alpha} \left(\frac{t - t_0}{1\ \mathrm{day}}\right)^{\beta} e^{-\tau} \qquad (1)$$

$$\text{where } \tau = K_2 \left(\frac{\nu}{5\ \mathrm{GHz}}\right)^{-2.1} \left(\frac{t - t_0}{1\ \mathrm{day}}\right)^{\delta} \quad . \qquad (2)$$

This formulation assumes that both the flux density S and the optical depth τ are well described by power law functions of the supernova age

$t-t_0$, with powers β and δ, after an explosion date t_0. The absorption is assumed to be purely, thermal, free-free absorption in an ionized medium (frequency dependence $\nu^{-2.1}$) external to the emitting region and the intrinsic emission is assumed to be due to the nonthermal synchrotron process with an optically thin spectral index α. K_1 and K_2 are two scaling parameters for the units of choice.

The intrinsic emission from the RSN is almost certainly of nonthermal (synchrotron) origin because its spectral index at late epochs is typical of nonthermal sources and definitely much steeper than that obtainable from any combination of thermal processes. Also, with measured flux densities and reasonable assumptions for upper limits on emitting region sizes derived from optically measured expansion velocities or from VLBI measurements, the apparent brightness temperatures are far too high for thermal processes (see e.g. Weiler *et al.*, 1981; Bartel *et al.*, 1984, 1985).

An external, rather than internal, absorbing process is indicated by the very steep initial rise of the flux density observed at 20 cm for SN1979c and at 6 cm for SN1983n. Theoretical work by, e.g., Marscher and Brown (1978) has shown that mixed, internal absorbing and emitting processes cannot cause the initial radio light curve to rise faster than $\sim \tau^{-1}$, much slower than that observed, independently of the absorbing mechanism.

Correspondingly, an external location of the absorbing material only permits a thermal process since the two known nonthermal processes, Razin-Tsytovich effect and synchrotron self-absorption, require much higher densities than are conceivable in the regions external to a supernova (see also Marscher and Brown, 1978; Chevalier, 1982a,b; Bandiera, Pacini, and Salvati, 1984).

With these assumptions, five parameters (K_1, K_2, α, β, δ) have been used to obtain a simultaneous fit to the data. The resulting values are listed in Table 2 and the curves are shown as the solid lines in Figures 1 and 2.

3.2. Properties

Firm conclusions are impossible based on such a small sample of objects. This is even more hazardous since the general classes of optical supernovae, Type I and Type II, are gradually, on the basis of more modern and accurate observations, being broken into subclasses. SN1979c was a linear Type II_L supernova (Barbon, Ciatti, and Rosino, 1982; Barbon *et al.*, 1982) and SN1983n was a peculiar, subluminous Type I_{SL} supernovae.

However, with these limitations in mind, some conclusions may be drawn from inspection of the light curves and the values in Table 2:

1) Type II_L supernovae show a flatter nonthermal spectral index ($\alpha \sim -0.6$; $S \propto \nu^{+\alpha}$) than Type I_{SL} supernovae ($\alpha \sim -1.0$).

2) Type II_L supernovae show a slower decline of flux density with time ($\beta \sim -0.7$; $S \propto t^{\beta}$) than Type I_{SL} supernovae ($\beta \sim -1.6$).

TABLE 2: Fitting Parameters[a]

SN Name	SN Type	Estimated Date of Explosion t_0	Explosion Days Before Optical Max. $(t_{opt\ max}-t_0)$	K_1	α	β	δ	K_2	m $(-\delta/3)$	v_i 10^4 km s^{-1}	$\dot{M}$ $M_\odot$ yr^{-1}
SN1979c	II_L	79/04/04	≡15	b) $(8.1\pm1.0)*10^2$	-0.76 ± 0.08	-0.69 ± 0.04	-2.81 ± 0.20	$(1.4\pm0.1)*10^7$	0.94	0.92	$5.3*10^{-5}$
				c) $(9.3\pm1.2)*10^2$	-0.72 ± 0.05	-0.71 ± 0.08	-3.01 ± 0.06	$(5.1\pm0.3)*10^7$	1.00	0.92	$6.2*10^{-5}$
SN1980k	II_L	80/10/01	29 ± 7	b) $(6.9\pm1.1)*10^1$	-0.50 ± 0.10	-0.64 ± 0.07	-2.82 ± 0.27	$(2.4\pm0.3)*10^5$	0.94	0.57	$3.3*10^{-6}$
				c) $(7.3\pm1.2)*10^1$	-0.50 ± 0.06	-0.65 ± 0.10	-2.85 ± 0.10	$(2.8\pm0.3)*10^5$	0.95	0.57	$3.3*10^{-6}$
SN1981k	II?	81/07/31	≡15	d) $(7.3\pm0.6)*10^1$	-0.91 ± 0.07	-0.73 ± 0.06	-3.18 ± 0.09[e]	$<4.5*10^4$[f]	1.06	1.0[g]	$<3.0*10^{-6}$
SN1983n	I_{SL}	83/06/29	18 ± 1	c) $(4.4\pm0.6)*10^3$	-1.03 ± 0.06	-1.59 ± 0.08	-2.44 ± 0.05	$(5.3\pm0.3)*10^2$	0.81	1.7	$2.0*10^{-6}$
SN1984ℓ	I_{SL}	84/08/12	≡18	(See Panagia, Sramek, and Weiler, 1986)							

[a]The radio light curves are assumed to fit a curve of the form $S(mJy) = K_1 \left(\frac{\nu}{5\ GHz}\right)^{\alpha} \left(\frac{t-t_0}{1\ day}\right)^{\beta} e^{-\tau}$ where $\tau = K_2 \left(\frac{\nu}{5\ GHz}\right)^{-2.1} \left(\frac{t-t_0}{1\ day}\right)^{\delta}$

[b]Five parameter fit
[c]Four parameter fit according to the Chevalier (1984b) model (i.e. $\beta = -3+\alpha-\delta$)
[d]Three parameter fit for K1, α, and β assuming the optical depth to be negligible at all epochs where measurements are available
[e]Derived from the best fit values of α and β by using the Chevalier model (i.e., $\delta = -3+\alpha-\beta$)
[f]Three sigma (3σ) upper limit
[g]Assumed

3) There is apparently a difference in the environments surrounding the supernovae of the two types with Type II_L supernovae appearing later in the radio than Type I_{SL} supernovae.

Considering Point 3 further, the apparent tendency for Type II supernovae to "turn on" later in the radio is generally consistent with a model where massive (> 8 $M_\odot$; see e.g., Nomoto, 1982) red supergiant progenitors establish their own local environment through mass loss in the last stages of stellar evolution. A range of masses gives a variation in mass loss rates and a difference in free-free absorption effects, but all are relatively high. On the other hand, the progenitors for Type I supernovae are considered under some models to be white dwarfs in interacting binary systems. The companion star is insufficiently massive to become a Type II supernova ($2 < M/M_\odot < 8$; Arnett, 1982) but is in its red supergiant phase and losing mass rapidly. Through this mass loss, it provides both the mass for accretion to its white dwarf companion, which triggers the white dwarf as a Type I supernova, and the stellar wind environment where the radio absorption takes place (Sramek, Panagia, and Weiler, 1984; Chevalier, 1984b). This circumstellar shell is similar to that for Type II supernovae but less dense from this smaller mass star.

If the matter providing the initial thermal absorption is produced by a stellar wind from either the supernova progenitor or a binary companion of the supernova progenitor, then the required mass loss rate can be estimated. From Panagia and Felli (1975) and Weiler et al. (1986) we can write:

$$\dot{M} = 3.02 * 10^{-6}\, \tau_{5GHz}^{0.5}\, m^{-1.5} \left(\frac{w}{10\ \mathrm{km\ s^{-1}}}\right) \left(\frac{v_i}{10^4\ \mathrm{km\ s^{-1}}}\right)^{1.5} \left(\frac{t_i}{45\ \mathrm{days}}\right)^{1.5}$$
$$\times \left(\frac{t}{t_i}\right)^{1.5m} \left(\frac{T}{10^4\ \mathrm{K}}\right)^{0.68} \quad M_\odot\ \mathrm{yr}^{-1} \quad . \tag{3}$$

Since our observations have shown that $0.8 \lesssim m \lesssim 1.0$ and from Equation 3 $\dot{M} \propto t_i^{1.5(1-m)}$, the dependence of the calculated mass loss rate on the date t_i of the initial ejecta velocity measurement is weak. Thus, we can take the best optical velocity measurements available without worrying about the deviation of their exact measurement epoch from the assumed 45 days after explosion. We also adopt values of $T = 10^4$ K, $w = 10$ km s^{-1}, and, from the best fits to our data given in Table 2 for t = 1 day, τ_{5GHz}(1 day) = K_2 and $m = -\delta/3$. $\dot{M}$ can thus be calculated. These estimated values of the mass loss rate are listed in the last column of Table 2. It is apparent that in all cases the estimated mass loss rates are quite high with $\dot{M} > 10^{-6}\ M_\odot\ \mathrm{yr}^{-1}$, reaching almost $10^{-4}\ M_\odot\ \mathrm{yr}^{-1}$ for SN1979c.

3.3. Models

There are a number of models available to describe the origin and development of the nonthermal radio emission from supernovae. For the acceleration of the relativistic particles two mechanisms have been suggested: 1) shock acceleration in a region external to the supernova photosphere and 2) "pulsar" acceleration by the stellar remnant in the center of the supernova. This second option has two variations a) observing at a frequency ν above a certain critical frequency ν_b ($\nu_{obs} > \nu_b$) and b) its converse ($\nu_{obs} < \nu_b$). For simplicity, we have labeled these two main classes of relativistic particle acceleration mechanisms 1) "mini-shell" from the resemblance of the shock model to shell-type supernova remnants like Cassiopeia A, and 2) "mini-plerion" from the resemblance of the "pulsar" generation model to plerionic supernova remnants like the Crab Nebula.

The "mini-shell" model involves the external generation of the relativistic electrons (and/or positrons) and the enhanced magnetic field necessary for synchrotron radio emission by the shock wave from the explosion interacting with a high density "cocoon" established by mass loss in a stellar wind before the supernova explosion. Since such a shock interaction region is external to the supernova photosphere, there is little problem in allowing the radio radiation to escape after relatively little free-free absorption. This model has mainly been explored by Chevalier (1981a, b; 1984a, b) with work also by Marscher and Brown (1978) and Fedorenko (1983). From the work of Chevalier (1984b) its observable properties are given by

$$S \propto \nu^{\alpha} t^{\beta} \quad \text{(mini-shell)} \tag{4}$$

$$\text{with} \quad \alpha = (1-\gamma)/2 \ , \quad \beta = -(\gamma+5-6m)/2 = -(3-3m-\alpha) \tag{5}$$

where α is the optically thin radio spectral index and m and δ ($\delta = -3m$) are the same model parameters which were discussed in the previous Section. γ is the exponent of the power law of the relativistic electron energy spectrum.

The "mini-plerion" models involve a central source of generation of the relativistic electrons (and/or positrons) and the enhanced magnetic field necessary for synchrotron radio emission by the remnant of the supernova progenitor star which has become something like a rapidly rotating neutron star. This energy source builds up a volume of relativistic particles and magnetic fields, presumably resembling the very early Crab Nebula and other plerions. These models have been explored by several workers (Pacini and Salvati, 1973, 1981; Shklovskii, 1981; Marscher and Brown, 1978; Bandiera, Pacini, and Salvati, 1983, 1984) and, for the variation of interest here ($\nu < \nu_b$), predicts (Pacini and Salvati, 1973)

$$S \propto \nu^{\alpha} t^{\beta} \quad \text{(mini-plerion, } \nu < \nu_b\text{)} \tag{6}$$

$$\text{with} \ \alpha = (1-\gamma)/2 \ , \quad \beta = \alpha + 3(1-\alpha)(1 - m)/2 \tag{7}$$

where m and γ are as discussed above and ν_b is a discontinuity or critical "break" frequency at which adiabatic losses and synchrotron losses become equal (see Pacini and Salvati, 1973).

There is one obvious difficulty for the centrally-driven "mini-plerion" models. On theoretical grounds one expects from one to several solar masses of material to be in the vicinity of a supernova explosion and such a large amount of mass, even if only a small fraction of it were ionized, would totally preclude the observation of radio emission from centrally-driven RSN for many decades. Two models have been proposed to avoid this: 1) the matter in the supernova photosphere clumps very early into dense filaments which contain essentially all of the mass but block very little of the view into the center of the supernova system a la the Crab Nebula (Bandiera, Pacini, and Salvati, 1983); or 2) the magnetic field and relativistic particles generated by the central plerion "leak" through the dense photosphere of the supernova, originating their radio radiation in an external region surrounded only by low density matter (Shklovskii, 1981).

Comparison of these model predictions with the optically thin RSN data of a form $S \propto \nu^{\alpha} t^{\beta}$ given in Table 2 indicates the mini-plerion with $\nu < \nu_b$ and an accelerated boundary ($m > 1$) is not readily distinguishable from the mini-shell model with a decelerated boundary ($m < 1$) in its predicted behavior. In fact, for a constant velocity boundary ($m = 1.0$) both models predict exactly the same behavior for the spectrum and for the time evolution of the flux density. Thus, to the level of accuracy to which the data for these supernovae are described by the model curves chosen, both the mini-shell and mini-plerion models remain viable.

Observationally the long term behavior of the light curves for these supernovae may eventually allow effective choice between the mini-shell and mini-plerion models. Since the mini-shell model is dependent on the presence of a high density "cocoon" of matter established by a presupernova stellar wind wherein the shock particle acceleration and magnetic field generation takes place, the radio emission may be expected to drop sharply when the supernova shock wave reaches the outer boundary of the circumstellar envelope. This time can be estimated to be as short as ~ 15 years for SN1979c (see Weiler et al., 1986). On the other hand the centrally driven mini-plerion models should continue declining at a relatively constant rate until they become undetectably weak. Obviously, the patient collecting of further data will shed light on these questions.

4. CONCLUSIONS

Although the radio supernova (RSN) phenomenon is very new, having been developed essentially in its entirety since the detection of SN1979c in 1980 April (Weiler et al., 1981), its study has met with significant success. Detailed multiwavelength radio light curves have been obtained for a number of supernovae; the number of high quality upper limits on other supernovae has been increased significantly; the number of definite detections of modern optical supernovae and the study of optically unidentified but possible older supernovae has increased dramatically; and theoretical interpretation of the phenomenon has developed apace. The gross structures of the radio light curves have been determined and can be related to the properties of the supernova's progenitor star, its immediate environment, and its likely state in the last stages of evolution before exploding.

5. REFERENCES

Aksenov, E.P. 1984, IAU Circ., No. 3907.
Allen, R.J., Goss, W.M., Ekers, R.D., and de Bruyn, A.G. 1976, Astr. Ap. 48, 253.
Argyle, R.W. 1984a, IAU Circ., No. 3912.
Argyle, R. W. 1984b, IAU Circ., No. 3985.
Arnett, W.D. 1982, Ap.J. 253, 785.
Bandiera, R., Pacini, F., and Salvati, M. 1983, Astr. Ap. 126, 7.
Bandiera, R., Pacini, F., and Salvati, M. 1984, Ap. J. 285, 134.
Barbon, R., Ciatti, F., and Rosino, L. 1982, Astr. Ap. 116, 35.
Barbon, R., Ciatti, F., Rosino, L., Ortolani, S., and Rafanelli, P. 1982, Astr. Ap. 116, 43.
Barbon, R., Cappellaro, E. Ciatti, F., Turatto, M., and Kowal, C.T. 1984, Astr. Ap. Suppl., 58, 735.
Bartel, N., Gorenstein, M.V., Marcaide, J.M., Rogers, A.E.E., Shapiro, I.I., and Weiler, K.W. 1984, Bull. Am. Astr. Soc. 15, 954.
Bartel, N., Rogers, A. E. E., Shapiro, I. I., Gwinn, C., Marcaide, J. M., and Weiler, K. W. 1985, Nature, in press.
Brown, R.L. and Marscher, A.P. 1978, Ap. J. 220, 467.
Benvenuti, P., Dopita, M., and D'Odonico, S. 1984, private communication.
Chevalier, R.A. 1981a, Ap. J. 251, 259.
Chevalier, R.A. 1981b, Ap. J. 246, 267.
Chevalier, R.A. 1982a, Ap. J. 259, 302.
Chevalier, R.A. 1982b, Ap. J. (Letters) 259, L85.
Chevalier, R.A. 1984a, Ann. N.Y. Acad. Sci. 422, 215.
Chevalier, R.A. 1984b, Ap. J. (Letters) 285, L63.
Cowan, J.J. and Branch, D. 1982, Ap. J. 258, 31.
Cowan, J.J. and Branch, D. 1984, Bull. Am. Astr. Soc. 15, 954.
Cowan, J.J. and Branch, D. 1985, Ap.J. 293, 400.
de Bruyn, A.G. 1973, Astr. Ap. 26, 105.
Detre, L. 1970, IAU Circ., No. 2269.
de Vaucouleurs, G. 1979, Ap. J. 227, 729.
Evans, R. 1983, IAU Circ., No. 3835.
Evans, R. 1984, IAU Circ., No. 3979.
Fedorenko, V.N. 1983, Ap. and Sp. Sci. 96, 25.
Fomalont, E. B. and Geldzahler, B. J. 1984, private communication.
Gottesman, S.T., Broderick, J.J., Brown, R.L., Balick, B., and Palmer, P. 1972, Ap. J. 174, 383.
Hummel, E., van der Hulst, J.M., Davies, R.D., Pedlar, A., and van Albada, G.D. 1983, IAU Circ., No. 3803.
Ishida, K. 1984, IAU Circ., No. 3981.
Johnson, G.E. 1979, IAU Circ., No. 3348.
Kirshner, R. B. and Kwan, J. 1975, Ap. J. 197, 415.
Kosai, H. and Kozai, Y. 1983, IAU Circ., No. 3789.
Landolt, A. U. and Connolly, L. P. 1980, IAU Circ., No. 3556.
Marscher, A.P. and Brown, R.L. 1978, Ap. J. 220, 474.
Meikle, W. P. S., Graham, J. R., and Bode, M. F. 1984, IAU Circ., No. 3918.
Nomoto, K. 1982, in Supernovae: A Survey of Current Research, eds. M.J. Rees and R.J. Stoneham (Dordrecht, Reidel), p. 205.

Pacini, F. and Salvati, M. 1973, Ap. J. 186, 249.
Pacini, F. and Salvati, M. 1981, Ap. J. (Letters) 245, L107.
Panagia, N. 1985, private communication.
Panagia, N. 1982 in Proceedings of the 3rd European IUE conference, ESA SP-176, p.31.
Panagia, N. and Felli, M. 1975, Astr. Ap. 39, 1.
Panagia, N., et al. 1980, M.N.R.A.S. 192, 861.
Panagia, N., et al. 1986, in preparation.
Panagia, N., Sramek, R. A., and Weiler, K. W. 1985, Ap. J. (Letters), in press.
Pennington, R.L. and Dufour, R.J. 1983, Ap. J. (Letters) 270, L7.
Pennington, R.L., Talbot, R.J. Jr., and Dufour, R.J. 1982, A. J. 87, 1538.
Rafanelli, P. and Iijima 1984, IAU Circ., No. 3910.
Rosino, L., Ciatti, F., Rafanelli, P., and Ortolani, S. 1979, IAU Circ., No. 3355.
Rosino, L. 1984, IAU Circ., No 3908.
Sandage, A. 1961, The Hubble Atlas of Galaxies, Carnegie Institution of Washington.
Sandage, A. and Tammann, G.A. 1975, Ap. J. 196, 313.
Schweizer, F. 1980, Ap. J. 237, 303.
Shklovskii, I.S. 1981, Sov. Astr. Lett. 7(4), 263.
Sramek, R.A., van der Hulst, J.M., and Weiler, K.W. 1980, IAU Circ., No. 3557.
Sramek, R.A., Weiler, K.W., and van der Hulst, J.M. 1982, in Extragalactic Radio Sources, eds. D.S. Heeschen and C.M. Wade (Dordrecht, Reidel), p. 391.
Sramek, R.A., Panagia, N., and Weiler, K.W. 1984, Ap. J. (Letters) 285, L59.
Ulmer, M.P., Crane, P.C., Brown, R.L., and van der Hulst, J.M. 1980, Nature 285, 151.
van der Hulst, J.M., Hummel, F., Davies, R.D., Pedlar, A., and van Albada, G.D., 1983, Nature 306, 566.
Weiler, K.W. and Sramek, R.A. 1980, IAU Circ., No. 3485. (Circular confuses the position of the SN with that of a nearby radio source 1220+160).
Weiler, K.W., van der Hulst, J.M., Sramek, R.A., and Panagia, N. 1981, Ap. J. (Letters) 243, L151.
Weiler, K.W., Sramek, R.A., van der Hulst, J.M., and Panagia, N. 1982, in Supernovae: A Survey of Current Research, eds. M.J. Rees and R.J. Stoneham (Dordrecht, Reidel), p. 281.
Weiler, K.W., Sramek, R.A., van der Hulst, J.M., and Panagia, N. 1983, in Supernova Remnants and Their X-Ray Emission, eds. J. Danziger and P. Gorenstein (Dordrecht, Reidel), p. 171.
Weiler, K.W., Sramek, R.A., Panagia, N., van der Hulst, J.M., and Salvati, M. 1986, Ap. J. 301, .
Wheeler, J. C. 1984, IAU Circ., No. 4008.
Wheeler, J. C. and Levreault, R. 1985, Ap. J. (Letters), 294, L17.
Wild, P. 1980, IAU Circ., No. 3532.
Wild, P. 1983, IAU Circ., No. 3803.
Wischnjewsky, M. 1980, IAU Circ., No. 3548.
Wood, R. 1981, IAU Circ., No. 3589.

ON THE ASSOCIATION BETWEEN SUPERNOVA REMNANTS AND PULSARS

D. Bhattacharya
Joint Astronomy Program
Department of Physics
Indian Institute of Science
Bangalore-560012, India

and

G. Srinivasan
Raman Research Institute
Bangalore-560080, India

ABSTRACT. From the extremely poor association between pulsars and supernova remnants, and the paucity of centrally filled synchrotron nebulae, presumably produced by active pulsars, we conclude that the majority of neutron stars must be born spinning rather slowly, with initial rotation periods $\gtrsim$ 100 milliseconds.

1. INTRODUCTION

The idea that neutron stars are born in supernovae (SN) dates back to Baade and Zwicky (1934). Subsequent theoretical studies of stellar evolution have confirmed that neutron star formation is associated with at least type II supernovae. Young neutron stars function as pulsars, and supernovae leave behind supernova remnants (SNRs). Thus one would expect an association between them. Interestingly, the various rates given below are consistent with each other lending support to this expectation: (i) SN II rate $\sim$1 in 50 yr (Tammann 1974), (ii) pulsar birthrate $\sim$1 in 40 yr (Taylor and Manchester 1977; Vivekanand and Narayan 1981) and (iii) SNR formation rate $\sim$1 in 30 yr (Higdon and Lingenfelter 1980; Srinivasan and Dwarakanath 1982; Mills 1983). However even though there are $\sim$400 pulsars and $\sim$150 SNRs known in our galaxy, the number of known pulsar-SNR association is only three. The rarity of directly observable associations may be due to several selection effects (low pulsar radio fluxes, pulse smearing due to ISM, beaming etc). However, Radhakrishnan and Srinivasan (1980) pointed out that though one may not be able to see the pulsar in an SNR, one should be able to detect the synchrotron nebulae (plerions) that active pulsars would produce by pumping relativistic particles and magnetic field within SNR cavity. Unfortunately most SNRs are found to have hollow interiors, i.e., there is no detectable plerionic emission within them. Radhakrishnan and Srinivasan (1980) interpreted this as possible evidence of neutron stars not functioning as pulsars. An alternative explanation may be that perhaps most neutron stars are born spinning slowly, thus not being able to produce bright plerions which can be easily detected (Radhakrishnan and Srinivasan 1983). In what follows we shall

F. Pacini (ed.), High Energy Phenomena Around Collapsed Stars, 235–242.

examine this viewpoint in greater detail.

2. IS CRAB NEBULA A PROTOTYPE PLERION?

Even though there is a tendency in the literature to regard the Crab Nebula as a prototype, clearly it is not so. For otherwise one must expect to see about 25 such nebulae born in last 1000 years; these would be at least as luminous as the Crab! One expects two factors to play important role in determining the luminosity of a plerion:- the parameters of the pulsar and the expansion velocity of the nebula. It is certainly true that the expansion velocity of the Crab nebula is almost an order of magnitude smaller than typical velocity of ejecta in supernovae. We also know that the Crab pulsar is not a prototype of pulsars. The statistics of pulsars reveals a fairly wide distribution of magnetic fields. Perhaps the same is true for initial periods, although according to conventional point of view, all pulsars should be spinning rapidly at birth.

The main purpose of this paper is the following: given the statistics of observed plerions, we wish to put some constraints on the periods of pulsars at birth (Srinivasan, Bhattacharya and Dwarakanath 1984). This we do in two different scenarios for the expansion of the nebular boundary, in both cases using the distribution of derived magnetic field (B_*) for the known pulsars. Following Radhakrishnan (1982) we shall assume that most pulsars are born with magnetic fields in the range 10^{12}-$10^{13.5}$ gauss.

3. PULSAR DRIVEN REMNANTS

One possible scenario is that the ejecta is accelerated by the pressure of the relativistic wind from the pulsar, as we know to be the case for the Crab nebula. After the acceleration phase is over, the kinetic energy of the ejecta will be directly proportional to the initial stored rotational energy of the pulsar and hence the velocity of expansion will be inversely proportional to the initial period P_o.

For simplicity we shall assume that magnetic fields of pulsars at birth is uniform in log B_*.

We shall go about deriving constraints on the initial periods of pulsars in the following way. To start with we shall assume that the initial periods are less than 20 milliseconds (i.e., all pulsars are born spinning at least as fast as the Crab pulsar). Given the above assumptions we shall estimate the number of expected plerions above a certain luminosity limit and compare it with observations. This will enable us to test the goodness of the hypothesis that all pulsars are born spinning faster than the Crab pulsar. For the luminosity limit we choose a value of 0.1 L_{Crab}. The flux from such a nebula will be 1 Jy at 1 GHz even if placed at an extreme distance of 20 Kpc. There is a general feeling in the literature that the catalogue of plerions is reasonably complete above this flux level.

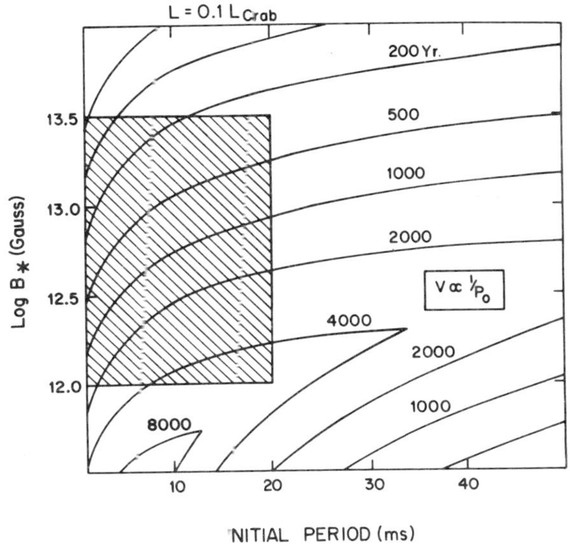

Fig. 1. Contours of constant lifetimes for pulsar-driven plerions. B_* is the magnetic field and P_0 the initial period of the central pulsar. Lifetime is defined as the time for which the plerion is more luminous than $0.1L_{Crab}$ (in radio). All pulsars born on a given contour will produce plerions of the same lifetime as indicated. In estimating the expected number of plerions with luminosities greater than $0.1L_{Crab}$ we have assumed that pulsars are born anywhere in the hatched region.

In Fig. 1 we have plotted constant lifetime contours for plerions. Pulsars powering the plerions have been assumed to be born within the hatched area in the B_*-P_0 plane. By "lifetime" we mean the duration for which the plerion is more luminous than $0.1L_{Crab}$. The lifetimes were calculated using luminosity evolution scheme developed by Pacini and Salvati (1973). Given these contours now it is a simple matter to calculate the average lifetime of the nebulae by weighting the indicated lifetimes appropriately. The expected number is obtained by multiplying the average lifetime by the plerion birthrate, which is nothing but the pulsar birthrate (assumed to be 1 in 40 yr). We find, that in this pulsar-driven scenario, there should be 35 plerions in our galaxy with luminosities larger than $0.1L_{Crab}$. However recent compilation by Weiler (1985) shows that there are only 3 known plerions above this luminosity limit. From a recent VLA search, Helfand et al (1984) have also concluded that there are unlikely to be many unknown plerions above this luminosity limit.

Given such a gross discrepancy between the expected number of plerions and the observed number we are forced to conclude that if the present scenario for the expansion of the nebula is relevant, then the majority of pulsars must be born outside the hatched region in fig. 1. Since the majority of pulsars having fields outside the interval indicated would be inconsistent with the statistics of pulsars, we conclude that most of them must be born spinning rather slowly compared to the Crab pulsar. In fact to reconcile with plerion statistics one will have to conclude that most pulsars are born with initial rotation periods as large as $\sim$200 milliseconds.

4. PULSARS IN RAPIDLY EXPANDING SHELLS

There is an alternative scenario which might be considered, namely, pulsars are born in rapidly expanding cavities ($v \sim 10^4$ km s^{-1}). At first sight it would appear that in this scenario it may not be necessary to deviate from the conventional wisdom that pulsars should be born spinning rapidly. For, after all, due to severe adiabatic losses the central plerion would not be as bright compared to one confined to a slowly expanding cavity. However, if we repeat the same exercise as in the previous section we predict that there should be 16 plerions above the chosen luminosity limit. This number would be even larger if we allowed for the deceleration of the ejecta due to interaction with the interstellar matter. Such a deceleration will also result in a prominent radio emitting shell so that such objects will have a hybrid shell-plerion morphology. Although there are a few remnants with such morphology (e.g. G326.3-1.8), none of them have a plerion with luminosity exceeding $0.1L_{Crab}$.

Thus, even in this more favourable scenario, one is forced to conclude that the majority of pulsars must have been much slower rotators at birth than the Crab pulsar. Even allowing for a substantial incompleteness in the sample of SNRs with hybrid morphology, we arrive at a lower limit to the initial period of $\sim$50 milliseconds.

5. THE HISTORICAL SHELLS

A similar limit on initial rotation periods of pulsars may be obtained from a consideration of the Historical Shell SNRs. According to the prevalent view, Kepler, Tycho, SNR1006 may be remnants of Type I supernovae and perhaps there are no neutron stars in them. However, we shall assume that there <u>are</u> pulsars in them and ask what kind of initial periods and magnetic fields would they have had? The published maps of these remnants show almost <u>no</u> emission near the centre. However, even from an optically thin shell one expects <u>some</u> emission from the central region. We shall make the reasonable assumption that if there was a central plerion whose surface brightness was above 1/5th the average surface brightness of the shell, we would easily have been able to detect it. Taking this as an <u>upper limit</u> we can draw a diagram showing "allowed" and "excluded" regions for pulsars in each of these remnants (Fig. 2). The meaning of these contours shown is the following. Let us take Kepler's remnant, for example. If there <u>is</u> a pulsar in Kepler, it could

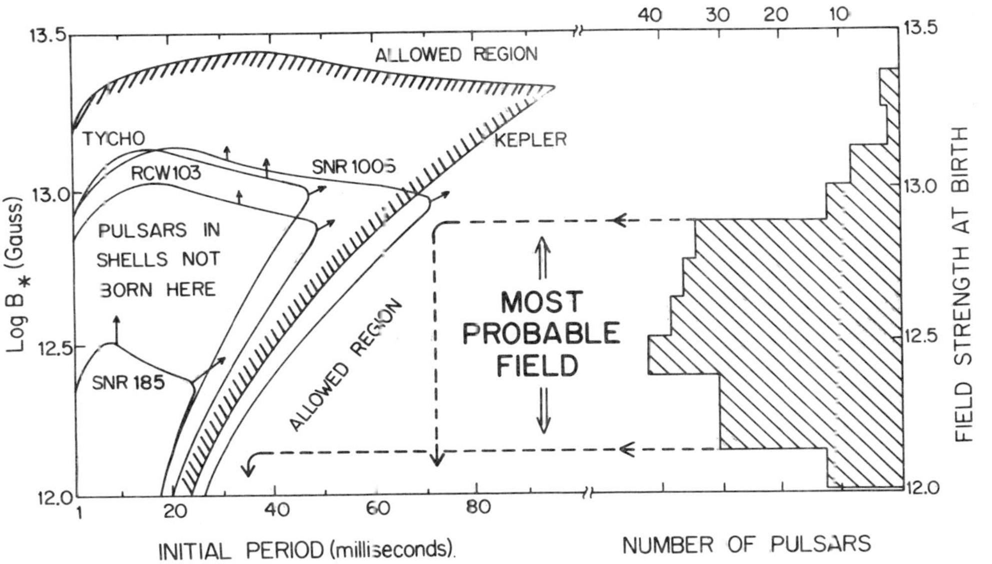

Fig. 2. Pulsars inside the historical shells. The contours correspond to surface brightness of an assumed central plerion equal to 1/5th the the average surface brightness of the SNR; the appropriate ages and expansion velocities inferred from their sizes were used. If pulsars in these shells were born inside the region enclosed by the contours, then the plerions produced by them should have been easily detected. The arrows on the contours indicate that the excluded region is likely to be much larger. Also shown is the histogram of pulsar fields at birth. Although pulsars in these remnants could have been born anywhere outside the region enclosed by the contours, the histogram suggests that for the majority of them the initial periods will be greater than 30 to 70 ms.

not have been born with its parameters anywhere within the region enclosed by the contour; otherwise it would have produced a distinct plerion. Similar excluded regions are shown for other historical remnants also. Not surprisingly, the constraint is stronger for the younger remnants. If the historical shells may be taken as typical representatives of galactic shell remnants, then what we can conclude is that most pulsars must be born in the region marked "allowed" in the figure. Pacini (1972) surmised long time ago that hollowness of the historical shells is consistent with the presence of very high field pulsars in them. This may indeed be so in some specific cases. But it cannot be true for the majority of shells. The histogram shown in fig. 2 is the distribution of pulsar magnetic fields at birth as deduced from observational data assuming a field decay timescale of $\sim$2 million years, as suggested by Radhakrishnan (1982). It is seen that very few pulsars have fields greater than 10^{13} gauss, or less than 10^{12} gauss. This distribution implies that their periods at birth must be longer than 35-70 milliseconds. This should be viewed as a lower limit. Since shell SNRs constitute nearly 90 percent of the total population of supernova remnants, this conclusion should apply to the majority of all pulsars.

6. THERMAL FIELD GROWTH

As an alternative to slow initial rotation periods, following Blandford et al (1983) one may like to suggest that association between supernova remnants and pulsars are rare because the magnetic fields of neutron stars are very low at birth, and it takes $\sim 10^5$ years before the typical surface magnetic fields $\sim 10^{12}$ gauss are built up by thermally driven battery mechanism. In the meantime the supernova remnant would have dissipated away. But, obviously, the known associations - Crab, Vela, 1509-58 and 0540-69.3 will be embarrassments to this picture. They have suggested that if the pulsar is rotating very rapidly at birth, then the field growth also can be very rapid, and this may explain the known associations. Conversely, the poor association implies, once again, that most pulsars are slow rotators at birth.

7. HIGH FREQUENCY EMISSION

The conclusions drawn so far are based on the radio emission from plerions. However, for all known plerions the major fraction of the luminous power is emitted in high frequency bands, for example, as X-rays. It will, therefore, be interesting to see what kind of constraints one obtains from considerations of X-ray luminosities of plerions. Adopting the same procedure as before, one can ask how many plerions would one expect to see with luminosity greater than, say, 10^{37} erg s^{-1} (in the Einstein observatory X-ray band). Once again assuming that pulsars are born in the hatched region in fig. 1, one would predict $\sim$75 X-ray plerions in the scenario of sec. 3 and 30 in the scenario of sec. 4! It might appear at first that the gross discrepancy with the statistics of X-ray plerions might give us much stronger constraints on the pulsar parameters than deduced above (see, for example, Helfand in this volume). Unfortunately this is not the case. The prediction of large number of very luminous X-ray plerions is, in part, a consequence of the simplifying assumption made that the spectrum of the injected particles is a simple power law extending all the way from radio emitting electrons to X-ray emitting electrons. However, one knows that there is a break in the injected spectrum of Crab nebula (Rees and Gunn 1974). It is likely that there is a break in the spectrum of particles in other cases too. In fact the observed ratio of X-ray to radio luminosities of plerions strongly suggest a break in the injected spectrum (Reynolds and Chanan 1984). Unfortunately, however, reliable spectral indices in the X-ray region are not yet available. In the absence of these, and with the lack of clear predictions from pulsar theories in this regard, it is not possible to obtain any useful constraints from the X-ray data.

7. CONCLUSIONS

We summarise below our main conclusions.

1. If the expansion of the nebular boundary is due to the pressure of the pulsar wind in all cases, then the paucity of plerions strongly suggest that the initial rotation period of most pulsars is $\gtrsim$150 milliseconds, or,

   such pulsar driven SNRs are very rare.

2. If pulsar-driven SNRs are rare, those like the Crab are even rarer. Such plerions are likely to be bright and long-lived only when the central pulsar has initial period $\sim$ 20 milliseconds and magnetic field $\sim 10^{12}$ G. The uniqueness of the Crab nebula should be understood in terms of its pulsar having just these required characteristics.

3. The hollow interiors of shell remnants implies that if there are pulsars in them, they must have been born as slow rotators.

4. The above remarks lend strong support to the conclusion arrived at by Vivekanand and Narayan (1981) that the majority of pulsars are "injected" with long periods.

REFERENCES

Baade, W., Zwicky, F. 1934, Phys. Rev., 45, 138
Blandford, R.D., Applegate, J.H., Hernquist, L. 1983, Mon. Not. R. astr. Soc., 204, 1025
Helfand, D.J., Chance, D., Becker, R.H., White, R.L. 1984, Astr. J., 89, 819
Higdon, J.C., Lingenfelter, R.E. 1980, Astrophys. J., 239, 867
Mills, B.Y. 1983, in IAU Symp. 101: Supernova Remnants and Their X-ray Emission, Eds. J. Danziger and P. Gorenstein, D. Reidel, Dordrecht, p. 421
Pacini, F. 1972, in The Physics of Pulsars, Ed. A.M. Lenchek, Gordon and Breach, London, p. 119
Pacini, F., Salvati, M. 1973, Astrophys. J., 186, 249
Radhakrishnan, V. 1982, Contemp. Phys., 23, 207
Radhakrishnan, V., Srinivasan, G. 1980, J. Astrophys. Astr., 1, 25
Radhakrishnan, V., Srinivasan, G. 1983, in IAU Symp. 101: Supernova Remnants and their X-ray Emission, Eds. J. Danziger and P. Gorenstein, D. Reidel, Dordrecht, p. 421
Rees, M.J., Gunn, J.E. 1974, Mon. Not. R. astr. Soc., 167, 1
Reynolds, S.P., Chanan, G.A. 1984, Astrophys. J., 281, 673
Srinivasan, G., Dwarakanath, K.S. 1982, J. Astrophys. Astr., 3, 351
Srinivasan, G., Bhattacharya, D., Dwarakanath, K.S. 1984, J. Astrophys.

Astr., 5, 403

Tammann, G.A. 1974, in Supernovae and Supernova Remnants, Ed. C.B. Cosmovici, D. Reidel, Dordrecht, p. 155

Taylor, J.H., Manchester, R.N. 1977, Astrophys. J., 215, 885

Vivekanand, M., Narayan, R. 1981, J. Astrophys. Astr., 2, 315

Weiler, K.W. 1985, in The Crab Nebula and Related Supernova Remnants, Ed. M. Kafatos and R.B.C. Henry, Cambridge, p. 265

THE PROGENITORS AND PRODUCTS OF SUPERNOVAE

David J. Helfand*
Astrophysics Laboratory, Columbia University

Robert H. Becker
Department of Physics, University of California, Davis
and
Lawrence Livermore National Laboratory

ABSTRACT. We review much of the recent radio and X-ray observational work relevant to the questions of the progenitor populations and explosion products of supernovae. Several illustrative examples are given of how X-ray data on shell-type supernova remnants, coupled with analytic and numerical hydrodynamic models, can constrain the mass and evolutionary state of the stellar precursor. We go on to examine the population of Crab-like remnants, pointing out the restrictions imposed on explosion models by the statistics of collapsed remnant cores.

I.1. Introduction

The first supernova explosion of which we have a written record occurred 1800 years ago this December** when Chinese astronomers recorded a bright new star which appeared suddenly in the constellation Centaurus. Only five other such events have been observed in our Galaxy since that time, none in the last 375 years. Despite their rarity, supernovae (SNe) have been recognized over the past few decades as a major contributor to the energy density of the interstellar medium (both in cosmic rays and turbulent energy) and as the driving force behind the chemical evolution of the Galaxy. Their remnants are among the most luminous galactic radio and X-ray emitters and have, in a few cases, been shown to harbor young neutron stars. The significance of SNe as markers of stellar death was first noted just over 50 years ago by Walter Baade and Fritz Zwicky (1934). Yet as we embark on our second half century of SN research, a number

*Alfred P. Sloan Fellow

**on the first author's birthday

F. Pacini (ed.), High Energy Phenomena Around Collapsed Stars, 243–259.

of rather fundamental questions remain regarding the percursor population, the mechanism of the explosion, and the evolution of the supernova remnant (SNR). In these lectures we will review a number of recent results which are illustrative of approaches to be taken in bettering our understanding of the progenitors and products of supernovae.

I.2 Supernova Facts

Since the appearance of S Andromeda was noted just 100 years ago, over 500 SNe have been observed in external galaxies, and it is from these that we have obtained virtually all of our knowledge of the explosions themselves and the early evolution of their remnants. The standard dogma divides the explosion events into two types, although more recent work suggests that this may represent a gross and misleading oversimplification. We list the salient characteristics of the two types in Table Ia. The defining distinction is the presence or absence of hydrogen lines in the spectrum near maximum light: Type II explosions have spectra dominated by hydrogen Balmer lines, while Type I objects show no evidence of hydrogen. Recent analysis of Type I spectra suggests that Ni lines, as well as features due to Silicon, Magnesium, and other heavy elements are present, although a definitive spectral synthesis has yet to be produced.

Only a small fraction of the recorded SNe have measured spectra and, as a result, classification of most events rests on the shape of the light curve. The canonical picture holds that Type I events rise rapidly to maximum with a relatively narrow peak in the optical band (declining by an order of magnitude in ~25 days) followed by an exponential decay with a half life of 56 days; Type II's exhibit a somewhat slower rise and a more erratic decline which is often interrupted by a "plateau" phase lasting ~50 days. The peak absolute magnitudes of Type I explosions were thought to be narrowly distributed around $M_B = -19.4$, whereas the maximum magnitudes of Type II events range from -16 to -20. More recent work, however, suggests considerable deviation from these norms: type I "peculiars" have been identified which peak ~2 magnitudes fainter than standard type I's and show different light curves, spectra (although still lacking H), and early-time radio emission. Type II "linear" and "plateau" subclasses are distinct in several characteristics, while a number of SNe which do not fit any of these types have been discovered. In particular, Branch has shown (see Pangia, this volume) that the historical light curves of Tycho's and Kepler's SNe are consistent with those of standard Type I's, of Type I peculiars, or of Type II linears. Clearly, it is only when spectrophotometric observations spanning >100 days are available for a large sample of events that quantitative classification will become meaningful, and it is our opinion that the current simple notion that two and only two fundamental types of SNe exist may well be incorrect.

Nevertheless, for the sake of completing the current catechism, we record in the remainder of Table Ia the other commonly held beliefs regarding Type I and Type II events. That uncertainties in even the standard picture are large is evident from the ranges of progenitor masses and the uncertainty regarding the existence of core remnants; more about such matters is to be found in the lectures by Hillebrandt, van den Heuvel, and Woltjer in this volume. It is our thesis in what follows, however, that study of the remnants of SN events is essential if the uncertainties in our current picture are to be narrowed.

Table 1

a) The Cannonical Wisdom on Supernova Types

Property	Type I	Type II
Spectrum	No H (possibly Ni)	H-dominated
Peak Magnitude	-19.4	-16 to -20
Light Curve	Narrow peak; exponential decay	Slower rise and fall
Expansion velocity	~ 14000 km s^{-1}	~ 8000 km s^{-1}
Progenitor Mass	1-9 $M_{\odot}$	8-60 $M_{\odot}$
Ejecta	1 $M_{\odot}$ Ni	$\gtrsim 5$ $M_{\odot}$ CNO, Si, etc.
Core Remnant	Rarely	Frequently (?)
Parent Galaxy Type	All	Never in ellipticals
Fraction in Galaxy	$\sim 50\%$	$\sim 50\%$

b) The Cannonical Wisdom on Supernova remnants

Property	Shell-like	Crab-like
Radio spectrum	$\langle\alpha\rangle$ -0.5	$\langle\alpha\rangle \sim 0$
Radio morphology	Limb-brightened	Centrally peaked
Radio polarization	$\lesssim 5\%$	$\sim 10\%$
X-ray spectrum	Thermal with strong He-like lines	Nonthermal power law
X-ray morphology	Any	Centrally peaked
X-ray point source	< 10%	> 50%
Number in Galaxy	140	6

I.3 Supernova Remnant Facts

For some years now, it has also been common to divide the observable remnants of supernovae into two categories -- shell-like and Crab-like -- and we list the properties of these two classes in Table Ib. Most of the 150 Galactic SNRs were first recognized by their radio emission: any extended radio source close to the galactic plane which exhibits a non thermal spectrum is a likely candidate for inclusion in remnant catalogs. It should be noted that the current census of galactic remnants is far from complete, and that even the defining criteria given above introduce selection effects into the catalogues. For example, Crab-like remnants have flat radio spectra and are thus indistinguishable from HII regions unless polarimetric mapping is available (Helfand et al. 1984). In addition, the requirement of an extended source means an angular diameter $\gtrsim$ few arcminutes, leaving out many small diameter objects (e.g., Green and Gull 1984, but see Chanan et al. 1985). Finally, nonthermal emission which arises from other causes may be mistakenly included (e.g., Becker and Helfand 1985). It is only with spatially resolved spectral and polarimetric mapping that definitive classification of radio SNRs can be made.

The essential difference between the two remnant classes is the source of the energy powering the emission. For shell-like objects, the radiation originates in the kinetic energy of the explosion itself: the radio synchrotron emission results from the compression of ambient magnetic fields and the acceleration of elections at the supernova shock front, while the thermal X-rays arise from plasma heated by the passage of this shock. In Crablike remnants, however, both the X-ray and radio emission is nonthermal, driven by the rotational kinetic energy loss of the central neutron star. At least half of the known Crab-like remnants show direct evidence for the central neutron star, either as an observable pulsar (e.g., the Crab, Vela, MSH 15-52, 0540-693) or as an unresolved X-ray point source at the remnant's center (e.g. CTB80, 3C58). The fraction of such remnants containing collapsed objects is quoted as >50% in Table 1b because the upper limits to point-like components in the remaining Crab-like SNRs are higher than many of the detected cases; i.e., the current data are consistent with all Crab-like nebulae containing a young neutron star. For the shell-like remnants, however, fewer than 10% show evidence for a compact remnant in any spectral regime (Helfand and Becker 1984) -- more on this below.

Recently it has become clear that, as with the simple division of SNe into two types, the binary classification of remnants is also incomplete. There are, in fact, a number of SNRs which show clear evidence of both shell-like and Crab-like components and, as of this writing, the number of such composite remnants actually exceeds the number of purely Crab-like objects. While it is possible that the underlying population represents a continuum from very weak to totally dominant Crab-like cores embedded in expanding shells (e.g., Manchester and Durdin 1983), it is our view that only a relatively modest fraction of shell-type remnants contain active neutron stars, a proposition we defend in Section III below.

Although various attempts have been made to associate the two classes of remnants with the two types of SNe (e.g., Type I's leave shells while Type II's produce Crab-like objects), we feel that such a procedure, in light of the questions raised above concerning the standard classification schemes, is at best premature. Instead, we present below two of the many approaches required to advance our understanding of the origin and evolution of SNe and their remnants. In section II, we outline the types of constraints which can be brought to bear on SN progenitors through the detailed study of the X-ray emission from their remnants, while in the following section we summarize the data on the fraction of SNe which leave neutron stars, a question relevant to the explosion mechanism itself.

II. CONSTRAINTS ON THE SN PROGENITOR

II.1. Introduction

Although virtually all catalogued Galactic SNRs were first recognized as a result of their radio emission, such radiation typically represents only a few percent of the emitted energy over the life of a remnant. The spectral regime in which the dominant radiative losses occur for at least the first 10^4 yrs is the 0.1-10 keV band and, with the advent of X-ray imaging detectors such as those on the *Einstein* Observatory, the information which the X-rays provide on the dynamics of the remnant's evolution and the composition of the material it contains is being exploited for the first time. In external galaxies where interstellar absorption is not too severe, X-ray surveys may well provide the best way of discovering new remnants; in the Magellanic Clouds, for example, the *Einstein* observations more than tripled the size of the remnant population known from previous radio and optical surveys to 43 (Helfand 1984 and references therein).

We outline here the general procedure for constraining the properties of a SN progenitor through the analysis of its X-ray emission. Following an introduction to the types of data available and the general theoretical framework within which they are interpreted, we present a brief account of several specific attempts to deduce progenitor parameters from *Einstein* Observations.

II.2 The Data

For a dynamically evolving remnant, a key parameter in any model calculation is the time since the explosion. Unfortunately, SNR ages are only available for remnants of the six SNe recorded in historical times -- RW86 (SN185), SN1006, Crab Nebula (SN1054), 3C58 (SN1181), Tycho (SN1572), and Kepler (SN1604) -- plus the young remnant Cas A which is thought to have occured in ~1667. For these seven remnants, age is an important and precisely defined input; for the remainder of the galactic and extragalactic populations, there exist essentially no age estimates independent of model calculations.

The most fundamental parameter obtained from an X-ray observation of a SNR is its intensity (in photons per second) in a specified X-ray band. In the case of the *Einstein* imaging instruments, these energy bands were ~0.15 to 4.5 keV and ~0.1 to ~3.0 keV in the imaging proportional counter (IPC) and high resolution imager (HRI), respectively; typical accuracies, at least for the brighter remnants, are $\lesssim$ 1%. Translation of these observed counting rates to intrinsic source luminosities includes several uncertainties which degrade substantially this accuracy. The count-rate-to-*detected*-flux conversion involves the convolution of the source spectrum with the detector-telescope response matrix and can introduce uncertainties of from ~10-30%. A determination of <u>intrinsic</u> source flux must then take account of the substantial photoelectric absorption suffered by soft X-rays in the intervening interstellar material. Even when the neutral atomic hydrogen column density to a remnant is well determined from HI absorption measurements, uncertainties in the amount of ionized and molecular material, the chemical composition of the absorbing gas (most of the integrated absorption cross section is contributed by oxygen and other metals), and the atomic cross section data can lead to errors of up to a factor ~2 in the detected-to-intrinsic flux conversion. Finally, the largest uncertainty arises from the unknown distances to most remnants. Even for the well-studied historical remnants, distance errors of a factor of 2 cannot be excluded, while a factor of ~3 uncertainty for many of the other galactic objects leads to intrinsic luminosity errors of up to an order of magnitude. The only class of remnants with substantially better distance determinations are those in external galaxies where errors of $\lesssim$ 20% are more typical, at least within the Local Group.

The most fundamental prediction of a dynamical SNR model is remnant size as a function of time, making the measured diameter a most important constraint. While frequently better determined by interferometric radio observations, angular diameters accurate to $\lesssim$ 5" to 10" were determined for ~50 SNRs with the <u>Einstein</u> HRI. The accuracy of linear diameters is, of course, again limited by distance uncertainties of a factor of ~2 to 3, excepting the Magellanic Cloud sources for which a very clean number-diameter relation can be constructed (see below).

In addition to allowing the accurate determination of X-ray sizes for the first time, the HRI allowed us to construct detailed surface brightness profiles of the brighter remnants. Basic parameters such as the degree of limb-brightening for shell-like objects and the presence or absence of point-like central components in remnants of all types are clearly of value in constraining a remnant's evolution (§II.4) and the properties of the progenitor (§III). In addition, the structure of shock front -- e.g., the presence or absence of blast-wave and reverse shock enhancements, and shock thickness -- provide additional clues as to the dynamical state of the remnant.

X-ray spectra at various resolutions were also provided by the <u>Einstein</u> instruments and are of the utmost importance in characterizing the thermal history and chemical composition of a remnant's shock-heated plasma and non-thermal electron component. The IPC provided a resolution of $\Delta E/E \sim 100\%$ in the 0.1 to 4.5 keV band, yielding a first approximation for the average plasma temperature or power law exponent, as well as the best available empirical determination of the absorbing column density. With the ~230 eV resolution between 0.5 and 4.0 keV provided by the solid state spectrometer (SSS), strong features corresponding to emission from the He-like ions of silicon, sulfur, magnesium, calcium, and neon were resolved in several of the young remnants. Such data provide valuable information on the ionization state and compositon of the plasma, which in turn leads to estimates of the ambient interstellar density and ejected mass so crucial to the dynamical evolution of the remnant. As will be seen below, however, such resolution is insufficient to resolve definitively ambiguities arising from competing density, temperature, and abundance effects. The Einstein focal plane crystal spectrometer (FPCS) with its ~10 eV resolution was capable in principal of sorting out these effects, and its beautiful spectrum of the intermediate-aged remnant Puppis A (Winkler et al. 1981) demonstrated the value of being able to resolve intercombination and resonance lines from various ionic species in determining the state of the emitting gas. Its small effective area and low observing efficiency severely limited its applicability, however, and it remains for future missions to exploit fully the rich plasma diagnostic potential of high resolution ($\Delta E/E \sim$ 100 to 3000) X-ray spectroscopy.

All of the Einstein data was limited to X-rays below ~4-5 keV by the telescope mirrors. At least for the moderately young remnants, however, substantial emission occurs at higher energies. Spectra in the 5-20 keV band from non-imaging detectors on OSO-8, HEAO-1, Tenma and EXOSAT (for a recent review, see Aschenbach 1985) have begun to contribute to remnant modelling by constraining the high energy continuum and sampling the ~6.5 keV iron line; the continuum slope and cutoff speak to the question of electron-ion temperature equilibration behind the shock, while the iron line strengths and energy are useful in determining the iron yield of the SN event as well as an additional temperature and density diagnostic for the plasma. The greatly improved high-energy sensitivity and imaging capabilities of the planned AXAF and XMM missions will provide a major step forward in this energy regime. Other desirable observational capabilities awaiting future development include X-ray polarimetry which could separate thermal and nonthermal components in shell type and composite objects as well as mapping the magnetic fields in Crab-like remnants, and spatially resolved moderate- to high-resolution spectroscopy which would add enormously to the constraints on remnant evolution models.

II.3 The Models

II.3.1 Philosophy. A supernova remnant is the observable manifestation of the conversion of the kinetic energy produced in a SN explosion to electromagnetic radiation. The dynamical evolution of the remnant depends on such parameters as time since the explosion, ejected mass, ejecta velocity, and ambient density. The evolving remnant emissivity in the various spectral regimes depends additionally on the elemental abundances in the ejecta and the ambient material, and on the detailed ionization and electron-ion equilibration mechanisms of the plasma. The problem of inferring the properties of the SN progenitor from its remnant, then, is to produce a radiative hydrodynamical model constrained by the fundamental observational inputs described above. The ultimate goal of such an exercise is to define the mass of the progenitor, the nature of the surrounding medium (e.g., is it dominated by mass lost prior to the explosion?), and the composition of the ejected material, along with details of the explosion such as the fraction of the energy which coupled to the environment and the fate of the stellar core (neutron star, black hole, or disruption).

II.3.2 Analytical models. Prior to the large increase in both the quantity and quality of X-ray data accumulated over the past decade, a relatively simple analytical hydrodynamic model was sufficient to describe the gross features of observed remnants such as diameter and integrated X-ray luminosity. As outlined by Woltjer (1972), the evolution of a remnant is describable as a progression through four distinct phases: the free expansion phase in which a blast wave is formed by the ejecta as it streams outward ballistically, with the radius increasing linearly with time and the temperature falling as $R^{-3(\gamma-1)}$ owing to adiabatic cooling; the Sedov or adiabatic stage when radiation is relatively unimportant, the sum of kinetic and thermal energies is regarded as conserved, and R increases as $t^{2/5}$; the radiative phase commencing when the temperature falls to $\lesssim 10^6$ K and radiative losses begin to effect substantially the remnant's dynamical evolution; and finally, the stage at which the shock velocity falls to $\lesssim 20$ km s^{-1}, comparable to the random turbulent velocity of the interstellar medium, and the remnant ceases to be a distinct entity. A recent refinement on this picture is the inclusion of a so-called reverse shock phase between free expansion and the Sedov regimes, (McKee 1974): as the ejecta begins to decelerate, a shock is driven into it, propagating inward in a coordinate system moving with the expanding blast wave. At times between a few hundred and a few thousand years after the explosion (depending on the explosion energy and ambient density), emission from this reverse shock can dominate that from the original blast wave.

Analysis of the *Einstein* imaging data for the remnant of SN185 (RCW86) by Pisarski, Helfand, and Kahn (1984) is illustative of this analytic approach used in conjunction with observations in the radio, optical, and X-ray bands to constrain the SN progenitor. These authors find the remnant's size, X-ray luminosity, X-ray temperature, and absorption column density to be consistent with the identification of the object with the 185 AD event. That age, plus recent optical data on extinction toward the filaments, requires a distance of $\lesssim 1$ kpc, 2-3 times smaller than previous estimates. Using this value, Pisarski *et al.* find that measurements of the plasma temperature (T ~ 1 keV) and the diameter (13 pc), plus the assumption that the remnant is in the Sedov phase, lead to reasonable and self-consistent values for the explosion energy (~ 2×10^{50} ergs), mean expansion velocity (~3500 km s^{-1}), ambient interstellar density (~0.3 cm^{-3}), ejected mass (~1 $M_\odot$), swept up mass (~ 5 $M_\odot$) and mean line-of-sight ISM density (~0.2 cm^{-3}). The energy, velocity and mass estimates coupled with the estimate of the absolute magnitude of the event (for d = 1 kpc and 0.8 magnitudes of extinction, $M_V \sim -19$) are all typical of the canonical Type I SN explosion. Type II parameters, are discrepant with the data. However, the authors go on to note that they cannot conclusively rule out the remnant's being the product of a Type II SN which is still in the free expansion phase. The observations coupled with the use of Type II parameters lead to a distance of ~1.4 kpc, a diameter ~18 pc, a total mass range of 7-14 $M_\odot$, and a local ISM density range of 0.15-0.3 cm^{-3}.

II.3.3 Numerical models. The advent of high-speed computers and the increasing detail of the X-ray observations (5" resolution surface brightness profiles and high resolution energy spectra) make it both possible and desirable to construct more sophisticated models of remnant evolution with which to compare the data. Increasingly, 1-D and even 2-D hydrodynamics codes are being developed to model both the SN explosion itself and the subsequent evolution of its remnant. Such SNR models have a number of significant advantages over the analytic approach. The ejected mass density and velocity profiles, as well as a non-uniform density profile for the ambient medium, can be included explicitly. In addition, radiative losses from the expanding remnant can be tracked throughout its evolution rather than being confined to a single phase, and the abundances in both the ejecta and the ambient material can be varied. Most importantly, however, the rate at which the elections and the ions in the plasma reach equilibrium and the time-dependent ionization process itself can be followed in detail. Several authors, including Itoh (1977), Gronenschild and Mewe (1982), Shull (1982), and Hamilton, Sarazin and Chevalier (1983), have shown that X-ray luminosities calculated for models of relatively young remnants ($\lesssim$ 3000 yr) can be underestimated by as much as an order of magnitude if non-equilibrium ionization (NEI) effects are ignored.

As an example of the application of such numerical modelling for remnant evolution, we refer (not surprisingly) to the work of Hughes and Helfand (1985). By employing a novel solution to the problem of time-dependent ionization in a shock heated plasma and incorporating it into a standard spherically symmetric hydrodynamic shock code, these authors produced two rather narrowly constrained models of the X-ray emission from Kepler's SNR. Observational inputs included the age, angular diameter, X-ray luminosity in the 0.1-3.0 keV and 0.15-4.5 keV bands, the detailed surface brightness profile and the line strengths of Mg, Si, S, and Ar; the distance was allowed to vary by a factor of two about a preferred value of 5 kpc. By using an eigenvalue method of solution for the matrix formed from the coupled system of rate equations expressing the time development of the ionization structure, Hughes and Helfand were able to calculate a self-consistent ion distribution in the plasma for a series of models with different explosion energies, ejected masses, and ambient densities. After evolving each model to an age of 376 years (the time since the SN at the epoch of the observation), they could then calculate the emission from each species, integrate along each line of sight through the remnant, and produce both a surface brightness profile and an integrated X-ray spectrum. The results for two best-fit models are shown below; further details may be found in the referenced paper.

The above cited work concludes that, barring the unlikely circumstance of a rather dense interstellar cloud surrounding the site of Kepler's SN 600 pc above the galactic plane, the progenitor star of this event had a mass of $\gtrsim 6.5\ M_\odot$, at odds with the view of Kepler as a Type I event originating in a Pop II system. It is not our point here to defend such a conclusion; indeed, Hamilton et al. (1985 preprint) have modelled the SSS spectrum of Kepler assuming the deflagration of an O Ne Mg white dwarf and obtain an equally acceptable fit (although their model does not simultaneously match the diameter and surface brightness constraints). This numerical evolution model is cited as but one illustration of the sorts of constraints on remnant progenitors which modern X-ray observations can provide.

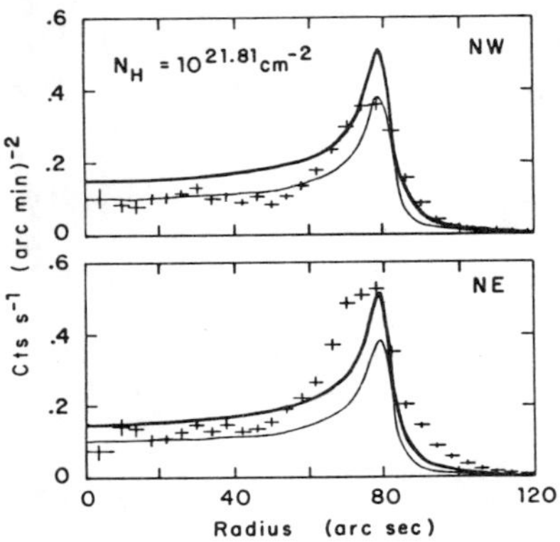

Figure 1:

Fits for two remnant models in which the strong limb brightening seen in the data (+'s) is due to the outward moving blast wave (thin lines) or the reverse shock (thick line).

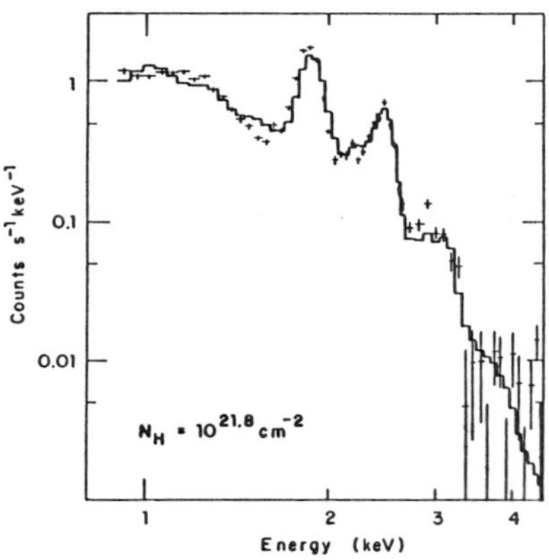

Figure 2:

The model X-ray spectrum for the reverse shock model with modified elemental abundances (solid histogram) compared to the *Einstein* SSS spectrum of Kepler's SNR (+'s).

See Hughes and Helfand 1985 for further details.

III.3.4 SNR Population Studies. As noted above, one of the principal limitations of galactic SNR studies is the large uncertainty in remnant distances. In addition, a drawback to modelling one remnant at a time is that individual peculiarities (e.g. local ISM irregularities) may mask more general features (e.g., are there just two well-defined progenitor mass ranges?). An obvious solution to both problems which has recently become practical is to study an entire population of remnants in an external galaxy of known distance. As noted above, the *Einstein* Magellanic Cloud surveys have turned up over 40 remnants, with many more likely to emerge from continuing analysis of the X-ray data. By measuring such parameters as X-ray temperature, luminosity, and diameter for the entire sample, one can constrain remnant models by requiring that they reproduce the observed parameter distributions as well as the combination of parameters appropriate to each individual remnant. Hughes, Helfand and Kahn (1984) and Fusco-Femiano and Preitze-Martinze (1984) have begun such a program by analyzing the number-diameter relation for LMC objects from which they draw conclusions concerning the distribution of SN explosion energies, interstellar environments, and stages of evolution characteristic of the population. Additional work along these lines may be found in Hughes (1984). Extension of the surveys to other Local Group members by the next generation of X-ray detectors will allow a tremendous advance in our understanding of remnant origins and evolution.

III. CONSTRAINTS ON THE EXPLOSION MECHANISM

III.1 Introduction

While X-rays dominate the radiative energy output of a SNR and can be used to infer much about the substantially greater kinetic energy in the explosion, it must be borne in mind that the entire remnant is but a 1% effect: the collapse of a stellar core releases $\sim 10^{53}$ ergs, most of which is radiated away in neutrinos and gravitational radiation. Thus, in order to study the physics of the explosion itself, it behooves us to find observational constraints more intimately tied to the dynamics of the event. An obvious choice is the fate of the stellar core: does it collapse to a black hole, form a neutron star, or disrupt completely? What are the mass, angular momentum, and magnetic field of any collapsed remnant? The answers to such questions, which are beginning to emerge from the search for compact objects in SNRs, are the subject of this section.

III.2 Review of Observational Results

The Crab Nebula is the prototype for the class of supernova remnants which clearly leave behind an active neutron star. It is also one of the rare examples where a historical SN can be directly associated with both a SNR and a rotation-powered pulsar. The Crab was the topic of lectures presented by Woltjer (this volume) so our discussion will be brief. The explosion occurred approximately 900 years ago. Today we observe a synchrotron nebula surrounding the pulsar and powered by the pulsar. The radio, optical, X-ray, and γ-ray properties of the Nebula are distinctive. The radio spectrum is relatively flat compared to that typical for shell remnants (but is steeper than spectra for most other Crab-like remnants). The nebula is brightest towards its center and shows no evidence for any limb brightening or surrounding shell structure. The radio emission is strongly polarized demonstrating its nonthermal nature. The energy spectrum of the radiating relativistic electrons extends to very high energies as evidenced by the presence of polarized diffuse X-ray emission. The X-ray spectrum is a power law, which combined with the polarization, proves that the diffuse emission at all wavelengths is nonthermal. In addition, the presence of X-ray emission requires the continued injection of relativistic particles into the Nebula, and there is a general consensus that the Crab pulsar is the source of these particles. Approximately 10% of the neutron star's rotational kinetic energy loss is radiated by the Nebular particles. The Nebula, then, serves as a marker of a SN event in which core collapse resulted in the formation of a rapidly spinning, magnetized neutron star.

With the Crab Nebula and its pulsar as our template, we can search for analogs to this remnant both within and outside of our Galaxy and, perhaps, learn something about the statistics of core collapse and neutron star formation. This search started slowly, hindered by the limited sensitivity and resolution of our instrumentation. In the past few years, however, the pace has quickened. In Tables 2 and 3 we summarize the current status of the search, listing those sources which seem most certain. In particular, Table 2 records the radio properties of the Crab-like nebulae; all but one were first detected in the radio band. We have included a total of fourteen sources. Note that this list is not exhaustive in the sense that at least half a dozen other sources have been suggested by other researchers (e.g. G27.8+0.6, G308.7+0.0, G328.4+0.2) which in time may prove to be Crab-like. In fact, the completeness, or lack thereof, of Table 2 is an important question to which we will return.

Table 2a:
Radio Properties of Crab-like Nebulae

Source	Distance (kpc)	Size (arcmin) Core	Size (arcmin) shell	Spectral Index Core	Spectral Index Shell
G0.9+0.1	?	2	8	0.0	-0.4
G20.0-0.2	?	12	-	0.0	-
G21.5-0.9	5	1.3	-	0.0	-
G24.7+0.6	?	5	13	?(-0.2)	?
G29.7-0.3	19	0.5	3	0.0	-0.7
G68.9+2.8 (CTB80)	3	6×10	40	0.0	-0.6
G74.9+1.2	>12	6×9	-	-0.24	-
G130.7+3.1 (3C58)	>3	6×10	-	-0.09	-
G184.6-5.8 (CRAB)	2	5×7	-	-0.26	-
G263.9-3.3 (VELA)	0.5	110×210	260	-0.1	-0.6
G291.0-0.1	8	7	-	-0.3	-
G320.4-1.2	4	-	30	-	-0.3
G326.3-1.8	5	8×15	36	-0.1	-0.4
0540-693 (LMC)	50	?	?	0.35	

Table 2b:
Radio Properties of Crab-like Nebulae (cont.)

Source	S(1 GHz) Core	S(1 GHz) Shell	L_{Radio} (10^{34} erg s^{-1}) Core	L_{Radio} (10^{34} erg s^{-1}) Shell	References
G0.9+0.1	6	8	?	?	Helfand & Becker (1985)
G20.0-0.2	10	-	?	-	Becker & Helfand (1985)
G21.5-0.9	6.4	-	1.5	-	Becker & Szymkowiak (1981)
G24.7+0.6	? (20)	?	?	?	Reich et al. (1984)
G29.7-0.3	.3	10	1.6	5	Becker & Helfand (1984)
G68.9+2.8 (CTB80)	2	100	.2	1.4	Angerhofer et al. (1981)
G74.9+1.2	8.6	-	≥5.4	-	Wilson (1980)
G130.7+3.1 (3C58)	33	-	≥2.1	-	Wilson and Weiler (1976)
G184.6-5.8 (CRAB)	1000	-	16	-	Wilson (1972)
G263.9-3.3 (VELA)	1100	650	2.0	0.3	Weiler & Pangia (1980)
G291.0-0.1	20	-	6	-	Wilson (1985)
G320.4-1.2	-	70	-	3.6	Manchester & Durdin (1983)
G326.3-1.8	40	100	7	6.6	Clark et al. (1976)
0540-693			4	7	Mills et al. (1984)

Of the 14 sources in Table 2, six are classic Crab-like SNR in that there is no discernible shell surrounding the filled-center nebula. The absence of shells in these objects is a serious difficulty with the standard picture as there is no reason to believe that the original SN did not eject high velocity material. In fact, serious searches for shells have been reported for only two of these six objects, the Crab nebula and 3C58 (Reynolds and Aller 1985). The remaining eight objects in Table 1, the composite remnants (Helfand and Becker 1985), consist of a central Crab-like nebula surrounded by the more usual remnant shell.

The radio data can in many cases be supplemented by X-ray observations, although substantial absorption by the interstellar medium below several keV limits the distance to which they can be observed. Since the X-ray telescopes flown to date have been insensitive above 3-4 keV, detections are difficult beyond 5 kpc. The current status of the X-ray observations are summarized in Table 3. All the images reported have been made with the *Einstein* Observatory IPC and HRI. The spectra were obtained with either the *Einstein* SSS experiment or the EXOSAT ME detector except in the cases of the Crab Nebula and the Vela pulsar.

Table 3
X-ray Measurements

Source	Spectal Index	$L_x(10^{35})$ ergs s^{-1}	Compact Source	L_x/L_R
G21.5-0.9	0.7±.1	3		20
G29.7-0.3	0.8±.2	40		350
G68.9+2.8 (CT80)		0.2	yes - ?	1
G74.9+1.2		0.9		1
G130.7+3.1 (3C58)	1.25±.25	0.1	yes - ?	0.5
G184.6-5.8 (CRAB)	1.1±.05	230	yes - pulsar	150
G263.9-3.3 (VELA)	1.2±.1	.05	yes - pulsar	.3
G291.0-0.1		0.5		1
G320.4-1.2		2	yes - pulsar	750
0540-693	0.8±.2	80	yes - pulsar	200

III.3 Implications of Crab-like Remnant Statistics

The question of whether shelled and unshelled Crabs are the result of different kinds of SN events is interesting to consider. Likewise, we can ask whether the size or luminosity of a Crab-like nebula depend on whether or not a shell is present, or, contrariwise, whether the shell around a Crab-like nebula differs from an empty shell. Examination of Table 2 reveals two trends. Sources without shells tend to have steeper radio spectra and tend to be more radio luminous. It is easy to imagine how both of these distinctions could result from selection effects. A steeper spectrum core would be harder to distinguish from the shell emission. On the other hand, a low luminosity Crab-like nebula without a steep spectrum shell would be harder to discover. If these selection effects are real, it implies the existence of a significant number of as yet unrecognized Crab-like objects.

We discern no obvious trends in the X-ray luminosities or spectra. Alternatively, one can search for correlations between the X-ray and radio luminosities by considering the radio of L_x/L_R. The values in Table 3 show no obvious dependence on the presence or absence of a shell. In fact, there is no convincing evidence that any of the properties of a Crab-like remnant depend on whether or not a shell is present. Therefore, we shall treat all the objects as members of a single class of which now includes 14 members.

We begin by considering sample selection effects, two of which have already been mentioned: the difficulty of detecting weak shelless Crabs against the galactic plane and our inability to distinguish a relatively steep-spectrum member of the class from a surrounding shell. In fact, any central component containing $\lesssim$ 1% the flux of the shell would be difficult to detect. All these effects could call into question any estimate of the percentage of SNR which have Crab-like components. Add to this the premises that 1) some pulsars may be born rotating slowly, never developing synchrotron nebulae, and 2) that a synchrotron nebula, once formed, may fade more rapidly than the associated shells (Battacharya, this volume), and one sees good reason to be cautious. However, let us proceed.

While only four active young pulsars have been detected within the Crab-like components in Table 2, the upper limits on the existence of such objects in the other members of the class are very weak, and we feel justified in assuming that each Crab-like nebula signals a pulsar birth. The converse of this statement, that each pulsar birth is heralded by a bright synchrotron nebula is less certain, although we note that the presence of such nebulae surrounding seven out of eight of the pulsar with energy loss rates $>10^{33.5}$ ergs s^{-1} (Helfand 1983) is supportive of such a contention. In principle, then, Crab-like SNR could provide a reasonable estimate of the pulsars' birthrate, as well as of the space distribution and nature of their progenitors.

Table 4 gives the z distances of the Crab-like SNR for which distances can be estimated. We find an average value of ~140 pc, significantly greater than that for all remnants and greater than the scale height usually given for pulsars at birth or for their presumed OB star progenitors.

One possible explanation for this z distribution is that Crab-like SNR originate from the second explosion in a runaway binary system which has acquired a high velocity in the first explosion (see van den Heuvel, this volume). If the high z values result from selection effects against discovering a low latitude Crab-like SNRs, one would expect the combination SNR to have a significantly lower average z value, since steep spectrum shells are easier to find than the naked Crab-like nebula; no such effect is present, however. The z distribution is remarkably flat out to z = 250 pc. To turn it into an exponential distribution with a scale height of 60 pc would require dozens of low latitude sources which a careful survey of the Galactic plane could reveal.

The current studies of the Crab-like SNR have interesting implications for pulsar beaming factors. Of the 14 sources in Table 1, six contain compact X-ray sources, most discovered by X-ray images. It seems likely that better X-ray images will reveal additional compact sources. If we identify all these with pulsars, then it would appear that the beaming factor is $\leqslant 2$. This is consistent with the general absence of pulsars or pulsar tracers in most SNR, since a significant decrease in the beaming factor reduces the birthrate of pulsars below that of Type II SN, (Lyne, this volume) suggesting that most of these SN do not leave behind pulsars.

Table 4
Spatial Distribution of Crab-like SNR

Source	Distance	z(pc)
Crab	2	200
G21.5-0.9	5	75
3C58	3	160
G74.9+1.2	7l2	>250
Vela	.5	40
G29.7-0.3	21	110
G326.3-1.8	5	160
MSH15-52	4	85
CTB80	3	150

IV. CONCLUSION

In conclusion, it would appear that a careful examination of the statistics of Crab-like and composite remnants, supplemented by high quality maps of known SNRs and a systematic search of the Galactic plane for heretofore overlooked members of this class, can tell us a great deal about the fraction of SNe which produce active neutron stars. The method of distinguishing among the remaining remnants -- which left slowly rotating or unmagnetized neutron stars, black holes, or nothing at all -- is less clear. Nonetheless, constraints on the parent stellar populations from study of the thermal shells, coupled with progress in delimiting the explosion mechanism through analysis of core remnant statistics, promises to stimulate rapid progress in the coming years toward an understanding of the progenitors and products of supernovae.

The authors would like to thank Professor Pacini for his organization of the NATO Advanced Study Institute and for partially supporting their attendance. The work described here was funded under NASA contracts NAG 8-501 (RHB) and NAG 8-497 (DJH), NSF grant AST83-13685 (RHB), Air Force grant AFOSR82-0014 (DJH), and the Alfred P. Sloan Foundation (DJH).

References

Angerhofer, P.E., Strom, R.G., Velusamy, T., and Kundu, M.R. 1981, Astron. Astrophys. **94**, 313.
Aschenbach, B. 1985, Sp. Sci. Rev. **40**, 447.
Baade, W. and Zwicky, F. 1934, Proc. Nat. Acad. Sci. **20**, 254.
Becker, R.H. and Helfand, D.J. 1985, Nature **313**, 115.
Becker, R.H. and Helfand, D.J. 1985, Ap.J. (Letters) **297**, L25.
Becker, R.H. and Helfand, D.J. 1984, Ap.J. **283**, 154.
Becker, R.H. and Szymkowiak, A.E. 1981, Ap.J. (Letters) **248**, L23.
Chanan, G.A., Helfand, D.J., Spinrad, H., and Elsneter, K. 1985, Nature (submitted).
Clark, D.H., Green, A.J., and Caswell, J.L. Astr. J. Phys. Ap. Suppl. **37**, 75.
Fresco-Femiano, R. and Preitze-Martinez, A. 1984, Ap.J. **281**, 593.
Green, D.A. and Gull, S.F. 1984, Nature **312**, 527.
Greenschild, E.H.B.n. and Mewe, R. 1982, Astr. Ap. Suppl. **48**, 305.
Hamilton, A.J.S., Sarazin, C.L., and Chevalier, R.A. 1983, Ap.J. Suppl. **51**, 115.
Helfand, D.J. 1983, Proceeding of IAU Symposium No. 101 "Supernova Remnants and Their X-ray Emission", P. Gorenstein and I.J. Danzinger, ed., (Dordrecht: D. Reidel), p. 471.
Helfand, D.J. 1984, PASP **96**, 913.
Helfand, D.J. and Becker, R.H. 1985, in preparation.
Helfand, D.J. and Becker, R.H. 1984, Nature **307**, 215.
Helfand, D.J., Chance, D., Becker, R.H. and White, R.L. 1984, A.J. **89**, 819.
Hughes, J.P. 1984, Ph.D. dissertation, Columbia University.
Hughes, J.P. and Hefand, D.J. 1985, Ap.J. **291**, 544.
Hughes, J.P., Helfand, D.J., Kahn, S.M. 1984, Ap.J. (Letters) **281**, L25.
Itoh, H. 1977, Publ. Astr. Soc. Japan **29**, 813.
Manchester, R.N. and Durdin, J.M. 1983, IAU Symposium 101, p. 421.
McKee, C.F. 1974, Ap.J. **188**, 335.
Mills, B.T., Turtle, A.J., Little, A.G., and Durdin, J.M. Astr. J. Phys. **37**, 321.
Pisarski, R.L., Helfand, D.J., and Kahn, S.M. 1984, Ap.J. **277**, 710.
Reich, W., Furst, E., and Sofue, Y. 1984, A.A. **133**, L4.
Reynolds, S. and Aller, H. 1985, A. J. **90**, 2312.
Shull, J.M. 1982, Ap.J. **262**, 308.
Weiler, K. and Panagia, N. 1980, A.A. **90**, 269.
Wilson, A. 1985, preprint.
Wilson, A. 1980, Ap.J. **241**, L19.
Wilson, A. 1972, M.N.R.A.S. **157**, 229.
Wilson, A. and Weiler, K. 1976, A.A. **49**, 357.
Winkler, P.F. et al. 1981, Ap.J. **245**, 574.
Woltjer, L. 1972, A.R.A.A. **10**, 129.

PLERIONS IN THEORY AND IN PRACTICE

M. Salvati
Istituto di Astrofisica Spaziale
C.P. 67, I-00044 Frascati, Italy (permanent address) and
European Southern Observatory, Garching bei München, FRG

ABSTRACT. The filled-centre SNRs (plerions) are the only observed connection between supernovae and pulsars, and are a key element in any unified picture of the end products of stellar evolution. Here we review the main features of the currently accepted plerion model, including some recent variants. The model allows a quantitative study of some individual objects, and of the statistical properties of the class.

1. INTRODUCTION

The filled-centre supernova remnants, or plerions, as proposed by Weiler and Panagia (1978), are a well established sub-class of the general SNR population. The definition, and a discussion of the individual cases - including several whose identification is uncertain - can be found in a series of papers by Weiler and co-workers (e.g., Weiler, 1983, 1984).

Briefly, however, a plerion differs from a normal, shell-type remnant in having the radio brightness maximum at the centre, rather than at the edges; a higher linear polarization and a flatter spectrum are additional identifying properties. Various plerions have been detected also at higher frequencies, namely in the X-rays and - in a few cases - in the optical. The brightness distribution is analogous to the radio one, and where spectral information is available it is indicative of non-thermal radiation processes. At a plerion's centre one often finds a point-like X-ray source (Helfand and Becker, 1984), which is sometimes seen to pulse in the X-ray themselves and/or in other frequency ranges (Seward and Harnden, 1982; Seward et al., 1984; Middleditch and Pennypacker, 1985).

The shell remnants are obviously interpreted as a signature of the shock wave, which propagates from the site of the supernova explosion. The filled ones, on the other hand, are thought to be produced and sustained by a central pulsar, with a high enough rate of rotational energy loss.

F. Pacini (ed.), High Energy Phenomena Around Collapsed Stars, 261–270.

Roughly one plerion out of two is in fact of a mixed type, and is associated with a more or less pronounced shell having normal properties. Including the composites, the entire class comprises only a small fraction of the total known SNRs. This scarcity has been interpreted as evidence that the plerion phenomenon is intrinsically rare, when compared with the production of shell remnants (see below).

Even if this were the case, however, one should remember that the known associations between pulsars and SNRs - without exception - are in fact associations between pulsars and plerions. They are, in order of discovery, PSR 0531+21 and the Crab nebula, PSR 0833-45 and Vela X, PSR 1509-58 and MSH 15-52 (Seward and Harnden, 1982), PSR 0540-69 and the homonymous remnant in the LMC (Seward et al., 1984). All pulsars and all SNRs are believed to have a common origin in SN explosions, with perhaps a discrimination between Type I and Type II SNe. The fact that the expected relation is observed only within the plerion subclass is certainly a clue to the formation processes and justifies a closer investigation of the plerion phenomenon.

In the following, we will review the model proposed by Pacini and Salvati (1973), with a brief description of the relevant formalism; the latter can then be applied to a quantitative analysis of filled-centre SNRs and analogous objects.

2. SOME THEORY

The model plerion is a bubble of relativistic fluid (relativistic particles plus magnetic field) which uniformly fills a spherical cavity. The cavity boundaries are provided by the SN ejecta and are assumed to be transparent to the radiation, on the one hand, and impermeable to the fluid, on the other. The assumption of uniformity within the bubble is partly justified by the smallness of the sound crossing time - comparable to the light travel time - with respect to the other time scales relevant to the problem. In principle, new and old particles are tied to central and peripheral lines, respectively, so that, even if the field is uniform, one should expect anisotropies and radial gradients arising from the conservation of the adiabatic invariant and from differential radiation losses. It is assumed that pitch angle and space diffusion are able to smooth out any large deviation.

The relativistic fluid is supplied by a central pulsar. The detailed mechanisms by which this is achieved are not known, and we must parametrize both the efficiency of conversion of the rotational energy into fluid energy and the shape of the injected electron distribution. The same situation arises in modelling the radio emission from shell-type remnants since it is not known how to describe the conversion from kinetic energy to turbulence, and from turbulence to the relativistic components needed for synchrotron emission.

After injection, we assume that no energy exchange occurs between the magnetic field and the relativistic particles, so that their evolution can be computed sequentially (for a different approach, see Reynolds and Chevalier, 1984). The field is taken to behave adiabatically, the electrons to suffer both from adiabatic and from synchrotron losses. The latter is perhaps an oversimplified description, since further loss mechanisms, or even acceleration mechanisms, could be in operation.

The quantitative set-up of the model proceeds as follows. The rate of rotational energy loss from the pulsar is described by a softened power law

$$- \dot{E}_{rot} = L = \frac{L_o}{(1 + \frac{t}{\tau})^{\alpha}} \quad (1)$$

This is equivalent to the usual assumption that the time derivative of the angular velocity, $\dot{\Omega}$, is proportional to Ω^n with $\alpha = (n+1)/(n-1)$. The initial luminosity, L_o, and slowing down time scale, τ, can be expressed as functions of the pulsar parameters if a specific braking mechanism is assumed. Note that several such mechanisms could be at work simultaneously, in which case the time behaviour of L would still be approximated by Eq. (1) with different parameters in different time intervals.

A certain fraction of the pulsar output is stockpiled within the plerion in the form of field and electron energy. We describe these two input terms as pL and qL, respectively, with $p+q < 1$. Bandiera et al. (1984) have considered the gravitational torque together with the usual electromagnetic torque, and in their case $p+q < L_{em}/L_{tot}$, which is <<1 close to the pulsar birth. If the injected electron spectrum is a flat power law up to a maximum energy E_{max}, then

$$j = K\,E^{-\gamma} \qquad E < E_{max} \qquad \gamma < 2$$

$$\int_o^{E_{max}} j\,E\,dE = \frac{K\,E_{max}^{2-\gamma}}{2-\gamma} = q\,L \quad (2)$$

All the parameters describing the conversion efficiency or the spectral shape could be arbitrary functions of time. It is plausible, for instance, that the maximum injected energy E_{max} varies as the pulsar slows down and the electromotive force declines (Bandiera et al., 1984). To date, however, most computations have not taken this possibility into account.

The bubble radius, R, must be known as a function of time. One might assume a kinematical law, usually a linear expansion, or an approximation to the Sedov regime (Panagia and Weiler, 1980).

Alternatively, one has to solve the proper dynamical problem, including the pressure of the relativistic fluid and the ram pressure of the swept-up matter (Maceroni et al., 1974; Reynolds and Chevalier, 1984). If the former approach is adopted, i.e., if the function $R(t)$ is known explicitly, the entire problem reduces to quadratures.

The evolution of the magnetic field is given by

$$W = \frac{R^3 B^2}{6}$$

$$\frac{dW}{dt} = -\frac{W}{R}\frac{dR}{dt} + p\,L \qquad \frac{d(WR)}{dt} = p\,R\,L \tag{3}$$

$$W = \left(p \int_0^t R\,L\,dt\right)/R$$

Knowing the field, one can compute the evolution of a single electron

$$\frac{dE}{dt} = -\frac{E}{R}\frac{dR}{dt} - a\,B^2 E^2 \qquad a = \text{constant}$$

$$\frac{d}{dt}\left(\frac{1}{ER}\right) = a\,\frac{B^2}{R} \tag{4}$$

$$\frac{1}{ER} - \frac{1}{E_i R_i} = f\,(t,\ t_i) = a \int_{t_i}^{t} \frac{B^2}{R}\,dt$$

The electrons which have energy E at time t were injected at various t_i's, but their injection energy had to be E_i as given by Eq. (4), so that they could evolve exactly into E from t_i to t. Thus

$$n(E,t) = \int_0^t K_i\,E_i^{-\gamma}\,\frac{\partial E_i}{\partial E}\,d\,t_i \qquad \frac{\partial E_i}{\partial E} = \frac{R_i\,E_i^2}{R\,E^2} \tag{5}$$

The synchrotron output of the plerion follows easily from Eq. (5) in the monochromatic approximation.

$$L_\nu \propto n(E,\ t)\,E\,B \qquad E \propto \left(\frac{\nu}{B}\right)^{1/2}$$

Sample solutions computed along these lines, or according to the variants discussed above, may be found in Pacini and Salvati (1973), Bandiera et al. (1984), and Reynolds and Chevalier (1984). The plerion can follow many different evolutionary paths, and it is difficult to give a compact, comprehensive description. The reason for this is to be looked for in the several time scales inherent to the problem: one has the slowing down time of the pulsar, the transition time between free expansion and the Sedov phase, the inverse rates of adiabatic and of synchrotron losses, proportional to t and to $B^{-2}\ E^{-1}$, respectively. By equating the two loss times one defines a break energy E_b - proportional to $B^{-2}\ t^{-1}$ -, which divides the expansion dominated and the synchrotron dominated regions of the spectrum. The time when $E_b = E_{max}$ is yet another characteristic time of the problem.

The various possible orderings of the characteristic times correspond to different domains of the parameter space, and give qualitatively different solutions. In most cases, the magnetic field evolves in such a way that E_b grows monotonically until it reaches E_{max}. Then, at early times, the particle distribution is convex, with the vertex moving together with E_b. The total distribution - i.e., the specific one multiplied by the plerion volume - grows faster at the larger energies, so that it appears to migrate up and right in the n-E plane. At late times the pulsar has already spent most of its energy, and the particle distribution diminishes. Even if the radiation losses are negligible, the high energy population declines faster, since it is sustained only by the declining pulsar input. Thus, the distribution is still convex, and its movement in the n-E plane is directed down and left.

A general consequence is that the ratio L_x/L_r of the X-ray to the radio emission reaches a maximum at about $t = \tau$.

3. COMPARISON WITH OBSERVATIONS

3.1. Individual objects

With the formalism developed in Eqs. (1) through (6) a quantitative study of plerions can be attempted. The obvious starting point is the Crab nebula (Bandiera et al., 1984), the class prototype, for which many of the model parameters can be determined independently. This applies to the pulsar properties, to the expansion velocity, and - to a lesser degree of approximation - even to the conversion efficiency. When the model is fitted to the radio data, the optical and the X-rays are discrepant, in the sense that the predicted luminosities are too large. The observed continuous bend in the Crab nebula spectrum cannot be explained entirely with evolutionary effects, instead, it must reflect some intrinsic bending already present at injection. The break frequency lies in the far infrared. Naively, by analogy with the classical solution of the impulsive injection problem ($\gamma < 2$, Kardashev, 1962), one could think that a bump develops there at $t \gtrsim \tau$. This is not found in the exact computations, though, and the bump observed by IRAS must be attributed to reprocessing by dust (Marsden et al., 1984). Finally, the secular decrease of the radio luminosity measured by Aller and Reynolds (1985) nicely fits the prediction of the theory. Note that the effect is quantitatively different from that expected in shell-type remnants (Shklovskii, 1960), because of the presence of the pulsar.

The association of PSR 1509-58 with MSH 15-52 has been questioned because the Σ-D relation implies a greater age for the remnant. If, instead, the remnant is given the same age as the pulsar, the expansion velocity becomes $\simeq 10^4$ km s^{-1}, which is high but still reasonable in a low density medium. With this velocity and with the observed pulsar parameters there is no difficulty in accounting for the X-ray measurements and for the radio upper limits (Srinivasan et al., 1984).

The pulsar-plerion association discovered in the LMC (0540-69.3, Seward et al., 1984) is extraordinarily similar to the Crab under all aspects. It is no surprise, then, if the predicted L_x/L_r ratio turns out to be higher than the observed one. The X-ray brightness distribution is known to be plerion-like, the radio observations instead lack the necessary angular resolution so that a mixed type cannot be excluded. It is then possible to attribute the observed excess radio emission to an unresolved shell component (Srinivasan and Battacharya, 1984).

The tendency of the model to overestimate the X-ray emission is confirmed when one considers the entire plerion class, rather than individual members (Reynolds and Chanan, 1984). In theory, the L_x/L_r ratio could be made arbitrarily small by tuning E_{max} so as to have the exponential synchrotron tail precisely in the X-ray window. Also, a quickly fading pulsar inside a slowly expanding plerion would produce a steep high-energy spectrum. This would be limited to a very narrow range, though, and again fine-tuning would be required. The discrepancy appears to be real and suggests more complicated injection and/or diffusion processes. Further class properties are elusive, since the plerions form a rather inhomogeneous class. The problem is well exemplified by the Σ-D diagram: Crab and Vela define an evolutionary path which is consistent with the theory (Bandiera et al., 1984), but the addition of other objects - even if one restricts oneself to the best known cases - blurs the correlation to almost nothing. The scatter is perhaps not surprising, given the great variety of environments and of pulsar properties which a plerion is likely to face. It casts however doubt on the reliability of population syntheses. There one has to model not only the plerion evolution, but also an initial distribution, widely spread in the parameter space.

3.2. Statistical considerations

The only straightforward issue about plerion statistics is the small number of objects. The plerions amount to about 10% of the total observed SNRs. Such a large discrepancy cannot be attributed solely to a difference in lifetimes. On the one side, the shell remnants appear to have ages T not much in excess of 10^4 years; even if one adopts the lowest birthrate ever proposed (150 years, Clark and Caswell, 1976), a population size of 150 implies T = 2 10^4 years. More recent analyses tend to bring T to even lower values, around a few times 10^3 years (Mills, 1983; Srinivasan and Dwarakanath, 1982). On the other hand, Vela has also T = 10^4 years, and the other plerions of known age would fall below the completeness limit in $T = 3\ 10^3$ years if they were to follow the model evolution. The argument can be refined by means of population synthesis (Srinivasan et al., 1984). Assuming a plerion birthrate comparable with the SNe, one estimates several tens of detectable objects, that is, an order of magnitude discrepancy as before.

A counter-argument has been put forward by Weiler (1984), who noted that plerions are more concentrated near the Sun than ordinary remnants, as if they were selected against in the existing surveys. The size of the selection effect - with perhaps some contribution from a small difference in the lifetimes - would be sufficient to remove the discrepancy.

Since the issue is not settled yet one has to live with two possible alternatives. In the first one, the plerion birthrate is comparable to either the SNR birthrate or the SN rate or the birthrate of pulsars. For the sake of the following discussion, it is important to note that agreement with the latter is obtained if a beaming factor of 5 is assumed (Lyne et al., 1985). Then a consistent picture can be put together. All Type II SNe leave behind powerful pulsars, which in turn produce plerions. Conversely, all pulsars are born very powerful, and go through the plerion phase at the beginning of their life. The majority of the observed pulsars are slow and weak because of evolution. They are pencil-beamed now, but beam evolution is suggested (Narayan and Vivekanand, 1983), and young, plerion-supporting pulsars could well be visible from all directions.

Apart from the initial assumption, this picture conflicts with the conclusions of Vivekanand and Narayan (1981). They analyze the pulsar distribution in the $P - \dot{P}$ plane and find evidence of very large initial periods, incompatible with the making of a plerion.

In the second alternative, plerions are indeed born at larger intervals than ordinary remnants. By the same token, the birth of fast, powerful pulsars is also a rare event. But the ratio of the high to the low birthrate is comparable with the canonical beaming factor of slow pulsars. It is still consistent to say that all pulsars are born fast and make plerions; only, they are born in 10% of the SN events, instead of 50% or 100% (Helfand, this volume). Then even slow pulsars would have fan beams, and the total number of neutron stars in the Galaxy would be lowered by about an order of magnitude.

The orthodox view holds that 50% or 100% of SNe make pulsars, only, most of them are born slow and weak, with just 10% being fast, powerful and Crab-like (Srinivasan et al., 1984). At least half of the non-plerionic remnants contain a non-plerionic pulsar, and the complete lack of associations outside the plerion sub-class - already stressed at the beginning - becomes here a crucial issue. We feel that a sensible solution cannot be found in beaming statistics. One must rather accept the possibility that the majority of neutron stars do not even function as pulsars just after birth. A suitable scheme has been proposed by Blandford et al. (1983). Here the neutron star is born unmagnetized, and the field needed for the pulsar mechanism is built up by an instability with a growth time much longer than the remnant lifetime.

Note that orthodoxy itself is flawed with contradictions. One is in fact concluding that, at birth, fast pulsars have strong fields, and vice-versa, which is plainly contrary to the intuitive picture of angular momentum removal during collapse.

3.3. Radio Supernovae

We come finally to the possible relevance of plerion theory to the phenomenon of radio SNe. Extensive reviews exist on the subject (e.g., Weiler et al., 1985; Weiler, this volume), and we will not duplicate them. We will only recall that, if the measured luminosities are extrapolated with their observed time laws, or - which amounts to the same - if a light curve is built up with radio SNe of different ages, one connects smoothly with the luminosities and ages of the plerions. The evolutionary law implied by this exercise varies inversely with time, with a slope somewhat smaller than 1. Immediate objections are the sudden disappearance of SN 1970g, and - perhaps - the incipient break in the light curve of SN 1979c. At any rate, the plerion theory can produce evolutionary curves which go through the same regions of the age-luminosity plane (Bandiera et al., 1984). A crucial issue is whether the SN envelope can break sufficiently fast to let the radio radiation out just after the explosion. If this were not the case, a newly born plerion would not be visible until about 10 years later.

The competing model avoids the problem by assuming the formation of a mini-shell remnant outside the envelope (Chevalier, 1982). Here the radio emission is expected to switch off in 10 to 30 years due to exhaustion of the circumstellar material, and to switch on again - after a poorly constrained interval of time - following the formation of the normal remnant (Gull, 1973).

All Type II SNe - and all Type I peculiar - appear to be radio SNe. Hence, if the connection between radio SNe and plerions were proven, it would be established that plerions occur frequently. A direct verification seems difficult - unless a pulsar is discovered at a SN site -, but a negative experiment is in principle possible: under the hypothesis of continuity between radio SNe and plerions, one expects in the Galaxy from a few to several intermediate age objects (10 to 30 years) more luminous than Cas A. If this were found to be untrue, the mini-shell model would prevail and the rarity of plerions would be established.

Searches at a fraction of a Crab luminosity would detect more aged objects. The brightness distribution would be needed in order to discern the shell remnants which would have formed by then. At this level, the radio SN statistics merge with the statistics of plerions, and we find anew the same order of mangitude discrepancy between expectations and observations.

4. CONCLUSIONS

The basic question to be answered is, whether the plerion list is or is not curtailed by selection effects. Also, it would be important to put firm limits on the pulsed emission from recognized plerions. Since one generally admits the presence of a pulsar there, this would amount to a measure of the beaming factor, at least in very active objects. And, finally, one should be reminded that, by general consensus, both pulsars and SNRs are the consequence of SNe, but we see together only the rare plerion-supporting pulsars and the equally rare plerion-type remnants. We still do not understand why we miss most of the action.

The author is grateful to R. Bandiera, F. Pacini, and N. Panagia for interesting discussions.

REFERENCES

Aller, H.D., and Reynolds, S.P.: 1985, in "The Crab Nebula and Related Supernova Remnants", eds. M.C. Kafatos and R.B.C. Henry, Cambridge University Press, p. 75.

Bandiera, R., Pacini, F., and Salvati, M.: 1984, Astrophys. J. 285, 134.

Blandford, R.D., Applegate, J.H., and Hernquist, L.: 1983, M.N.R.A.S. 204, 1025.

Chevalier, R.A.: 1982, Astrophys. J. 259, 302.

Clark, D.H., and Caswell, J.L.: 1976, M.N.R.A.S. 174, 267.

Gull, S.F.: 1983, M.N.R.A.S. 161, 47.

Helfand, D.J., and Becker, R.H.: 1984, Nature 307, 215.

Kardashev, N.S.: 1962, Sov. Astron. AJ 6, 317.

Lyne, A.G., Manchester, R.N., and Taylor, J.H.: 1985, M.N.R.A.S. 213, 613.

Maceroni, C., Salvati, M., and Pacini, F.: 1974, Astrophys. Space Sci. 28, 205.

Marsden, P.L., Gillett, F.C., Jennings, R.E., Emerson, J.P., de Jong, T., and Olnon, F.M.: 1984, Astrophys. J. 278, L29.

Middleditch, J., and Pennypacker, C.: 1985, Nature 313, 659.

Mills, B.Y.: 1983, in IAU Symp. 101 "Supernova Remnants and their X-ray Emission", eds. J. Danziger and P. Gorenstein, Reidel, Dordrecht, p. 551.

Narayan, R., and Vivekanand, M.: 1983, Astron. Astrophys. 122, 45.

Pacini, F., and Salvati, M.: 1973, Astrophys. J. 186, 249.

Reynolds, S.P., and Chanan, G.A.: 1984, Astrophys. J. 281, 673.
Reynolds, S.P., and Chevalier, R.A.: 1984, Astrophys. J. 278, 630.
Seward, F.D., and Harnden, F.R.: 1982, Astrophys. J. 256, L45.
Seward, F.D., Harnden, F.R., and Helfand, D.J.: 1984, Astrophys. J. 287, L19.
Shklovskii, I.S.: 1960, Sov. Astron. AJ 4, 243.
Srinivasan, G., and Bhattacharya, D.: 1984, Curr. Sci. 53, 513.
Srinivasan, G., Bhattacharya, D., and Dwarakanath, K.S.: 1984, J. Astrophys. Astron. 5, 403.
Srinivasan, G., and Dwarakanath, K.S.: 1982, J. Astrophys. Astron. 3, 351.
Vivekanand, M., and Narayan, R.: 1981, J. Astrophys. Astron. 2, 315.
Weiler, K.W.: 1983, in IAU Symp. 101 "Supernova Remnants and their X-ray Emission", eds. J. Danziger and P. Gorenstein, Reidel, Dordrecht, p. 299.
Weiler, K.W.: 1984, in "The Crab Nebula and Related Supernova Remnants", eds. M.C. Kafatos and R.B.C. Henry, Cambridge University Press, p. 227.
Weiler, K.W., and Panagia, N.: 1978, Astron. Astrophys. 70, 419.
Weiler, K.W., and Panagia, N.: 1980, Astron. Astrophys. 90, 269.
Weiler, K.W., Sramek, R.A., Panagia, N., van der Hulst, J.M., and Salvati, M.: 1985, submitted to Astrophys. J.

THE BINARY ORIGIN OF PULSAR VELOCITIES

V. Radhakrishnan and C.S. Shukre
Raman Research Institute
Bangalore 560 080, India

ABSTRACT. Analysis of a sample of pulsar proper-motions leads to the conclusion that most pulsars were born in and released from binary systems.

A minority of pulsars belong to a class characterized by low velocities and low magnetic fields. These were significantly spun-up by accretion and released from the binary through a non-explosive disruption of the companion. The rest belong to the high velocity, high magnetic field class. Using relations between the speeds at present and those in the binary orbit prior to its disruption, we show that such binaries must have undergone a common envelope phase of stellar evolution. The limits obtained on the masses and separations of these binary systems conform closely with those from stellar-evolution theory.

The linear velocities of pulsars have not received much attention, perhaps because they can be determined only after several years of observations. For single pulsars, only the proper-motions can be observed while for pulsars in binary systems, only the radial velocities can be determined. But even though one knows pulsar velocities incompletely, as we shall show much can be learnt from them.

The proper-motion of the Crab pulsar was estimated soon after its discovery to be ~ 100 km/s (Trimble, 1968). From this value and the pulsar Z-distances Gunn and Ostriker (1970) concluded that all pulsars must have high velocities ~ 100 km/s and that these may be the runaway velocities from binary systems which were disrupted. Subsequent radio determinations of pulsar proper- motions confirmed that pulsars were indeed high velocity objects (see references in Lyne et al., 1982). The best measurements due to radio interferometry give a range of 10 to 370 km/s (Lyne et al., 1982).

Theoretically, two other mechanisms have been proposed to explain the pulsar velocities. They are, the asymmetric supernova explosion (Shklovskii, 1970) and the 'rocket' mechanism (Harrison and Tademaru, 1975). On the basis of observations, these seem unlikely to be operative (for details see Radhakrishnan and Shukre, 1985 a, b).

F. Pacini (ed.), High Energy Phenomena Around Collapsed Stars, 271–278.

As there is no other satisfactory explanation to date for the velocities observed, we are left with the only alternative that all pulsars had an origin in binary systems (see also van den Heuvel, 1981b). Other indications in this direction are as follows. The majority of evolving stars today occur in binary or multiple systems (Abt and Levy, 1976, 1978). The difficulty of getting rid of the angular momentum in gas clouds which are collapsing to form stars appears to strongly favour the formation of binary or multiple systems (Bodenheimer, 1981). Many O and B stars which could be expected to later explode as supernovae and leave behind a pulsar are in 'runaway' binaries which acquired their velocities on the explosion of their companions; indeed, there seems to be little or no evidence that single stars of this category exist (Stone, 1979). We shall therefore proceed with the assumption that all single pulsars are released from binaries.

Observations also show a correlation between proper-motions and derived magnetic dipole moments (Anderson and Lyne, 1983). Most of this correlation finds a satisfactory explanation on the basis that the sample of pulsars is a mixture of two classes, one with 'high' velocities and the other with 'low' velocities (Radhakrishnan and Shukre, 1985 a, b). There is a residual correlation in the high velocity class alone, and this can be attributed to observational selection effects (Stollman and van den Heuvel, 1986).

Pulsars with high velocities have high fields and the low velocity ones have low fields and $P/2\dot{P} > 10^7$ yrs. The low field pulsars can be naturally identified with 'recycled' pulsars proposed by Radhakrishnan and Srinivasan (1981) (also later by Backus, Taylor and Damashek, 1982). They were spun-up in binary systems through accretion, and later released as single pulsars on the disruption of the binary. Their fields would be low due to decay during the evolution time of the companion stars, but the neutron stars would still be observable as radio pulsars due to their anomalously short periods (for the field). A necessary but not sufficient condition, which is an excellent diagnostic for such recycled pulsars, is that they must have characteristic ages $(P/2\dot{P}) > 10^7$ yrs <u>independent</u> of the decay time scale for the magnetic fields (Radhakrishnan 1984). Figure 1 which is a Log B - Log P plot shows the field-period equilibrium condition or 'spin-up' line (taken from Srinivasan and van den Heuvel, 1982) for accretion in binaries. All low field recycled pulsars should lie on or below it.

Now if pulsar velocities are due to their release from binary systems which are disrupted by symmetric supernova explosions, what are the implied characteristics of those binary systems? We assume that all the supernova explosion did was to reduce the mass of the exploding star on a time scale very short compared to the orbital period, without affecting the positions and velocities of the component stars of the binary. We also assume that the resulting orbit is unbound. The speeds are very soon expected to approach their asymptotic values which correspond to negligible gravitational attraction between them. We need a relation between the velocities just after (or what is the same thing, just before) the explosion and the asymptotic velocities,to infer something about the binary systems.

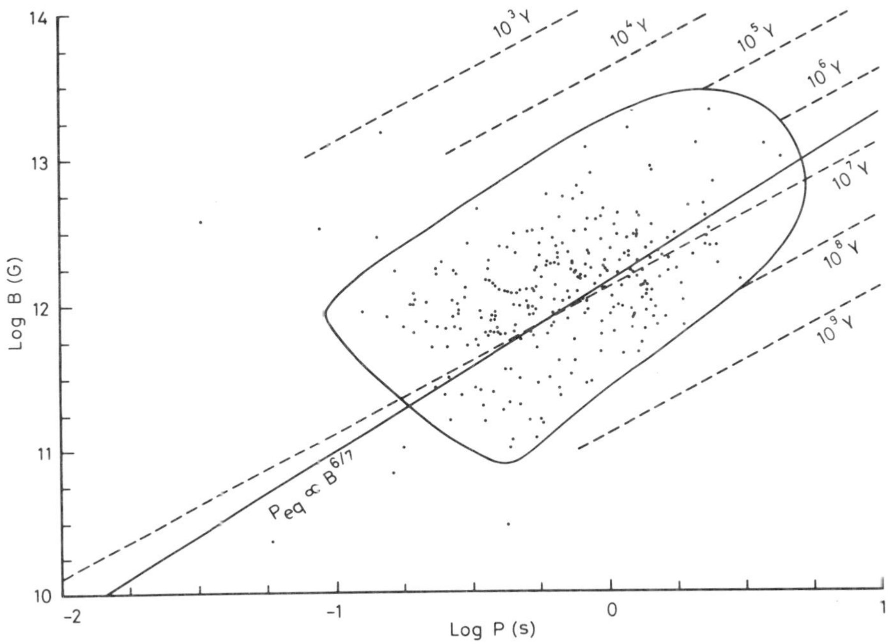

Fig. 1: Plot of derived magnetic field vs. period for pulsars. Solid line is the field-period equilibrium condition for spin-up given by Srinivasan and van den Heuvel (1982). Broken lines are the characteristic age $P/2\dot{P}$ = constant lines. [Figure adapted from Radhakrishnan (1982)].

Towards this aim, we must digress to mention a remarkable property of binaries which are disrupted by symmetric explosions. To an observer situated in the same frame of reference as the centre of mass of a binary system with a circular orbit, a sudden symmetric expulsion of mass from one of the stars leaves the speed of the remnant of this star unchanged for all time. Remarkably, this is true irrespective of the amount of mass lost, i.e., whether as a consequence the system remains bound, or is disrupted (for details see Radhakrishnan and Shukre, 1985b; Shukre, 1986).

Returning to our objective, the implication of this extraordinary property is that the progenitor of a pulsar had precisely the same speed in orbit (assumed to be circular) just before the explosion, as the pulsar has today. Ignoring for the moment the possibility of elliptical orbits, we can then conclude that the observed distribution of space velocities for the pulsars whose births disrupted the binaries was exactly the distribution of the orbital speeds of their progenitors. (This is not true for the other object in the binary system whose mass did not undergo any sudden change.) Here the motion of the binary system as a whole before the explosion is neglected as it is in any case a minor perturbation.

Below are the formulae for the observed speeds of the pulsar and its companion in terms of the progenitor speed in the circular orbit.

$$V_{psr} = V_{prog} \tag{1}$$

$$V_{comp} = [1 + (2\alpha - 1) (\frac{M_1 + M_2}{M_2})^2]^{1/2} V_{prog} \tag{2}$$

and

$$V_{prog} = \frac{438 M_2}{\sqrt{(M_1+M_2)a}} \text{ km/s} \tag{3}$$

with

$$\alpha > 1/2 \tag{4}$$

where M_1 is the mass of the pre-supernova object and M_2 that of the companion, in solar masses. The separation 'a' between M_1 and M_2 is in solar radii. The fraction of the original total mass lost in the explosion is α. Eqn. 4 just expresses the fact that the resulting orbit is unbound.

We can now use this to learn something about the binary systems from which the pulsars originated. A pulsar can be released as a single object either by the first or the second explosion in the binary system. There are many systems (e.g., X-ray binaries) containing a neutron star which must have clearly survived the first explosion, and at least one case (i.e., PSR 1913+16) which survived two. The conclusion that the first explosion will not disrupt the binary follows from the inevitability of mass transfer in such close binaries (Paczynski, 1971). We thus consider the release of pulsars due to the second explosion.

At this point it should be noted that binaries disrupted by the second explosion will produce two neutron stars. The younger one will have a lower velocity and a higher magnetic field which is exactly the opposite of what is observed. This clearly indicates that the velocities of our two classes require different scenarios to explain them. Indeed, the low velocities are too low to follow from eqn. (3) for masses appropriate for massive close binary systems. Consequently, we conclude that these binaries do not produce a pair of pulsars such that one of them belongs to the low velocity class and the other to the high velocity one.

We consider first the high velocity class. To estimate the space velocities we multiply the data of Lyne et al. (1982) by $\sqrt{3/2}$ to allow for the radial components in a statistical sense. The range of space velocities for the high-velocity pulsars thus obtained is 100-450 km/s with a mean of 270 km/s. As only the second explosion disrupts the system, the above formulae have the additional constraint that the mass of the companion is that of a neutron star, i.e., $M_2 \sim 1.4\ M_\odot$. We can go to

Figure 2 which depicts, for $M_2 = 1.4\ M_\odot$, the dependence of the separation between the stars on the total mass, as given by eqn. 3. Considering that speeds in the range 100-450 km/s are to be accounted for, we see that the separations have to be very small, and the dense cores of the collapsing stars have to be very compact indeed. This implies that the evolution of the binary system between the first and second SN explosions must go through a common envelope phase. The consequent tidal interactions would inescapably lead to circularisation of the orbits, fully justifying our earlier assumption to this effect.

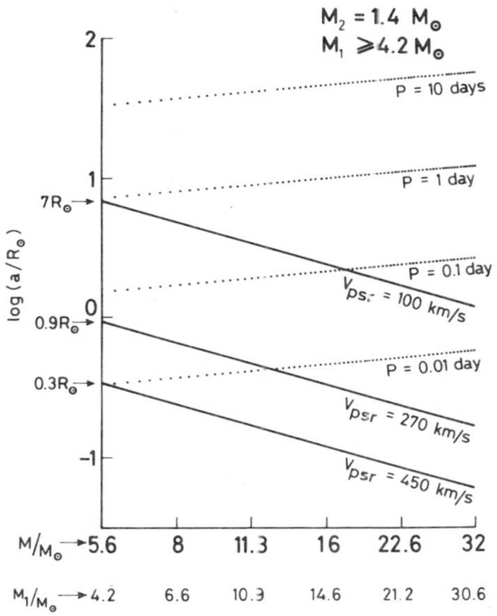

Fig. 2: Plot of separation in solar radii ($a/R_\odot$) between the stars vs. the total mass in solar masses ($M/M_\odot$) of the binary system, for different values of V(psr), as derived from eqn. 1. The companion neutron star mass M_2 is assumed to be $1.4\ M_\odot$. Numbers with arrows along the Y-axis indicate maximum separations appropriate for the observed range of velocities; they correspond to the minimum value of $M = 5.6\ M_\odot$ required to disrupt the binary. Dotted lines are of constant binary orbital period and are labelled by P_d, the period in days.

Our conclusion about the common envelope phase, drawn here purely on the basis of pulsar proper-motions, accords well with the conclusions from the point of view of stellar evolution in binaries (Tutukov, 1981; van den Heuvel, 1981a). However, it still remains to explain the magnitudes of the observed pulsar velocities for both classes. As already mentioned the low velocities must be related to some special property of binary systems which produce this class. These velocities could be explained naturally if originating in binary systems consisting of a neutron star and a companion which eventually disappears in a non-explosive manner. The single neutron star left behind will then have all the characteristics of a 'low-velocity' pulsar. The spiralling-in of the neutron star (Paczynski, 1976; Taam, Bodenheimer and Ostriker, 1978) during the common envelope phase will result in such an outcome if the core of the companion of the neutron star had not evolved to He-ignition at the start of the common envelope phase (Radhakrishnan and van den Heuvel, 1986). The velocity of the pulsar will be that of the system due to the kick received in the first explosion, i.e., ~ 30 km/s (van den Heuvel, 1983).

In most such binaries, however, the core would have evolved past the He burning stage. Thus, the more likely result of the spiral-in is an even closer binary than the initial one. Typically, the resultant binary will have a neutron star (~ 1.4 $M_\odot$) and a He (or more evolved) star. When the He star explodes as a supernova a pair of pulsars is released. Of this pair, the young pulsar will obviously have a high magnetic field. The old pulsar's field will also be high as the field decay time scale is of the order of the time interval between the two SN explosions, ~ 4 x 10^6 yr (van den Heuvel, 1977).

This outcome of the spiral-in thus leads to two high-velocity high-field pulsars after the second SN explosion. Our high-velocity class of pulsars is then a mixture of two subclasses. It should be pointed out here that half of these pulsars may be considered 'recycled' ones. However, because of the high value of their magnetic fields during the accretion phase the spin-up will not make their periods anomalously short.

While concluding from Figure 2 that binaries which produce pulsars must undergo a common envelope phase, we identified the velocities of high-field pulsars with the V(psr) of eqn. 1. However, if this class is indeed a mixture, the data on high-field pulsars would contain both V(psr) and V(comp) of eqns. 1 and 2, and it will not be possible to unscramble them.

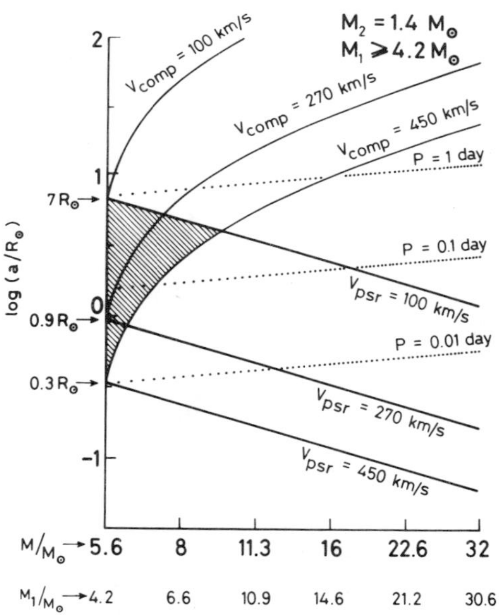

Fig. 3: Same as Figure 2 but including plots of separation a/$R_\odot$ vs. the total mass M/$M_\odot$ for different values of V(comp) as derived from eqn. 2. Shaded region indicates the values of 'a' and M allowed by the proper-motion data as explained in the text.

Figure 3 shows both V(psr) = const. and V(comp) = const. lines. In this plane, a binary system can be uniquely specified by a particular total mass and a particular separation prior to the second SN explosion, or alternatively by giving V(psr) and V(comp). The observational requirement that

100 km/s < V(psr), V(comp) < 450 km/s restricts the relevant binary systems to the intersection of the two bands. They are bounded by V(psr) = 100, 450 and V(comp)= 100, 450 km/s lines. In Figure 2 we had considered only the V(psr) band. We see now that the binaries in question must lie in the triangular overlap between these two bands. The true region would be still smaller if we could differentiate between V(psr) and V(comp). However, even the liberal ranges in Figure 3 put severe constraints on the binaries, e.g., if $M = 8\ M_\odot$, the separation must lie between $2.4\ R_\odot$ and $4.8\ R_\odot$. In any case, $3.3 < a/R_\odot < 6.7$ and $5.6 < M/M_\odot < 9.5$. We conclude that pre-supernova masses for the second neutron stars formed in binaries must lie in the range $4\ M_\odot$ to $8\ M_\odot$.

CONCLUSIONS

Pulsar proper-motion data indicate that all single pulsars were most likely born in and released from binary systems. If so, these binary systems must have undergone a common envelope phase of stellar evolution. The pulsars in the sample considered fall into two classes. One class consists of low-velocity pulsars which also have low magnetic fields and characteristic ages larger than 10^7 yr. These are old pulsars which were spun-up significantly during accretion from their non-explosively disrupted companions. Pulsars in the other class have high velocities and high magnetic fields. They must be produced in binary systems, which have separations less than $7\ R_\odot$ prior to the second explosion which disrupts the system. As one of the two stars was then already a neutron star, it implies limits of 4 to $8\ M_\odot$ for the pre-supernova mass of the companion. These conclusions drawn independently are totaly consistent with those from stellar evolution considerations. This provides compelling support in favour of a binary origin for all pulsars.

REFERENCES

Abt, H.A., Levy, S.G. 1976, Astrophys. J. Sup., 30, 277.
Abt, H.A., Levy, S.G. 1978, Astrophys. J. Sup., 36, 241.
Anderson, B., Lyne, A.G. 1983, Nature, 303, 597.
Backus, P.R., Taylor, J.H., Damashek, M. 1982, Astrophys. J. Lett., 255, L63.
Bodenheimer, P. 1981, in IAU Symp. 93: Fundamental Problems in the Theory of Stellar Evolution, Eds. D. Sugimoto, D.Q. Lamb & D.N. Schramm, Reidel, p.5.
Gunn, J.E., Ostriker, J.P. 1970, Astrophys. J., 160, 979.
Harrison, E.R., Tademaru, E.P. 1975, Astrophys. J., 201, 447.
van den Heuvel, E.P.J. 1977, Ann. N.Y. Acad. Sci., 302, 14.
van den Heuvel, E.P.J. 1981a, in IAU Symp. 93: Fundamental Problems in the Theory of Stellar Evolution, Eds. D. Sugimoto, D.Q. Lamb & D.N. Schramm, Reidel, p.155.
van den Heuvel, E.P.J. 1981b, in IAU Symp. 95: Pulsars, Eds. W. Sieber & R. Wielebinski, Reidel, p.379.
van den Heuvel, E.P.J. 1983, in "Accretion-Driven Stellar X-ray Sources", Eds. W.H.G. Lewin & E.P.J. van den Heuvel, Cambridge Univ. Press. p.303.
Lyne, A.G., Anderson, B., Salter, M.J. 1982, Mon. Not. R. astr. Soc., 201, 503.

Paczynski, B. 1971, Ann. Rev. Astron. Astrophys., 9, 183.
Paczynski, B. 1976, in IAU Symp. 73: Structure and Evolution of Close Binary Systems, Eds. P. Eggleton, S. Mitton & J. Whelan, Reidel, p.75.
Radhakrishnan, V. 1982, Contemp. Phys., 23, 207.
Radhakrishnan, V. 1984, in Proceedings of the Workshop held at 'NRAO', Green Bank, West Virginia, 6-8 June: Millisecond Pulsars, Eds. S.P. Reynolds & D.R. Steinbring, p. 130.
Radhakrishnan, V., van den Heuvel, E.P.J. 1986, (to be published).
Radhakrishnan, V., Srinivasan, G. 1981, in Proceedings of the Second Asian-Pacific Regional Meeting of the IAU, Bandung, Indonesia, 24-29 Aug.
Radhakrishnan, V., Shukre, C.S. 1985a, in Proceedings of the Third Asian-Pacific Regional Meetinq of the IAU, Kyoto, Japan, 1-6 Oct., 1984. Astrophys. Space Sci. 118, 329.
Radhakrishnan, V., Shukre, C.S. 1985b, in Proceedings of the Workshop on Supernovae and their Remnants, Bangalore, India, Oct. 29 - Nov. 2, 1984 (to be published by Indian Academy of Sciences, Bangalore).
Shklovskii, I.S. 1970, Astron. Zh., 46, 715.
Shukre, C.S. 1986, (to be published).
Srinivasan, G., van den Heuvel, E.P.J. 1982, Astron. Astrophys., 108, 143.
Stollman, G.M., van den Heuvel, E.P.J. 1986, (to be published).
Stone, R.C. 1979, Astrophys. J., 232, 520.
Taam, R.E., Bodenheimer, P., Ostriker, J.P. 1978, Astrophys. J., 222, 269.
Trimble, V. 1968, Astron. J., 73, 535.
Tutukov, A.V. 1981, in IAU Symp. 93: Fundamental Problems in the Theory of Stellar Evolution, Eds. D. Sugimoto, D.Q. Lamb & D.N. Schramm, Reidel, p.137.

BLACK HOLES IN OUR GALAXY

Martin J. Rees
Institute of Astronomy
Madingley Road
Cambridge CB3 0HA
ENGLAND

ABSTRACT. Some astrophysically-relevant properties of black holes are briefly reviewed. The physics of spherically-symmetric and disc-mode accretion flows is discussed, as well as the possibility of electromagnetic extraction of spin energy from Kerr holes. The results are relevant both to binary X-ray sources, and also to "lone" black holes which may abound in the Galactic Disc and Halo.

1. INTRODUCTION

Flow patterns around <u>neutron stars</u> are complicated by the effects of boundary layers, shocks and magnetospheres. When the central compact object is a <u>black hole</u>, these particular features are generally absent. However, there is a new complication: whereas the overall <u>efficiency</u> of accretion onto a neutron star (i.e. the amount of energy radiated for each unit mass that eventually comes to rest on the star's surface) is readily calculable, for black holes the situation is different. This is because radiation emitted deep in the potential well may itself be captured by (or advected into) the hole; moreover, a hole can, in some situations, swallow infalling material, along with its kinetic and internal energy, before more than a small fraction of this energy can be radiated.

After some order-of-magnitude considerations, I shall briefly review the main features of Schwarzschild and Kerr black holes. One would like, of course, to find some observational characteristic of X-ray sources that depends on the quantitative details of some inherently relativistic effect, but this still seems impossible even for the best-studied black accretion flows relevant to such objects, and conclude with some comments on the huge number of black holes (<u>not</u> in binary systems) that could exist in the galactic disc and halo.

F. Pacini (ed.), High Energy Phenomena Around Collapsed Stars, 279–296.

2. SOME GENERAL FEATURES OF RADIATION FROM ACCRETION FLOWS

The characteristic radius of a collapsed object is

$$r_g = \frac{GM}{c^2} = 1.5 \times 10^5 \frac{M}{M_\odot} \text{ cm} \qquad (1)$$

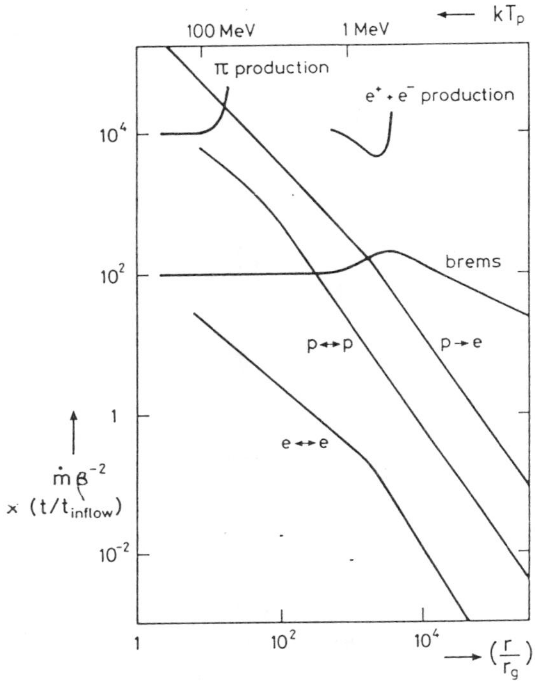

Figure 1. The time scales for various two-body plasma processes are here compared with the inflow time scale for an accretion flow. Processes shown are the self-equilibration time for electrons (e<>e) and protons (p<>p) (the latter includes nuclear as well as Coulomb effects at > 10 MeV); the time scale for transferring the proton thermal energy to the electrons (p→e); the bremsstrahlung cooling time for the electrons; and the effects of $e^+ + e^-$ and production. The proton temperature is taken as the virial temperature $KT = m_p c^2 (r/r_g)^{-1}$, but the electron temperature is assumed to vary as $r^{-\frac{1}{2}}$ when the electrons are relativistic. The diagram shows that for free-fall accretion at the "critical" rate ($\dot{m} = 1$), two-body cooling processes are inefficient, and the electrons and protons are not thermally coupled by Coulomb interactions when KT > 1 MeV. The ratio of the various time scales to t_{inflow} scales as $\dot{m}^{-1}$; for inflow at β times the free-fall speed, the ratio scales as β^2 for given $\dot{m}$ (adapted from Rees (1984).

It is a basic prerequisite for any modelling to know how the various cooling and microphysical time scales compare with the dynamical time. The latter can be written (for $r \gg r_g$) as

$$t_{inflow} = \beta^{-1} (r_g/c) (r/r_g)^{3/2} . \qquad (2)$$

The parameter β, equal to one for free-fall, is introduced at this stage because the numbers all scale straightforwardly to cases (with $\beta < 1$) where the inflow is impeded by rotation or by pressure gradients. (In deriving these characteristic numbers, the flow is treated as sperically symmetric: although this is roughly true for thick tori, further geometrical factors obviously enter for thin disks as discussed in section 4.)

The value of $\dot{M}$ needed to supply a luminosity L with efficiency ε can be written as $\dot{m}\dot{M}_E = (\frac{L}{L_E})\, \varepsilon^{-1}$, where $\dot{M} = L_E/c^2$. The particle density at radius r corresponding to an inflow rate $\dot{m}$ is

$$n \simeq 10^{19}\, \dot{m}\, \beta^{-1} (M/M_\odot)^{-1} (r/r_g)^{-3/2}\ \mathrm{cm}^{-3} \qquad (3)$$

Another quantity of interest is the Thomson optical depth at radius r, which is

$$\tau_T \simeq \dot{m}^{-1} (r/r_g)^{-\frac{1}{2}} . \qquad (4)$$

The "trapping radius", within which an accretion flow would advect photons inward faster than they could diffuse outward [i.e. within which $\tau_T > (c/v_{inflow})$] is

$$r_{trap} = \dot{m} r_g . \qquad (5)$$

Note that this depends only on $\dot{m}$ and not on β.

Figure 1 shows the ratios of various physically important time-scales to t_{inflow} for a radial free-fall with $\dot{m} = 1$, calculated on the assumption that the ions at each radius are at the virial temperature [i.e. $kT_i = m_p c^2 (r/r_g)^{-1}$]. This assumption is self-consistent because bremsstrahlung cooling and electron-ion coupling are indeed ineffective for $\dot{m}\, \beta^{-2} = 1$.

The magnetic field could build up toa value such that the magnetic and kinetic energy densities were comparable. This implies

$$B \lesssim B_{max} \simeq 4 \times 10^{8}\, \dot{m}^{\frac{1}{2}}\, \beta^{-\frac{1}{2}} (M/M_\odot)^{-\frac{1}{2}} (r/r_g)^{-5/4}\ \mathrm{Gauss} \qquad (6)$$

Note that these fields are far below those of neutron stars (the latter are, of course, anchored in the stars themselves, rather than requiring associated currents in the inflowing material).

A characteristic feature of the plasmas with $KT_i >> m_e c^2$ is that the electron-ion coupling time is so long that equality of the electron and ion temperatures is not guaranteed. For low $\dot{m}$, the collisional mean free paths for each species may exceed r (see Figure 1), though even a very weak magnetic field would suffice to make the inflow fluid-like.

A dynamically-significant magnetic field would be strong enough for synchrotron losses to be important. Suppose that the magnetic energy is q times the rest mass density of the plasma: we might expect $q \lesssim KT_i/m_p c^2$ for accretion flows. The ratio of the cyclotron cooling time (neglecting absorption) to the bremasstrahlung time for a sub-relativistic electron is $\alpha_f (m_e/m_p)\ q^{-1}\ (kT_e/m_e c^2)^{-\frac{1}{2}}$, which is << 1 for a plasma with $kT_i \gtrsim$ MeV with an equipartition field; for ultrarelativistic electrons the dominance of synchrotron losses over bremsstrahlung is even greater. Analogously, Compton losses can be very important; indeed, in any source where Thomson scattering on electrons (or positrons) yields $\tau_T > 1$, the requirement that the Compton y-parameter be $\lesssim 1$ implies that the electrons or positrons must be mostly subrelativistic.

The conventional distinction between thermal and nonthermal particles becomes somewhat blurred in these contexts where two-body coupling processes cannot necessarily maintain a Maxwellian distribution. Various acceleration mechanism (relativistic shocks, reconnection, etc.) may, moreover, boost some small fraction of the particles to high γ: such mechanisms operate in many contexts in high-energy astrophysics and should be even more efficient in an environment where the bulk velocities and Alfvén speeds are both ~ c. These particles would then emit synchrotron or inverse Compton radiation. This radiation would extend into the γ-ray band, in which case pair production (via $\gamma + \gamma \rightarrow e^+ + e^-$) could ensue if the "compactness parameter" L/r were high enough. These pairs would not only modify the spectrum of the radiation (cf. Guilbert et al. 1983), but their effect on τ would modify the effective Eddington limit and the trapping radius (equation (5)).

3. BLACK HOLES ACCORDING TO GENERAL RELATIVITY

3.1 The Metric

The gravitational potential is of Newtonian "1/r" form for $r \gg r_g$. However, the energy release, and the interesting dynamical effects, are concentrated in the inner parts of the flow, where the effects of general relativity must be allowed for.

If the angular momentum of a black hole is zero, its metric is given by Schwarschild solution, which can be written

$$ds^2 = -\left(1 - \frac{2r_g}{r}\right) c^2\, dt^2 + \left(1 - \frac{2r_g}{r}\right)^{-1} dr^2 +$$

$$+ r^2 (d\theta^2 + \sin^2\theta\, d\phi^2) . \qquad (7)$$

The coordinate radius r is defined so that the proper area of a sphere of radius r is $4\pi r^2$. The gravitational redshift factor for a body stationary at radius r is $(1 - 2\, r_g/r)^{-1/2}$; this becomes infinite as r approaches the event horizon (Schwarzschild radius) at $r_s = 2r_g$.

The more general Kerr metric, corresponding to a hole of mass M and angular momentum J, can be written

$$ds^2 = -\frac{\Delta}{\rho^2} (c\, dt - a \sin^2\theta d\phi)^2 + \frac{\sin^2\theta}{\rho^2} \left[(r^2 + a^2)\, d\phi - ac\, dt\right]^2$$

$$+ \frac{\rho^2}{\Delta} dr^2 + \rho^2 d\theta^2 , \qquad (8)$$

where

$$\Delta = r^2 - 2r_g r + a^2 ,\ \rho^2 = r^2 + a^2 \cos^2\theta ,\ a = J/r_g c .$$

The quantity a specifies the ratio of angular momentum to mass, and the Kerr metric is therefore a one-parameter family of solutions (with $J = 0$ reducing to the Schwarzschild case). Note that this metric has a dt dϕ term. This reflects the fact that it is no longer STATIC though it is still STATIONARY (in the sense that the metric components are themselves independent of t). The presence of the dt dϕ term reflects the fact that space near the hole is, in a sense, being "dragged around" with the hole. The event horizon in the Kerr solution occurs when = 0 and has a radius

$$r_E = r_g - (r_g^2 - a^2)^{1/2} . \qquad (9)$$

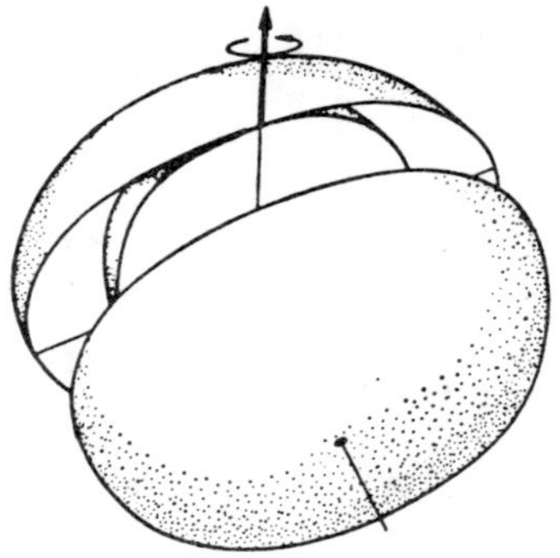

Figure 2. Schematic view of Kerr geometry. The region between the surface of stationarity (flattened figure of revolution) $r = r_{star}$ and the horizon $r = r_E$ is called the "ergosphere". A particle entering the ergosphere and emitting a disintegration product can emerge with more energy (rest-plus-kinetic) than it had when it entered.

This horizon disappears if $J = J_{max} = GM^2/c$ (corresponding to $a = a_{max} = r_g$). The so-called "cosmic censorship hypothesis" would imply that holes always form with $J < J_{max}$. The Kerr solution has a critical radius called the STATIC LIMIT, within which particles must rotate with the same sense as the hole, relative to nonrotating, distant observers, even though they can still escape - this is a consequence of the fact that the dragging of inertial frames can be so strong that even light cones necessarily point in the ϕ direction. This critical radius is

$$r_{stat} = r_g + (r_g^2 + a^2 \cos^2\theta)^{1/2} \tag{10}$$

it corresponds to the radius at which g_{oo} vanishes. The region $r_E < r < r_{stat}$ is called the ergosphere because one can in principle extract energy from it (Penrose, 1969): if a particle enters the ergosphere and splits in two, in such a way that one fragment falls into the hole, then the other may leave the ergosphere with more energy than the original particle (energy being extracted from the hole). A Kerr hole can thus be considered to have two kinds of mass-energy (Christodoulou, 1970): a fraction associated with its spin, which can in principle be extracted via the Penrose process, and an irreducible mass. The fraction of the mass which can be extracted is

$$1 - 2^{-1/2} \left\{ 1 + \left[1 - (r/r_g)^2 \right]^{1/2} \right\}^{1/2} \tag{11}$$

This is 29% for a "limiting Kerr" hole with $a = a_{max}$.

Accretion-powered sources introduce a characteristic timescale

$$t_E = \frac{\sigma_T c}{4\pi G m_p} \simeq 4 \times 10^8 \text{ years} \tag{12}$$

This is the time it would take an object to radiate its entire rest mass if its luminosity were L_E. The time scale over which an accretion-driven source would double its mass is $(L/L_E)^{-1} \times (\text{efficiency})^{-1} \times t_E$.

The dynamical and orbital time scales near black holes are vastly shorter than t_E, so the mass of material in any stationary flow pattern around the hole is generally negligible compared to the mass of the hole itself - this vindicates us in postulating an undistorted Kerr metric and neglecting the self-gravity of plasma around the hole. (Radiation of rest mass on a time scale r_g/c yields a stupendous mass-independent luminosity of $c^5/g \simeq 3 \times 10^{59}$ erg s^{-1}).

3.2 Orbits in Schwarzschild Metric

The binding energy per unit rest mass of a circular orbit of radius r around a Schwarzschild black hole (with Schwarzschild radius $r_s = 2r_g$) is

$$\frac{c^2 r_s}{2r} \left(\frac{r - 2r_s}{2r - 3r_s} \right). \tag{13}$$

For $r \gg r_s$, this reduces to $GM/2r$, which is just the Newtonian orbital energy. However, the binding energy has a <u>maximum</u> of 0.057c at $r = 3r_s$. Circular orbits closer in than this have <u>more</u> angular momentum and are less tightly bound (as for orbits in Newtonian theory when the effective force law is r^{-n} with $n > 3$: the orbits have zero binding energy for $r = 2r_s$; and for $r = 1.5r$ the expression goes infinite (implying that photons can move in circular orbits at this radius).

The qualitative forms of <u>non</u>-circular orbits in a Schwarzschild metric are as follows. Bound orbits near the hole are rosette-shaped (the periastron precession, a small effect in the Solar System where planetary orbits have radii $> 10^7\, r_g$, being important even in a single orbital period). The periastron of bound orbits must be outside $r = 2r_s = 4r_g$ (because a particle in a non-circular orbit has, at periastron, more energy than a circular orbit with the same r); and the apastron of such orbits must be outside $r = 3r_s = 6r_g$. Even unbound orbits must have periastrons outside $r = (3/2)r_s = 3r_g$.

As a curiosity, one notes that 29% efficiency is in principle possible for accretion onto a Schwarzschild hole. This could come about if a stream of material (of zero binding energy) fell towards the hole with angular momentum corresponding to a circular orbit at $r = 2r_s$. This stream would then, without loss of energy, follow such an orbit. If, however, it then collided with an equal and opposite stream, and the material came to rest, a fraction $(1 - 2^{-1/2})$ of the original mass - energy would be dissipated.

3.3 Orbits in the Kerr Metric

The energy per unit rest mass of a prograde circular orbit of radius r around a Kerr black hole (the orbit being in the equatorial plane) is

$$e = c^2 \left\{ \frac{r^2 - 2r_g r + a(r_g r)^{1/2}}{r[r^2 - 3r_g r + 2a(r_g r)^{1/2}]^{1/2}} \right\} \tag{14}$$

For corotating equatorial orbits in a Kerr mertric, the innermost stable orbit moves inward compared to the Schwarzschild case and becomes more tightly bound. As $a \to a_{max}$ the stable orbits extend inwards and the binding energy approaches $(1 - 1/\sqrt{3})c^2 \simeq 0.42c^2$.

These numbers determine the maximum theoretical efficiency of accretion disks. A related inference is that there are no stationary bound orbits whose angular momentum is less than a definite threshold value - this important qualitative feature of the orbits means the axisymmetric flow patterns cannot extend very close to the rotation axis.

An orbit around a spinning (Kerr) hole which does not lie in the equatorial plane will precess with an angular velocity $\approx 2(r/r_g)^{-3} r_g^{-1} \times c(a/a_{max})$ (Bardeen and Petterson (1975)). For $a \simeq a_{max}$ this precession has a time scale $t_{BP} \simeq (r/r_g)^{3/2} t_{Kep}$, when t_{Kep} is the orbital period. However, if material is spiralling slowly inwards (at a rate controlled by viscosity) on a time scale exceeding t_{BP}, then the precession effect can mount up. The important consequence follows that the flow pattern near a black hole may be axisymmetric with respect to the hole irrespective of the infalling material's original angular momentum vector. The Lense-Thirring precession, an inherently relativistic effect, thus guarantees that a wide class of flow patterns near black holes will be axisymmetric - an important simplification of the problem. The quantitative details of this effect are rather subtle (see, for instance, Kumar and Pringle (1985)).

3.4 Electromagnetic Effects

The rotational energy of a spinning black hole can in principle be extracted, as was first recognised by Penrose (1969). Astrophysically plausible mechanisms for extracting this energy depend on exploiting the remarkably close analogy between a black hole and an ordinary electric conductor. This analogy is most simply illustrated, for a Schwarzschild hole, by calculating the electric field due to a point charge held at rest near the hole (Hanni and Ruffini 1972). As the charge, with radial coordinate r_c, is moved closer to the Schwarzschild horizon, the field lines get progessively more distorted: they "wrap around" the hole so that as $r_c \to r_s$ they appear to emanate from $r = 0$, the field being essentially radial for $(r - r_s) >> (r_c - r_s)$. It is as though the charge has spread itself over the hole's surface. For a charge in free fall, the spreading happens in a time ~ (r_s/c).

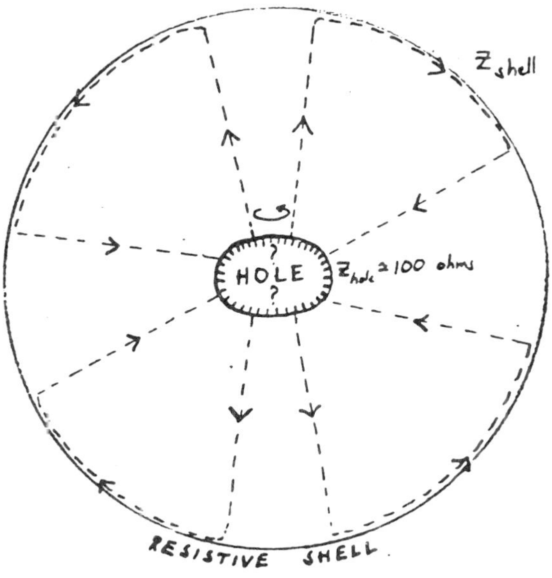

Figure 3. The unipolar inductor mechanism schematically depicted. A magnetic field (not shown) is applied to a spinning black hole surrounded by a much larger non-rotating conducting shell. A current system (dotted lines) is then induced, which dissipates energy in the hole (resistance $Z_{hole} \simeq 100$ ohms) and in the surrounding shell. The power generated in the shell (rather than going to waste down the hole) is maximised, for a given applied magnetic field, when $Z_{shell} = Z_{hole}$.

Comparing this with the "classical" estimate of the time $r_s^2/4\pi\sigma$ taken for charge to spread over a sphere of radius r_s and conductivity σ, we find that the effective resistance of a black hole is of order 100 ohms (cf. Znajek 1977).

A spinning (Kerr) black hole behaves like a spinning conductor (Blandford and Znajek 1977), in the sense that there are constraints on the orientations of any stationary electric and magnetic fields near the horizon. This analogy, spelt out in detail by Macdonald and Suen (1985), is sufficiently close that a "unipolar inductor" mechanism can indeed tap the spin energy of a hole.

Figure 3 depicts , very schematically, a unipolar inductor mechanism. For this to operate there must be:

(1) A magnetic field threading the hole (not shown in Figure 3).

(ii) Currents flowing in a circuit into and out of the hole.

(iii) A near-optimal impedance match, so that the currents dissipate a good fraction of their energy in the outer conducting sphere, rather than in the hole itself.

The maximum electric potential drop is ~ $B_o r_g((a/a_{max})$ where B_o is the imposed field. This can be very large, as it is when a similar argument is applied to spinning magnetised neutron stars in conventional models for pulsars. However, just as in this latter case, a realistic external current system and plasma distribution, though very hard to calculate, is likely to "short out" the field. General arguments show that, if currents are allowed to flow into the hole from larger radii the power dissipated can be up to $B_o^2\ r_g^2\ (a/a_{max})\ c$ (cf. an "inside out" pulsar light cylinder). If $e^+ - e^-$ plasma can be produced near the hole there could indeed be a current system flowing "into" and "out of" the hole. Power dissipated outside the hole comes, in effect, from the hole's spin energy or "reducible mass" .

The likely astrophysical contexts where this mechanism operates are further discussed in Section 4.3.

In all astrophysical situations the <u>gravitational</u> effect of charges and magnetic fields is negligible. When a gravitating system is electrically charged, the gravitational effect of the charge (and, more generally, of the electromagnetic field) modifies the Kerr metric to the more general Kerr-Newman form. The charge Q becomes significant if $Q^2 \simeq GM^2$, which requires a fractional deviation $\underline{f}$ from charge neutrality of order $\mathcal{N}^{-1/2}$, where $\mathcal{N} = e^2/Gm_p^2 \simeq 10^{36}$. Any gravitating plasma body (even the Sun) acquires a charge such that the electrical and gravitational binding energy of an electron are comparable: it is the resultant radial electric field which allows the electron pressure gradient (~½ the total pressure gradient) to be used to counteract the gravitational force on the ions (more by ~ m_p/m_e than that on the electrons). However the value of $\underline{f}$ needed to provide a surface potential of even 10^9 volts is $\simeq \mathcal{N}^{-1}$, and for this value, Q^2/GM^2 is itself only of order $\mathcal{N}^{-1} \simeq 10^{-36}$. Even if the hole is charged to the potentials $\sim 10^{16}$ eV discussed above, the value of Q^2/GM^2 is still only of order 10^{-23}. (A related statement is that the magnetic energy density in realistic accretion flows is $\lesssim 10^{-23}$ of $(M/r_g^3)c^2$). We are thus justified, by a large margin, in treating the space-time as given by the ordinary Kerr metric.

4. APPLICATION TO MODELS OF ACCRETION FLOW, ETC.

When "real" plasma surrounds a black hole (just as for a neutron star), we need to tackle the complexities of MHD, fluid-dynamical instabilities, and radiative transfer in order to calculate what is going on. I shall not go into this topic in any detail here, but will summarise some of the features of promising models. (Pringle (1981) gives an excellent review of accretion discs).

4.1. Thin Discs

Suppose that the central object is being fuelled steadily by the inflow of material (from, for instance, a companion star) which has a large amount of angular momentum in a well-defined direction. In the standard thin disc model, it is assumed that the gas at each radius is nearly in Keplerian orbit about the hole. Slow radial inflow occurs if there is a viscous stress which transfers angular momentum outward. The energy dissipated by the viscous stress is radiated locally at a rate three times the local rate at which gravitational energy is liberated ($GM \times \dot{M}dr/r^2$ between r and $r + dr$). The factor of 3 occurs because the viscous stress transports energy as well as angular momentum outward, this local imbalance being compensated in the innermost region of the disc, where the local release of binding energy exceeds the dissipation. For thin discs, slow inflow can be maintained down to the innermost stable orbit; within this radius, each element of gas can plunge into the hole without further dissipative energy release. The efficiency is then 5.7% for a Schwarzschild hole; it is higher for co-rotating orbits around a Kerr hole (cf. (12) and (13)).

The weak point in the modelling of thin discs is the specification of a viscosity law. Ordinary molecular viscosity is almost certainly too small to be relevant and one must appeal to magnetic or turbulent viscosity, the nature of which is not well understood. Although the radial dependence of the viscosity law will not change the energetics for a disc with given $\dot{M}$, it will affect the column density through the disc and may change the "vertical" structure and the emitted spectrum.

Suppose that the viscous stress has the form $t_{\phi r} = \alpha p$, where p is the pressure in the equatorial plane (Shakura and Sunyaev 1973). If $\alpha < 1$ then the inflow is subsonic with respect to the gas in the disc and the approximation of hydrostatic equilibrium normal to the disc plane is valid. The ratio of disc thickness h to disc radius r (for $h \ll r$) is

$$\left(\frac{h}{r}\right)^2 \simeq KT_{gas} \left(1 + \frac{p_{rad}}{p_{gas}}\right) \left(\frac{r}{r_s}\right) \frac{1}{m_p c^2} \qquad (15)$$

This is essentially the ratio of thermal energy to gravitational energy. To maintain a thick disc ($h \gtrsim r$) at distances of the order r_s, requires either $P_{rad} \gg P_{gas}$ (i.e. $L \simeq L_E$) or $KT_{gas} \simeq 100$ MeV.

Magnetic fields, as well as (maybe) determining the viscosity, can have a big effect on the radiation spectrum emerging from a realistic thin disk. Energy transported by magnetic buoyancy into a hot corona could dominate the (approximately blackbody) radiation from the dense part of the disk. Magnetic flares in the corona may accelerate relativistic electrons that radiate nonthermally. If the magnetic field were sufficiently well ordered, a coronal wind (rather than outward transfer via viscosity within the disk itself) could be the main sink for the angular momentum of accreted material (Blandford and Payne 1982).

The thermal balance of the outer parts is likely to be controlled by irradiation (causing photoionization, Compton heating, etc.) from the central region. Even where thin disks exist, they could be embedded in hotter quasi-spherical structures. There may thus be no sharp demarcation between thin discs, and the toroidal structures which I shall review next.

4.2 Thick Discs, "Donuts" or Tori

A radiation pressure-supported disc becomes geometrically thick in its inner parts (i.e. $h \simeq r$ in (14)) when the accretion rate exceeds the critical value $\dot{M}_E$ required to generate a luminosity L_E. A gas pressure-dominated disc would become thick in the opposite limit - that of low accretion rate - because the cooling time varies inversely with $\dot{M}$. The bulk of work to date has focused on the radiation-dominated case. However, virtually all of the arguments about structure and stability carry over to thick discs dominated by gas pressure.

As is true of thin accretion discs, the largest uncertainty in constructing a model of a thick disc is the nature of the viscosity. This uncertainty is not crucial to many of the qualitative features of thin disc theory (e.g. the energetics). However, in a thick disc one must deal explicitly with shear stresses in two directions. The stresses determine the distribution of both angular momentum and enthalpy, and therefore the shape of the isobars inside the disk. The isobars at the disk surface, however, depend on the surface distribution of specific angular momentum. This fact has enabled the Warsaw group (Abramowicz, Calvani and Nobile 1980 and references cited therein) to construct models of thick discs without reference to the viscosity. They specify, a priori the angular momentum distribution along the surface of the disc in such a way that it smoothly connects a cusp solution near the hole with a thin disc at sufficiently large radii. Although self-consistency and stability considerations constrain the acceptable forms of the angular momentum distribution to an extent which is still unknown (cf. Papaloizou and Pringle 1984, 1985), they may nevertheless allow one considerable freedom to specify the shape of the surface arbitrarily, and yield virtually no information about the interior. A typical example is shown in Figure 4.

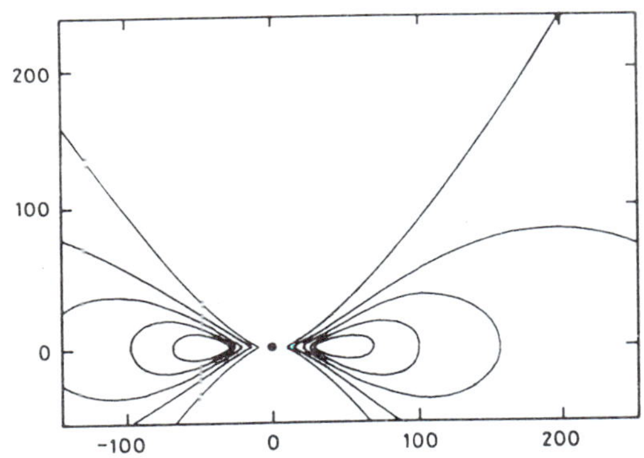

Figure 4. A typical disc or torus. The isobars are drawn for a case where the angular momentum in the equatorial plane varies as $r^{2/5}$ (i.e. it increases with r slower than the Keplerian dependence), and obeys the von Zeipel condition. The hole is assumed to have a Schwarzschild metric. The empty funnel along the rotation axis would be narrower for a Kerr hole, and for angular momentum distributions that increased still more slowly with r. The length scale is given in multiples of r_g.

Because radial pressure gradients are important, the rotation velocity in the equatorial plane will be Keplerian just at one radius - the radius at which the pressure attains a maximum. Outside this maximum it is sub-Keplerian; closer to the hole, the orbital speeds in the equatorial plane exceed Keplerian. Matter can flow into the hole from a cusp, lying somewhere between the orbit of maximum binding energy (innermost stable orbit for a test particle) and the circular orbit of zero binding energy. The shape of the surface depends on the angular momentum distribution. For the limiting (and highly unstable) case when the angular momentum is constant, the evacuated vortex is bounded by a funnel of roughly paraboloidal shape. In other cases, the funnel is wider. If the effective viscosity were very high (perhaps because of serious non-axisymmetric instabilities) the profile would resemble that shown in Figure 4 only in a time-averaged sense, and there could be large-amplitude variations on a dynamical timescale.

4.3 Inefficient Accretion and Electromagnetic Power Extraction

A generic feature of accretion tori is that they are less efficient - in the sense that they liberate less energy per gram of infalling matter - than thin disks. The efficiency is given by the binding energy of the material at the cusp; this depends on the angular momentum profile but can be very low for "bloated" tori.

In any torus with a strongly sub Keplerian rotation law, rotation is unimportant (gravity being essentially balanced by pressure gradients, and the isobars almost spherical) except near the funnel along the rotation axis. The situation resembles the quasi-spherical inflow discussed in section 2, the parameter β being equivalent to the viscosity parameter α . To avoid convective instability, the density must fall off with radius at least as steeply as the isentropic laws $n \propto r^{-3}$ for $\gamma = 4/3$ (e.g. radiation pressure support), and $n \propto r^{-3/2}$ for $\gamma = 5/3$ (e.g. ion pressure support).

If $\dot{m} \gg 1$, the expected luminosity would still be only $\sim L_E$. The efficiency would fall roughly inversely with $\dot{m}$: the inner cusp would move inward towards the circular orbit of zero binding energy, and the surface would "puff up" (with the consequence that the spectrum would soften as $\dot{m}$ increased). If the effective viscosity in such tori were low (and the peak density correspondingly high) internal H-burning could significantly augment the inefficient release of gravitational energy.

Accretion can also be inefficient in the opposite limit when $\dot{m} \to 0$. This is because the material may then be unable to radiate its binding energy on the inflow timescale; the energy dissipated via viscous friction is then retained as internal energy. In fact, for a given low $\dot{m}$ there may be two accretion modes - one would be a thin disc, supported by gas pressure; the other a thick ion-supported torus. Which of these flow patterns gets established would depend on whether the material is fed in via a stream in a well-defined plane, or whether the infall is more chaotic and isotropic. In the latter case, shocks would form, after which the gas would be unable (cf. section 2) to cool and establish a thin disc. A thick ion-supported torus yields an accretion-powered luminosity that is low for two (multiplicative) reasons - because $\dot{m}$ is small and because ε is low. The most conspicuous effect of the hot plasma around the hole may then be to catalyse extraction of the hole's energy via electromagnetic torques. In section 3.4 three requirements were outlined; how can these be met when the hole is surrounded by a thick torus of diffuse plasma at temperatures exceeding 1 MeV?

(i) Conducting plasma near the hole can carry the currents that maintain a magnetic field; even a very low level of accretion suffices for this.

(ii) This same plasma can indirectly supply the charges that carry electric current "into" the hole. Dilute plasma near a black hole would radiate inefficiently, and would be at a temperature of >1 MeV (the viral temperature for $r<10^3\ r_g$ being higher than this). Its low-level radiative emission would then include bremsstrahlung gamma rays. Some of these will interact very close to the hole, yielding a cascade of electron-positron pairs, with more than enough charge density to 'complete the circuit' and carry the necessary current - enough, indeed, to make the magnetosphere essentially charge-neutral, in the sense that $(n^+ + n^-) >> |n^+ - n^-|$, so that relativistic MHD can be applied.

(iii) The issue of the impedance match, and the consequent efficiency, is rather more subtle. Phinney (1983) has explored the physics of the relativistic wind whose source is the pair plasma created by $\gamma + \gamma \rightarrow e^+ + e^-$ in the hole's magnetosphere, and which flows both outward along the funnel and into the hole. He finds consistent wind solutions in which up to 0.2 of the hole's spin energy is transformed into Poynting flux and a relativistic electron-positron outflow.

Models along these lines have been applied primarily to extragalactic radio sources e.g. Rees et al. (1982). However, a scaled-down version could apply to stellar-mass holes that accrete magnetised plasma at a low rate; the strongest power output from such objects could be relativistic $e^+ - e^-$ winds energised by the black hole's spin.

5. HOW MANY BLACK HOLES MIGHT THERE BE IN OUR GALAXY ?

Massive stars may end their lives by collapsing to black holes. It is unclear how the formation rate of black holes compares with that for neutron stars (this depends on the minimum mass of a star that forms a black hole); but the ~ 3 X 10^8 supernovae that have exploded in our Galactic Disc since it formed could have left ~ 10^8 black holes of

~10 $M_\odot$. More speculatively it has been proposed that the Galactic Halo could consist of black holes of 10^2 - 10^6 $M_\odot$ - remnants of an early stellar generation (Population III).

The entire halo mass could not be in holes less massive than ~10^2 $M_\odot$ without overproducing heavy elements. These are expelled from massive stars in their terminal phases unless the stars are so massive that they end their lives not by exploding, but by collapsing to black holes after the pair-production instability (Truran and Cameron 1971; Woosley and Weaver 1982; Carr, Bond and Arnett 1984 and references cited therein). Collapse is thought to occur for core masses above 200 $M_\odot$. If the halo mass were in VMOs, then there must be a cutoff above 10^6 $M_\odot$, because dynamical friction would have increased the velocity dispersion of disc stars to an excessive degree (Carr 1978; Lacey 1984; Ipser and Semenzato 1985, Lacey and Ostriker 1985). Within the context of VMO theories, we have little evidence on whether the preferred mass is closer to 10^2 or to 10^6 $M_\odot$.

The upper mass limit could be pushed downward if we had a firmer understanding of what luminosity would result from accretion onto black holes passing through the Galactic Disc. Black holes in the mass range 10^2 to 10^6 $M_\odot$ (VMO remnants) could reveal their presence by accretion of surrounding gas. The accretion rate for supersonic motion at speed V through gas of density n is proportional to nV^{-3} M^2. The luminosity depends on the accretion rate, and also on the efficiency ε. For spherical accretion, where, as discussed in section 2, the efficiency is low because the radiative cooling time is long compared to the free-fall time, ε should scale with M, making the luminosity proportional to M^3. For disc-like accretion, ε may be as much as 0.1, independent of M. The spectrum of the emergent radiation is also uncertain. The case of spherical inflow has been considered by Ipser and Price (1977, 1982), who argue that the radiation emerges mainly in cyclotron harmonics from the thermal electrons whose temperatures are marginally relativistic. This emission would peak in the infrared. The Ipser/Price estimates may be rather conservative, because they do not allow for a possible non-thermal tail of relativistic electrons. Disc-type accretion could yield a high luminosity predominantly thermal radiation in the ultra - violet.

The most conspicuous halos would be those which were passing through dense gas clouds, and which had V much less than the mean velocity. Although, for a Maxwellian distribution, the fraction with speed $< V$ scales with V^3, the resultant higher $\dot{M}$ ($\propto V^{-3}$) makes these specially slow-moving holes more readily detectable.

Unfortunately, the distinctive signature of an accreting black hole is hard to estimate, and so one cannot at the moment place firm limits on the number (or mass) of putative halo objects of this kind.

(For fuller discussions, see Ipser and Price (1982), Lacey and Ostriker (1985) and McDowell (1985). Lower mass of faster-moving holes, accreting at a rate corresponding to very low $\dot{m}$, might release energy primarily via a low-level version of the electromagnetic process described in section 4.3.

Finally, I would like to mention one very special object in our Galaxy - the strange object that lurks at its centre. There is a compact (and variable) radio source at the Galactic Centre (Lo et al. 1985), with dimensions ~ 3×10^{14} cm or less. Another peculiar phenomenon towards the Galactic Centre is a strong electron-positron annihilation line, which there is reason to believe is variable. (The variability is claimed on the basis of obtaining different results on balloon flights six months apart). This can be modelled in terms of low-level accretion onto a $\lesssim 10^6 M_\odot$ hole (Rees 1982).

REFERENCES

Abramowicz, M., Calvani, M. and Nobili, L. 1980. Ap.J., 242, 772.
Bardeen, J.M. and Petterson, J.A. 1975. Ap.J. (Lett)., 195, L65.
Blandford, R.D. and Znajek, R.L. 1977. MNRAS, 199, 883.
Blandford, R.D. and Znajek, R.L. 1977. MNRAS, 179, 433.
Carr, B.J. 1978. Comm. Astrophys., 8, 161.
Carr, B.J., Bond, J.R. and Arnett, W.D. 1984. Ap.J., 277, 445.
Christodoulou, D. 1970. Phys.Rev.Lett., 25, 1596.
Guilbert, P.W., Fabian, A.C. and Rees, M.J. 1983. MNRAS 205, 593.
Hanni, R.S. and Ruffini, R. 1972. In 'Black Holes' ed. B. De Witt and C. De Witt, p. R75 (Gordon and Breach).
Ipser, J.R. and Price, R.H. 1977. Ap.J., 216, 578.
Ipser, J.R. and Price, R.H. 1982. Ap.J., 255, 651.
Ipser, J.R. and Semenzato, R. 1985. Astron.Astrophys., 149, 408.
Kumar, S. and Pringle, J.E. 1985. MNRAS, 213, 435.
Lacey, C.G. 1984. In 'Formation and Evolution of Galaxies and Large Structures in the Universe' eds. J. Audouze and J. Tran Tranh Van (Reidel, Dordrecht).
Lacey, C.G. and Ostriker, J.P. 1985. Ap.J. (in press).
Lo, K.Y., Backer, D.C., Ekers, R.D., Kellermann, K.I., Reid, M. and Moran, J.M. 1985. Nature, 315, 124.
MacDonald, D.A. and Suen, W-M. 1985. Phys.Rev.D., 32, 848.
McDowell, J.C. 1985. MNRAS (in press).
Papaloizou, J.L.B. and Pringle, J.E. 1984. MNRAS 208, 721.
Papaloizou, J.L.B. and Pringle, J.E. 1985. MNRAS 213, 799.
Penrose, R., 1969. Riv.Nuovo.Cim., 1, 252.

Phinney, E.S. 1983. Cambrdige Ph.D. Thesis (unpublished).
Pringle, J.E. 1981. Ann. Rev.Astr.Astrophys., 19, 137.
Rees, M.J. 1982. In 'The Galactic Centre' eds. R. Blandford and G. Riegler, p. 166 (Amer. Inst. Phys. N.Y.)
Rees, M.J. 1984. Ann.Rev.Astr.Astrophys., 22, 471.
Rees,M.J., Begelman, M.C., Blandford, R.D. and Phinney, E.S. 1982. Nature, 295, 17.
Shakura, N.I. and Sunyaev, R.A. 1973. Astron.Astrophys., 24, 337.
Truran, J.W. and Cameron, A.G.W. 1971. Astrophys.Sp.Sci., 14, 179.
Woosley, S.E. and Weaver, T.A. 1982. In 'Supernovae: a Survey of Current Research' eds. M.J. Rees and R. Stoneham (Reidel, Dordrecht).
Znajek, R.L. 1977. MNRAS, 179, 457.

"GAMMA - RAY SOURCES IN OUR GALAXY"

G.F. Bignami
IFC/CNR
15, via Bassini
20133 Milano, Italy

QUESTION: La poser c'est la resoudre
(G.FLAUBERT: "Dictionaire des idées reçues")

1. ON GAMMA-RAY ASTRONOMY GENERAL, AND ON THE OVERALL EMISSION FROM OUR GALAXY

Astronomy with gamma-rays in the tens of MeV-few GeV energy region was started in the sixties, with the availability of reasonable space vehicles, in the hope of opening a new window of observational astronomy which was to shed new "light" on a variety of energetic astrophisical processes in compact objects as well as in the interstellar medium (ISM).

The energy region chosen was that characterized by the pair production process for the photon detection, giving an important asset in reducing the background by exploiting the unique electron-positron pattern signature of the high energy gamma-ray. As soon as, in the late sixties, technology allowed it, it was also possible to use in space the spark chamber, a detector developped for elementary particle/cosmic-ray work, offering the great advantage of its pictorial display of the materialization event.

Astrophysically, the predictions on gamma-ray astronomy were connected mostly with the production of photons from the short-lived neutral mesons produced in the p-p interaction of CR and ISM, in the hope of solving the CR origin problem, and with high-energy, electromagnetic processes taking place in few special celestial sources. As it happens, such predictions turned out to be quite optimistic on the quantitative side, although qualitatively correct. This resulted in a very slow progress towards significant results, owing clearly to photon limitation. In fact, the two major spark chamber experiments on which most of todays's Gamma Ray Astronomy (GRA) is based (SAS-2 and COS-B) were designed to have precisely the same geometric surface (600 cm2) and placed on board similar missions for one-to-two year duration, in the hope of meeting with high enough fluxes. In retrospect, it is fair to say that COS-B owns its success to having lasted over three times more than planned (while SAS-2 only lasted about a third of its planned life), and having thus has a chance to collect nearly 200,000 gamma-ray

F. Pacini (ed.), High Energy Phenomena Around Collapsed Stars, 297–316.

photons, it established the current reference GRA database.

Fig. 1 shows, in a classic logN-A.D. plot, the actual number of gamma-ray photons collected per mission, starting with the first detection of 31 extraterrestrial gamma-rays by the Explorer XI. The eye is tempted to see a smooth line joining the 4 available mission yields, probably giving too much significance to the issue; it will be nevertheless interesting to see if the next missions in the same energy range (GAMMA-1 and EGRET on GRO) will follow the trend. It is also timely to note that the 200,000 COS-B photons have, at the time of writing (Oct. '85), just been rendered public by ESA to anyone willing to undertake more in-depth analysis than was done by the "Caravane" Collaboration, responsible for the COS-B satellite.

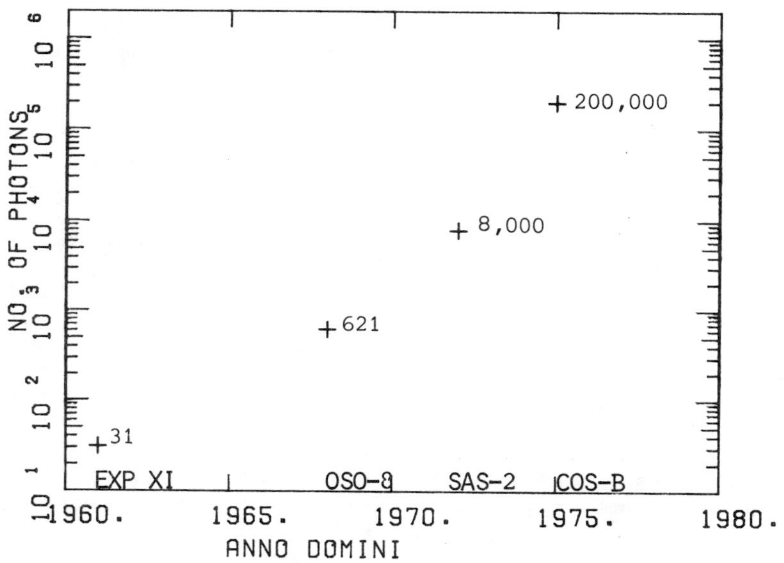

Fig. 1 - Number of gamma-ray photons collected per mission vs. the launch date.

How are these COS-B photons distributed in the sky? Fig. 2a shows, in an isophote plot, their flux distribution in galactic coordinates. As is known, the disc of our Galaxy stands well out above several tens of MeV owing to the diffuse emission from the CR-ISM interactions, also well visible at medium latitudes. One of the big results of COS-B has been to prove the nature of such emission by showing that cloud complexes can be well mapped in gamma-rays when, like in the case of Orion and Ophiucus, they are at high enough latitude. This point is in fact the focal one also for the

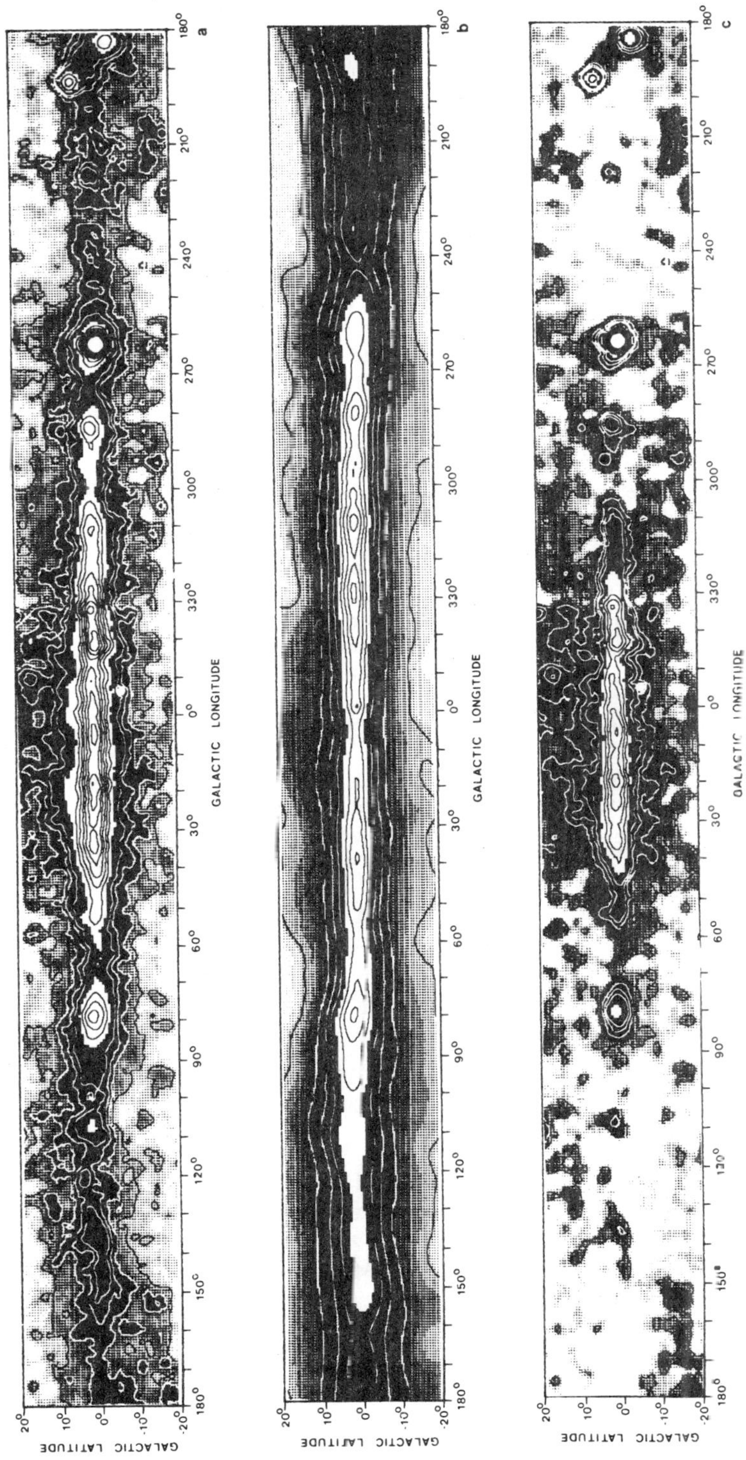

Fig. 2 - a) Gamma-ray intensity map for the energy range 70 MeV - 5 GeV.
b) Skymap of the expected gamma-ray intensity from atomic HI alone after convolution of the HI column densities with the COS-B Point Spread Function and with the average spectrum of the galactic gamma radiation in the same energy range.
c) Result of the 2a) - 2b) subtraction.

objects of major interest here, i.e. Gamma-Ray Soures (GRS). Clearly, their visibility improves dramatically when going away from the plane, and yet a glance to fig. 2a) shows that one must search for them at low-latitudes, i.e. against the intense diffuse emission from the plane, to be treated as a background. The standard approach to finding an unresolved source is the same in GRA as in all other astronomies: the available sky data are compared to the instrument's Point Spread Function, as determined by pre-and post-launch calibration, after having defined the level of the (more or less smooth) surrounding background. Those excesses that are, at the same time, significant above the background and compatible with the PSF are called GRS.

The problem is then shifted to the definition of such "background". Clearly the diffuse emission must be treated as such, and one could proceed to model it on the basis of the following factors: the gamma-ray yield per target atom when hit by a given CR spectrum, the distribution across the galaxy of the target IMS, plus, of course, some reasonable assumption on the large scale CR behaviour in the Galaxy. The gamma-ray yield (or emissivities) can be measured reasonably well from local (i.e. high-latitude) data, and found to be consistent with accelerator p-p interaction results; on the IMS distribution, however the data available until recently were the radio measurement of the 21cm HI line, only mapping the neutral hydrogen. An example of what one can do, on the basis of such observations, is given in fig. 2b), showing the result of the following process: the HI data have been folded with the gamma-ray emissivity, "degraded" in angular resolution to become comparable with GRA data and finally subtracted from fig.2a). The outcome (fig. 2c) is clearly those gamma-rays which cannot be accounted for by CR-HI interaction, and they represent a better data base in which to search for GRS. Historically, that is what happened, and that is basically how the COS-B GRS were discovered.

Presently, however, this is only part of the story, as will be seen in what follows, which shall deal with the successive generation of COS-B results on GRS. For recent reviews on both past and not so past work on GRS, including what discussed so far, as well as for a full list of references, the reader is referred to Bignami and Hermsen (1983, hereafter BH 83) and Bignami (1984).

2. FIRST GENERATIONS OF GRS, THE 2CG CATALOGUE AND SOME PROPERTIES OF THE GRS POPULATION.

At the end of the SAS-2 mission (see Fichtel et al. 75), the recognized GRS were few, but very important: the Crab pulsar and nebula (tentatively seen in previous balloon flights), the Vela pulsar, discovered by SAS-2, Cyg X-3, claimed on the basis of the orbital periodicity, and an unidentified but very significant excess in the anticenter, called "gamma 195+5". Armed with a complete coverage of the galactic disc, obtained well before the end of the mission, the COS-B team started on the task of preparaing a catalogue of sources.

The method chosen was that outlined above: to cross-correlate the galactic plane gamma-ray data, taking appropriately into account the instrumental and galactic backgrounds, with the instrument PSF. An idea of the PSF is given by the following analytical expression:

$$f(\Theta) \propto \exp\left[-\left(\theta/\theta_o(E_\gamma)\right)^{2c(E_\gamma)}\right]$$

where E_γ is the reconstructed photon energy and Θ is the incidence angle, and Θ_o and c have tabulated variations.

Such PSF corresonds to "error boxes" in the sky of the order of one degree radius (≥ 100 MeV), with small variations, depending on source photon statistics. After the first, partial CG (COS-B Gamma) source list of Hermsen et al (1977), the 2CG catalogue of Swanenburg et al. (1981) was constructed, and is shown in fig. 3.

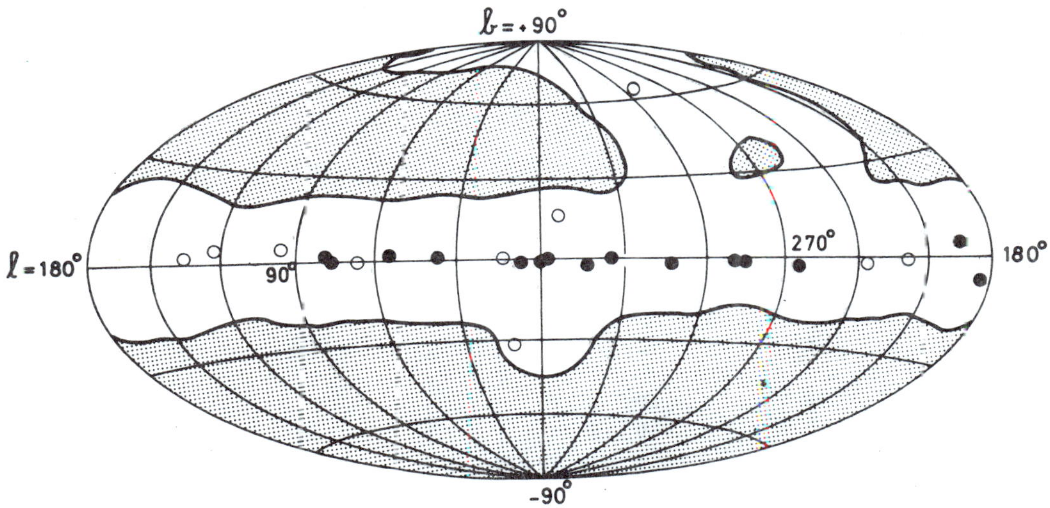

Fig. 3 - Distribution in galactic coordinates of GRS in the 2CG catalogue (Swanenburg et al., 1981). Open circles: sources with flux (> 100 MeV) ≤ 1.3×10^{-6} ph/cm^2 sec. Solid circles: sources with higher flux.

Before discussing the main body of the catalogue, i.e. the numerous (>20) unidentified sources shown, let us briefly comment on the few identified ones. The only high-latitude one (2CG289+64), has been identified with the local bight quasar 3C273 and will not be discussed in this galactic context. It is instructive, however, to note the procedure for its identification: in the 3°x3° error box it was noted that an X-ray source like 4U1226+02 (solidly identified with 3C273) had a ~2% probability of occurrence, hence 2CG289+64 ≡ 3C273, which was not the target of the COS-B pointing.

The gamma-ray emission from the Crab and Vela pulsars is extensively covered elsewhere in this book, and only some highlights from relatively recent COS-B data will be given here. Fig. 4 (taken fom Wills et al 82) shows several COS-B observations separately: an indication of variability in the shape of the light curve, namely the pulse/interpulse ratio, is present at the per-cent level. For the Vela Pulsar, fig. 5 shows an example of the detailed analysis possible with the ample COS-B statistics (see Kanbach et al 1980): the PSR 088-45 data

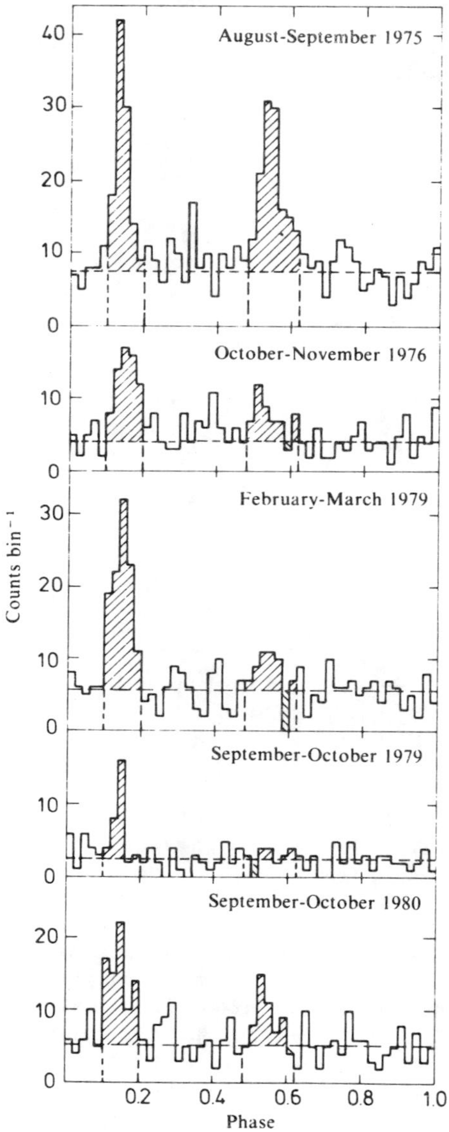

Fig. 4 - Time history of the COS-B observations of the Crab pulsar (see Wills et al 82)

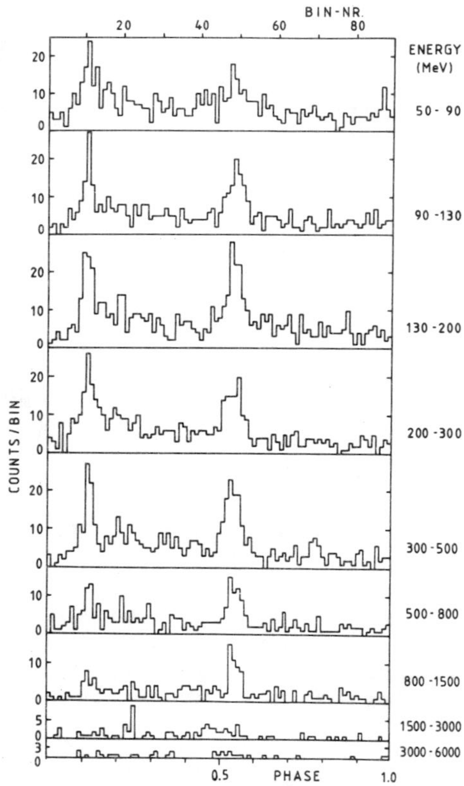

Fig. 5 - COS-B data of the Vela pulsar, divided in energy intervals. (see Kanbach et al,1980)

are broken down in energy intervals, permitting a study of the light curve spectrum.

The only remaining 2CG source for which an immediate identification was possible is 2CG353+16, positionally coincident with the local (~150 pc) cloud complex near the star ρ Oph, and for many years suggested as a GRS candidate (Black and Fazio, 1973). This source is interesting, because there may be associated to it the necessity for a localized CR increase (or acceleration process), since its gamma-ray flux appears somewhat greater than expected on the basis of the known mass estimated (see Bignami and Morfill, 1980; Morfill et al 1981). However, further accumulation of COS-B data has resulted in a much better detection of this cloud which has now to be dropped from a catalogue of PSF-compatible enhancements (see Hermsen and Bloemen, 1983).

For the other sources in fig. 3, no immediate identification is possible, in the sense that no peculiar or bright X-ray source is present in their error box, at least beyond what is expected by random coincidence, or any other "special" celestial object, in any way interesting. In order to at least understand these UGO's (Unidentified Gamma Objects) as a population, if not individually, it is interesting to look at their average properties, as deduced from the 2CG catalogue and from elementary astrophisical assumptions (see Table 1, from BH 83). Another similar approach is to study the number-intensity (logN-logS) curve for the sources (Bignami and Caraveo, 1980). If the sources are distributed uniformly in the disc, a slope ~ S-1 is expected, and appears reasonably followed for the anticentre data (fig. 6a). In the central region (fig. 6b), a significant steepening of the logN-logS implies, since one knows anyway that the sources are confined in a disc, a source density gradient towards the centre. This is another way of looking at the longitudinal anisotropy of the strongest sources, as shown by the solid circles in fig. 3.

In summary, the various forms of population property studies for the UGO yield only the result that they are a galactic population, probably young in the sense that they appear to be narrowly confined along the galactic disc, and probably with a centripetal density gradient. They can also contribute significantly to the overall galactic luminosity of 5×10^{38} ergs/sec (>100 MeV) (see e.g. Bignami et al 1978 and Riley and Wolfendale 1981) although fluxwise they are <10% of the whole galactic flux.

Further work along these lines is hampered by the fact that one is not at all sure that one is dealing with an homogeneous sample of objects, or that, for instance, one may be grouping together truly compact object (e.g. neutron stars, be they in the active radio pulsar phase or not) with small bright ISM cloud complexes, seen unresolved because of their distance.

3. NEW GENERATION OF COS-B SOURCES, AND STEPS TOWARDS A "FINAL CATALOGUE"

In the early eighties, with the advent of sistematic radio millimetric

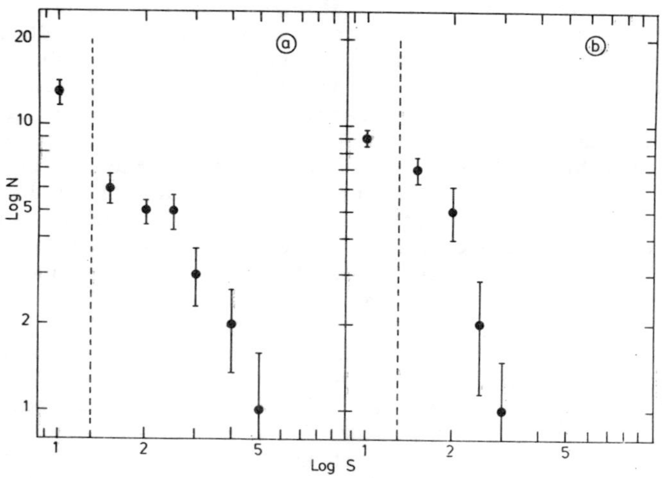

Fig. 6 - The LogN-LogS graph for the sources in the 2CG catalogue. 6(a): anticentre data; 6(b) central regions data (see Bignami and Caraveo, 1980). The vertical dotted line is the completeness threshold of the 2CG catalogue. (Figure from BH 83)

Table 1 - Average properties of the unidentified galacti gamma-ray sources (from BH 83)

Average b	1.5°
Angular size	0° - 2°
Photon-flux range (100 MeV<E<1 GeV)	~(1-5) x 10^{-6} ph cm^{-2} s^{-1}
Distance range	2 - 7 kpc
Scale height < z > (300°<l<60°)	< 130 pc
Spectral shape	different for individual objects, average dN/dE E^{-2}
Average photon energy	~ 250 MeV in the 100 MeV - 1 GeV range
Energy-flux range (100 MeV<E<1 GeV)	(4-20)x10^{-10} erg cm^{-2} s^{-1}
Luminosity range (100 MeV<E<1 GeV)	(0.4-5)x10^{36} erg s^{-1}

surveys of the Milky Way, it became possible to start mapping another elusive, yet probably abundant, component of the ISM, namely molecular hydrogen. This is normally done by mapping the 2.6 mm line of CO, and then use an appropriate formula to translate into H2 the CO columnar densities. Clearly, such a result is of paramount importance to GRA, since it permits a much better definition of the expected galactic background. Hence, a joint project between the caravane Collaboration and the Goddard Institute/Columbia Univ N.Y. based on H2 mapping and its yield in galactic gamma-rays. Unfortunately, the CO surveys of the galactic disc are not yet complete and the work is still in progress. For the 1st galactic quadrant, Pollock et al (1985a) were able, however, to use the new data for a better background definition and hence for a new GRS search. Several of the 2CG objects cannot be called sources any longer, in the sense that they are well well explained by the gas with standard CR irradiation, as had been suspected could be the case for several the 2CG "sources" (see e.g. Arnaud et al. 1982). Several others, however, remain as significant, unresolved excesses above the (now) complete gas background; of particular interest is 2CG54+01, clearly seen to vary between two different observations separated by many months. In this case, one has a rather convincing proof of an object with typical dimensions less then light-months. A first extention of this work (Pollock et al 1985b) to include the 2nd quadrant and part of the 4th in Carina has yielded similar results, as shown in Table 2. Clearly, the survey must be completed before conclusions can be drawn: it is to be expected, anyway, that the "final" COS-B source catalogue will be somewhat less rich that the 2CG, owing to the recognition that a part of the "sources" are well explained by the bumpy structure of the H2.

Table 2

Object	l°	b°	λ*
2CG078+01	78.1	2.3	60.9
2CG075+00	76.1	0.5	
284-01	284.7	-1.8	12.0
286+00	286.8	0.6	
083+03	83.0	3.1	8.4
2CG054+01	55.0	2.2	8.3
2CG125+01	135.1	1.0	7.7
2CG013+00	14.0	0.3	5.9
2CG095+04	95.5	4.2	2.5
2CG036+01	36.5	1.5	1.6
2CG121+04	121.4	4.0	0.9
2CG065+00	65.7	0.0	0.0

Summary of the search for point-source emission above the gas in the COS-B data above 300 MeV. l and b are galactic coordinates which for those objects markec c are taken from the 2CG catalogue. The positions are uncertain to between 0.5° and 1.5° . λ* is the likelihood statistics; only objects with *>5 can be considered as not explained by the CR-ISM interaction.

A region of particular interest, and one well studied in CO, is that around Cyg X3. There the situation was confused in the 2CG catalogue, while now, owing to a better background definition, improved statistics and a better analysis method, one can clearly see three bright sources standing well above the gas-explained emission. This is most visible in the energy range > 500 MeV (fig. 7), where the instrument resolution is at its best. As was the case in the 2CG data, here again none of the sources appear compatible with the position of Cyg X-3, a candidate of particular interest in view of the longstanding claims of VHE ($\gtrsim 10^{12}$ eV) gamma-ray detection from it (starting with Vladimirsky et al 1973), of the SAS-2 observations (Lamb et al 1977), and of the recent UHE ($\gtrsim 10^{15}$ eV) detections (Samorski and Stamm 1983). While the full story of the COS-B work on Cyg X-3 is given in Hermsen et al (1985) we summarize it here by saying that, in addition to the lack of a spatial enhancement mentioned above, also the temporal analysis done exploiting the 4.8 hrs. periodical signature of the source has given negative results for all the repeated (seven) Cyg X-3 observation by COS-B, as well as for the sum of the total data.
The COS-B upper limit is well below both the SAS-2 reported detection and of an extrapolation of the Samorski and Stamm spectrum back from the 10^{15} eV region. Further work is currently in progress on this point and on its implications for Cyg X-3 models.

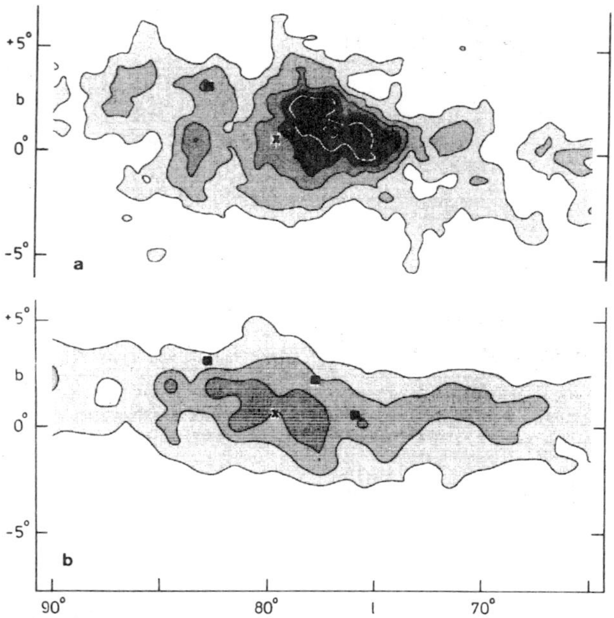

Fig. 7 - Gamma-ray intensity (>500 MeV) distributions in the Cygnus region. (a) Measured by COS-B; (b) estimated from the total gas distribution. Squares: positions of GRS of Pollock et al (85 a,b) Cross: position of Cyg X-3.

4. POSSIBILITIES OF ASTRONOMICAL IDENTIFICATION OF UGOs, AND WHAT WE LEARN IN THE PROCESS.

In contrast with the few, well discussed cases were GRS are identified, the majority of tentative models and astrophisical interpretations for USO's are necessarily vague, suffering from the lack of observational constraints. Thus, it is found that a variety of objects, e.g. from SNOBs to pulsars to special X-ray binaries, could produce the required 10^{35} ergs/sec or so in gamma-ray (not a dramatic requirement energetically, only 1/1000 of the typical Eddington luminosity for accreting systems), while not violating the observations. For a review of this important work from the beginning to 1983, the reader is referred to BH 83; for a more recent example of this "buckshot" approach, see the recent Morfill, Forman and Bignami, 1984.

On the other hand, the necessary effort towards astronomical identification of UGOs can at the moment only follow two ways: a) discovery, in the gamma-rays from a source, of time variabilities, in particular periodicities, identical to those of known objects in the error boxes, or, by exploration at other wavelengths in the hope of finding a "peculiar" object which, as for 3C273, can be associated to the gamma-ray source for obvious astrophisical reasons.

Time variability searches have been carried out on the COS-B data base in the search for possible gamma-ray emission from radio pulsars (Buccheri et al 1983), and the results have been negative for all the pulsars explored. Very similar results had previsioulsy been obtained by SAS-2 (Thompson et al 1983), so that one can convincingly say that known radio pulsars are not gamma-ray sources at the SAS-2/COS-B sensitivity level.

Other objects of potential interest, and with charcteristic time variabilities, are present in some COS-B error boxes, or are, anyway, in the disc at low latitudes. Searches for evidence of time variability at these timescale (e.g. binary orbits) have been carried out is the data base (Caraveo et al 1985), also with negative results. Among the objects investigated were some of the X-ray binaries recently reported to have VHE gamma-ray emission (e.g. Vela X-1, Protheroe et al, 1984) and others, like LSI+61.303, discovered as an X-ray source inside the error box of 2CG135+1 (Bignami et al, 1981). Fig. 8 shows an example of such search, in the case of 2CG135+01, where the data are plotted folded modulo the $\sim$ 26 day radio period of the radio X-ray star. The χ^2 test reveals no significant deviation of the source signal from the background (see Caraveo et al, 1985).

The method of exploring at other wavelenghts selected COS-B error boxes has been pursued in two directions: radio and X-ray. In radio, some sistematic work has been carried out from the Bonn (Sieber and Schlikeiser, 1982) and Westerbork telescopes (Spoelstra and Hermsen, 1984) at various frequencies in the continuum, and a lot of new radio sources were found, but none bearing very special marks. More interestingly, searches for new pulsars were undertaken from those COS-B error boxes visible by Arecibo (Boriakoff et al 1983) and, as expected,a couple of new pulsars were found, one of which is the second

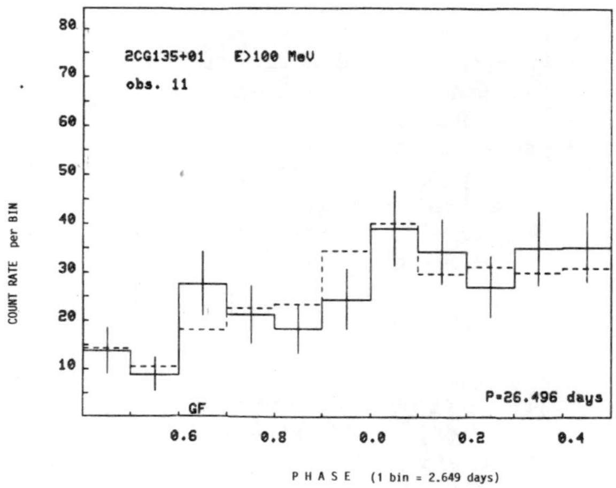

Fig. 8 - COS-B countrates (>100 MeV) for the observation period 11, including the variable X-ray star LSI+61.303. "Source" counts (full line) are seen to follow closely the backgraound distribution (broken line). Abscissa in fractions of orbital period. (Caraveo et al. 1985).

fastest known in the Galaxy,with a period around 6 msec. However, no evidence could be found, through timing, of association of COS-B source with these new pulsars, as for the old ones. In summary, one can say that, as a result of the various radio searches done, one has only learnt that UGOs do not have conspicuous radio counterparts, if they have any at all.

In the X-rays, the work has started already prior to the imaging detectors, leading to the identification with SAS-3's RMC of 4U0241+62 with the nearest QSO to our Galaxy, which became thus a very legitimate candidate for 2CG135+01 (Apparao et al, 1978). With the advent of th Einstein Observatory, a systematic coverage of several error boxes was possible through the Guest Observer Program, with a number of follow-up observations on some of the new and most interesting sources. Most of this work is described in Caraveo (1983), where the details of the program are given as well as the actual error box coverage. Exploiting the 1°x1° field of view of the IPC, the UGOs can be covered with a mosaic of few fields, as shown in fig. 9, displaying the 2CG135+01 region. The example is particularly interesting, because many sources are seen, and all of them have been identified, sometimes with follow-up HRI work. Well visible are the QSO 0241+622 and the radio/X-ray Be star GT0235(LSI + 61°.303), i.e. the two brightest sources in the region and both of which, as already mentioned, are reasonable counterpart candidates for 2CG135+01. The final result of the X-ray search exercise is similar, in most respects, to that of

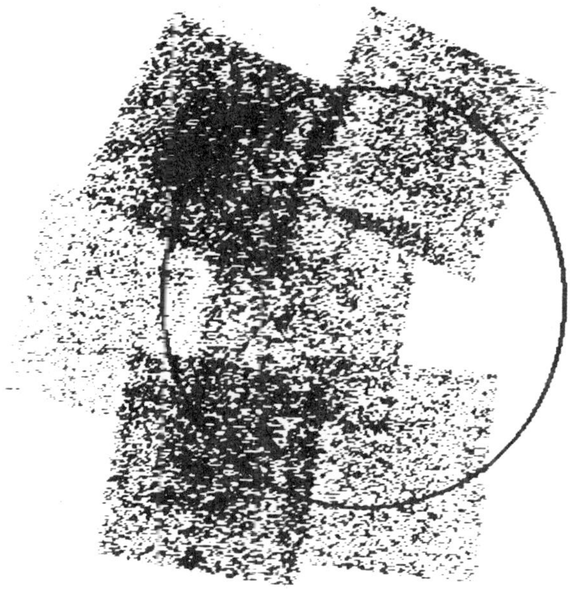

Fig. 9 - Eistein IPC X-ray coverage of the 2CG135+01 error box.
Bright source in the upper left is QSO 0241+622.
Also visible (low left, just inside error box)
is the radio/X-ray Be star LSI+61°.303.

the radio searches: a few new X-ray sources have been found which are interesting per se, but with no possibilities of being firmly associated to UGOs. However, in one important case, that of Geminga (2CG195+04), the X-ray search has on the contrary turned out what is definitely the best candidate so far, and in all probability the key to understanding at least one bona fide gamma ray source.

5. THE BEST CASE: GEMINGA - FACTS WITHOUT FICTION

Two overlapping Einstein IPC exposures, for a total of 10,000 secs were obtained in 78 and 81 of the field covering entirely the error box of 2CG195+04, the smallest of all UGOs with a radius of 0.4 square degrees, i.e. a big improvement with respect to the error box of SAS-2, which had, however, discovered the source (Thompson et al 78). Fig. 10 shows a composite picture of the region, including the follow-up HRI exposure the brightest new source found in the field, 1E0630+178 (Bignami et al 83). Not clearly visible in the figure, there are however three other faint sources in the field, which also need to be discussed: two of them have been identified on the basis of the IPC data with a normal star and a radio galaxy (Bignami et al 1983, Sieber

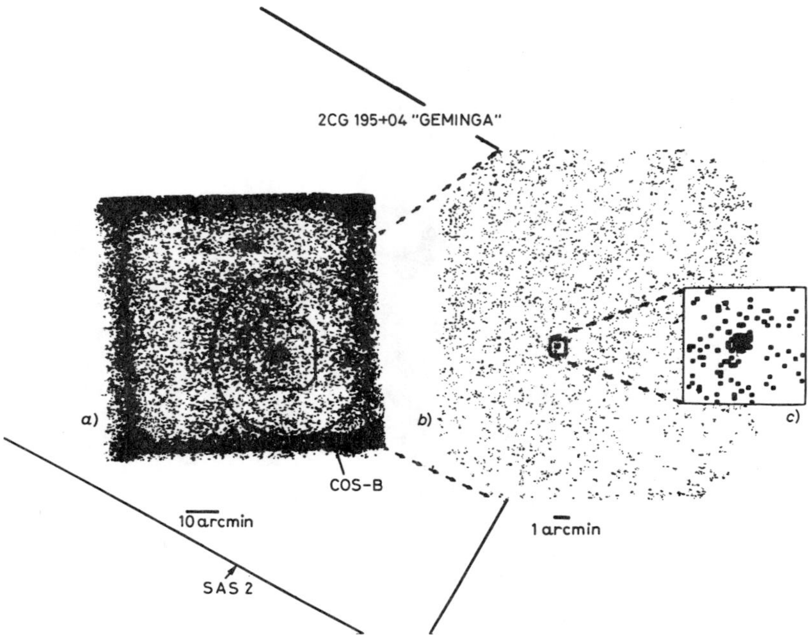

Fig. 10 - Composite picture of the Geminga field, showing the SAS-2, and COS-B > 100 MeV gamma-ray error boxes (to scale), the IPC field a), with the HRI field around 1E0630 + 178, the evident IPC source. Blown-up b) is the HRI field, showing a single source, again blown-up in c) to show its pointlike nature.

and Schlickeiser 1982, Spoelstra and Hermsen 1984), while for the third one it was necessary to wait for the CMA instrument on the EXOSAT mission. This has recently allowed the accurate (∼10") positioning of the IPC source, and only one object appears compatible with its position on the POSS; starlike and of mv∼17, it implies and Fx/Fv of 1, typical, for example, of many extragalactic sources.

Of the 1,000 photons from 1E0630+178 (which has a flux of $2x10^{-12}$ ergs/sec in the IPC range), both a spectral and temporal analysis were performed, and a similar work was done on the photons from the source observed by EXOSAT in 1983 (see Bignami et al 83 and Caraveo et al 84). Only a summary of the results will be presented here. The limited IPC spectral information prevents the accurate determination of the two-parameter which characterize an X-ray source spectrum: the actual spectral shape and the columnar density of absorbing material (NH). Nevertheless, a best fitting procedure over a reasonable range of spectral types and NH values has yielded the following results:

1) no fit is acceptable with NH > $2x20^{20}$ cm^{-2}
2) the region were the best fits are found is that of steep power laws or thermal black bodies with NH∼ 10^{19} cm^{-2}
3) no lower limit can be set to NH, which could be very small

4) the best fit of all is a thermal b.b. with $T = 9\times10^5$ °K and $NH = 10^{19}\ cm^{-2}$ (fig. 11).

Nothing more precise can be extracted from the IPC data, and practically no spectral information is available from EXOSAT. The spectrum of fig. 11 is, however, indicative of a soft X-ray source which cannot be located at a distance greater than 100/200 pc, but which could also be much nearer.

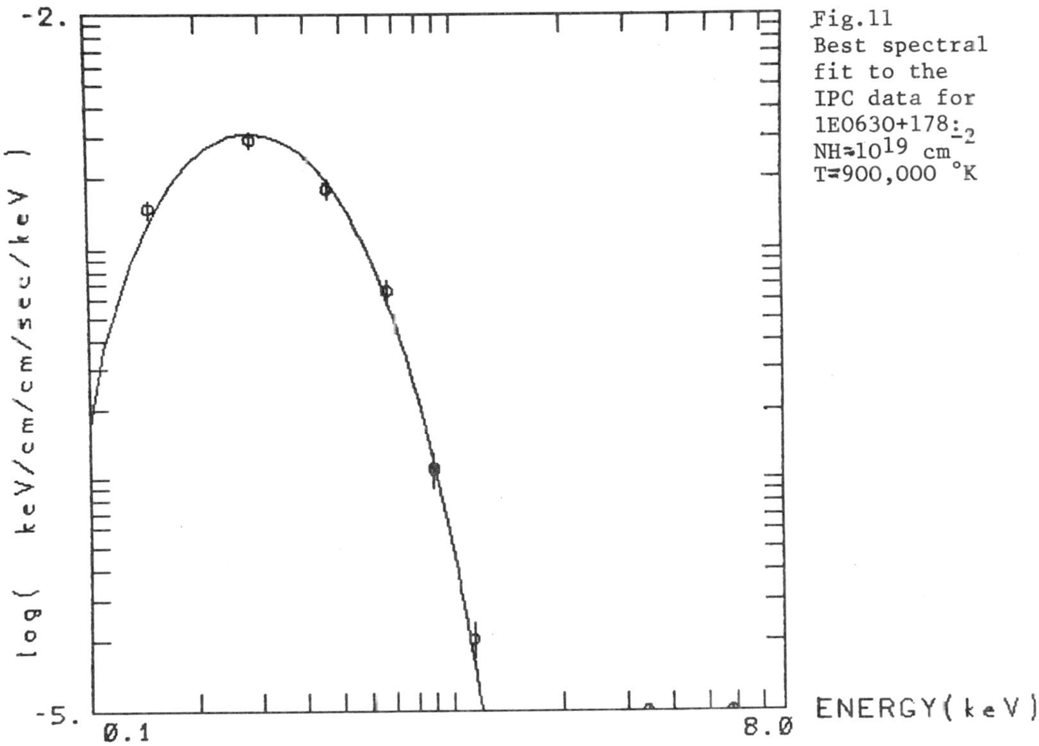

Fig.11 Best spectral fit to the IPC data for 1E0630+178: $NH \approx 10^{19}\ cm^{-2}$ T=900,000 °K

The temporal analysis of the photon arrival times resulted in the discovery of a 59 sec periodicity in both the Einstein and EXOSAT data. Such periodicity was searched for in the data because of previous reports, both by SAS-2 and COS-B, of such an effect at gamma-ray wavelengths. More recently, ground based VHE ($>10^{12}$ eV) gamma-ray observations by the Crimean group (Zyskin and Mukanov, 1983) supported such finding. On the other hand, a reanalysis of the COS-B data for such periodicyt yielded a negative result (Buccheri et al, 1985), with, however, varying degree of significance depending on the light-curve shape assumptions. The X-ray data, in any case, yield evidence for the effect: Fig. 12 gives the light curve for the Einstein March 81 observations (IPC+HRI) folded modulo 59.73 secs, the period value obtained after a χ^2 search of periodicities in the range predicted by the available gamma-ray observations. The reduced χ^2 of the light

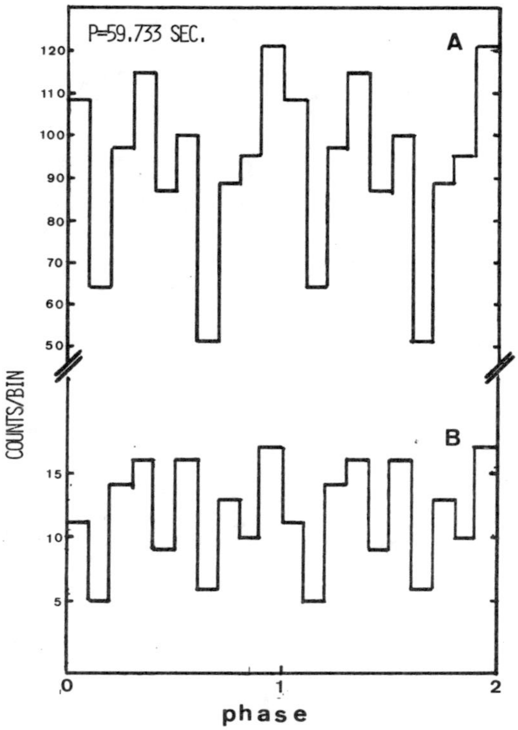

Fig. 12 - Light curves for the Einstein March 81 observations of 1E0630+178, foldeld modulo P - 59.733 secs. Upper curve (A): IPC+HRI data Lower curve (B): HRI data only

curve of fig. 12a is 5.06 over 9 d.o.f., which corresponds to a very low chance probability, after correction for the number of independent trials performed. Further support to the result is given by the HRI-only light curve of fig. 12b, where the $\sim$ 100 HRI photons are folded at the same period and exactly at the same phase, and are seen to be well in phase with the data of fig. 12a, which is dominated by the photons taken in th IPC observation some 12 hours earlier. All arrival times were reduced to solar system barycenter, to correct for the satellite and the Earth orbital motion. As an a posteriori check of the result, a period scan was performed covering the very ample range 10 to 210 seconds, with about 50,000 steps and covering, in period and phase, the maximum value at 59.73. As shown in fig. 13, this period value stands well out above the normal random fluctuations expected. Further details on this Einstein '81 result, and similar work on the Einstein 79 and EXOSAT 83 data are given in Bignami et al

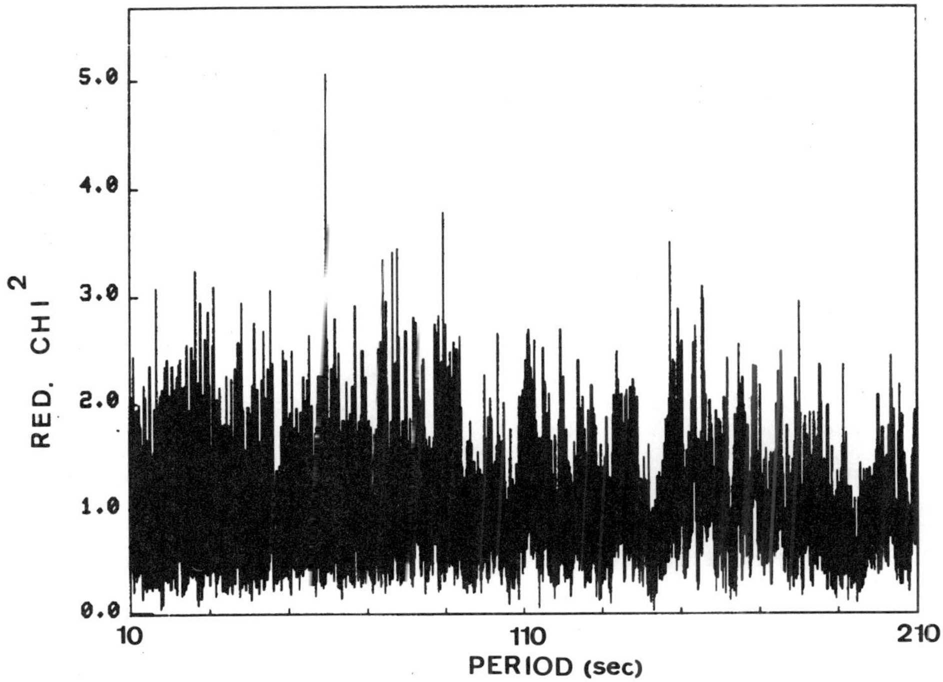

Fig. 13 - A posteriori period scan for the Einstein 1981 data of 1E0630+178. The 59.733 value is seen to stand well out, with X^2-5.09 (9 d.o.f.).

84. The overall picture so far available is very puzzling: a similar periodicity effect is present (albeit at a lower significance value due to the lower statistics) also in these data, but the actual best period values are at a 59.46 and 60.06, respectively, thus implying a very high period derivative, in the region from 2 to 4×10^{-9} sec/sec. If the observed period is taken as the rotational one of the X-ray source, such a P implies, for instance, a "spindown" age of < 1,000 years. Clearly, this period and period variation are very important aspects for models of the source.

As mentioned in the X-ray source discovery paper, 1E0630+178 does not show an optical counterpart on the red or blue POSS plates within the 4" radius HRI position. This was a puzzling point per se: except for regions of very high galactic obscuration (not so here, in the anticentre and at +4° latitude), all Einstein sources of comparable strength have counterparts on the POSS, and, in fact, so do all but extremely few even of the weakest ones. The following optical search (Caraveo et al, 1984) from the CFH 3.6 m instrument led to an mv ~ 21 object (also observed from ESO and reported in Sol et al, 1985) which was at 4.2" from the best HRI position.

While the positional compatibility with the HRI error box was not excellent, the object was proposed as a counterpart because 1) nothing else was present in the HRI error box down to mv $\sim$ 23, and 2) its B-V (0.6) rated it as the "bluest" of objects of comparable magnitudes, both in the photometry of Caraveo et al (84) and in that of Halpern et al (85). However, as pointed out by Halpern et al (85), the chance appearance probability of a mv 21 object in a 4.2" circle is high, 15%; this, coupled with spectral data (of both Halpern et al (85) and Djorgoski & Kulkarni (86) and with the lack of optical pulsations at the 59 sec X-ray period (Kulkarni and Djorgovski, (86))) led to the suspicion that the mv$\sim$21 object was not the true counterpart. A more cogent point in the same direction comes from some recent IR measurements kindly taken for us by M. Lebofski at the MMT, showing that the photometry of the object can be well understood from K to B with a black body emission at 5500-6000 °K with, however, an interstellar absorption $A_v > 1$; this translates in a columnar density $> 10^{21}$ cm^{-2}, in strong contrast with the X-ray data absorption.

If the object considered so far is not the optical candidate, the Lx/Lv of the source is >1,000, since nothing brighter than 24.5 is present in the HRI box. Two objects are seen there in the sum of the Lick CCD plates, one called G' (24.5), possibly the same suggested by Sol et al (85) and a fainter one, G", (> 25,5) at the edge of the error box. Work is actively in progress at the time of writing (Oct 85) on the understanding of these objects; while nothing much can be said on G', which has, of course, a very high change occurrence probability, an interesting result seems to appear on G". Summing the 1984 January CFHT observations, Bignami et al (1986) also reveal what appears to be G", but at 1.6" arcseconds away. Figs. 14 (a) and (b) show the result of the CFHT 84 and Lick 85 data, where the G" proper motion seems apparent. If confirmed by furhter observations, this has the disquieting implication that, for transverse velocities of G" between 30 and 300 Km/sec, its distance is less then 40 pc, and thus its absolute magnitude M_R is between, +22 and +27. When compared to the optical magnitude of the Vela pulsar (M_B +15), one sees that this could certainly be the faint optical emission of a very local neutron star. This, in turn, would be a natural explanation for 1E0630+178 and for Geminga, the brightest (nearest?) UGO, with Lx/L$\sim$ 2,000.

Confirmation of this proper motion result is needed soon; if it comes, beyong Geminga we may have a clue on the existance of "defunct" neutron stars, not active in radio. Several models proposed recently (e.g. Nulsen and Fabian, 1984; Bisnovathy-Kogan 84) rely on a collapsed binary system to explain Geminga; however isolated neutron stars (Katz, 1985) have also been considered. In summary, one hopes that COS-B results, and subsequent identification work, have led us to the evidence for something long suspected and postulated but never actually seen. Whether this has much to do with other UGO's remains a challenge for the future.

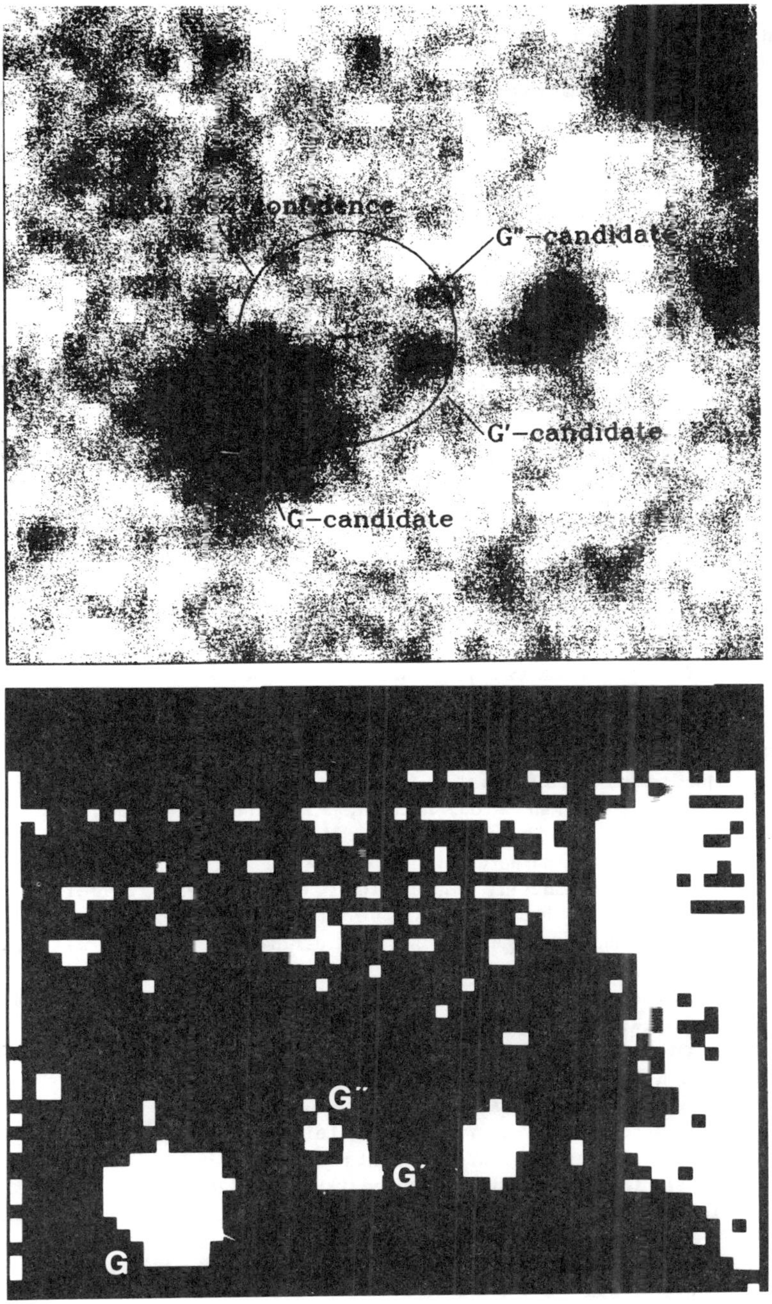

Fig. 14 - (a, lower part) : CFHT January 84 data, result of sum of 12 separate exposures
(b, upper part) : Lick March 85 data, also the result of summation of plate stack
(Djorgovski and Kulkarni, 1986).

REFERENCES

Apparao, K. et al., 1978 Nature 273: 450
Arnaud, K., et al. 1982 M.N.R.A.S. 201: 745
Bignami, G.F., Caraveo, P.A., Maraschi, L., 1978 A. A. 67: 149
Bignami, G.F. and Caraveo, P.A., 1980 Ap. J. 241: 1161
Bignami, G.F. and Morfill, G.E., 1980 A. A. 87: 85
Bingami, G.F., Caraveo, P.A., Lamb, R.C., Markert, T.H., Paul, J.A. 1981 Ap. J. 247: L85
Bignami, G.F. and Hermsen, W., 1983 Ann. Rev. A. A. 21: 67
Bignami, G.F., Caraveo, P.A., Lamb, R.C. 1983 Ap. J. 272: L9
Bignami, G.F., 1984 Rivista del Nuovo Cimento 7: 1
Bignami, G.F., Caraveo, P.A., Paul, J.A. 1984 Nature 310: 464
Bignami, G.F. et al., 1986, in preparation
Bisnovathy-Kogan, G.S., 1985 Nature, 315, 555
Black, J.H., Fazio, G.G., 1973 Ap. J. 185: L7
Boriakoff, V., Buccheri, R., Fauci, G.G., 1983 Nature 304: 417
Buccheri, R., et al. (Caravane Collaboration), 1983 A. A. 128: 245
Buccheri, R., et al. (Caravane Collaboration), 1985 19th I.C.R.C. San Diego OG 2.4-3
Caraveo, P.A., 1983 Space Science Rev. 36: 207
Caraveo, P.A., Bignami G.F., Vigroux, L., Paul J.A., 1984 Ap. J. 276: L45
Caraveo, P.A., et al. (Caravane Collaboration), 1985 19th I.C.R.C. San Diego OG 2.5-9
Fichtel, C.E., Hartman, R.C., Kniffen, D.A., Thompson, D.J., 1975 Ap. J. 198: 163
Halpern, J., Grindlay, J., Teitler, D., 1985 Ap. J. (1 Sept.)
Hermsen, W. and Bloemen, J., in "Surveys of the Southern Galaxy" Eds. W. Burton and F.P. Israel, Reidel, 1983 p. 65
Hermsen, W., et al. (Caravane Collaboration), Nature 296: 494
Hermsen, W., et al. (Caravane Collaboration), 1985 19th I.C.R.C. San Diego OG 2.2-2
Kamback, G., et al. (Caravane Collaboration), 1980 A. A. 90: 163
Katz, J.I., 1985 Astrophys. Lett. 24: 183
Kulkarni, S. and Djorgovsky, S., 1986 A. J. (Jan.)
Lamb, R.C., Fichtel, C.E., Hartman, R.C., Kniffen, D.A., Thompson, D.J., 1977 Ap. J. 212: L63
Nulsen, P.E. and Fabian, A.C., 1984 Nature 312: 48
Morfill, G.E., Volk, H.J., Forman, M., Bignami, G.F., Caraveo, P.A., Drury, L., 1981 Ap. J. 245: 810
Morfill, G.E., Forman, M. and Bignami, G.F., 1984 Ap. J. 284: 856
Pollock, A.M., et al. (Caravane Collaboration), 1985 A. A. 146: 352
Pollock, A.M., et al. (Caravane Collaboration), 1985 19th I.C.R.C. San Diego OG 3.1-9
Protheroe, R.J., Clay, R.W., Gerhardy, P.R., 1984 Ap. J. 280: L47
Riley, P.A., WOlfendale, A.W., 1981 Rivista del Nuovo Cimento 4: 1
Stamorski, M. and Stamm, W., 1983 Ap. J. 268: L17
Sieber, W., Schlickeiser, R., 1982 A. A. 113: 314
Sol, H., Tarenghi, M., Vanderniest, C., Vigroux, L., Lelievre, G., 1985 A. A. 144: 109
Spoelstra, T. and Hermsen, W., 1984 A. A. 135: 135
Swanenburg, B.M., et al. (Caravane Collaboration), 1981 Ap. J. 243: L69
Thompson, D.J., Fichtel, C.E., Harman, R.C., Kniffen, D.A., Lamb, R.C., 1977 Ap. J. 213: 252
Thompson, D.J., Bertsch, D.L., Hartman, R.C., Hunter, S.D., 1983 A. A. 127: 220
Vladimirsky, B.N., et al., 1973 13th I.C.R.C. Denver 1: 456
Wills, R., et al. (Caravane Collaboration), 1982 Nature 296: 723
Zyskin, Y.U. and Mukanov, D., 1983 Sov. Astron. Lett. 9: 117

COSMIC GAMMA RAY BURSTS

K. Hurley
Centre d'Etude Spatiale des Rayonnements
CNRS-UPS
B.P. 4346
31029 Toulouse Cedex
France

ABSTRACT. The observational properties of gamma ray bursts - their time histories, energy spectra, statistical properties, and counterpart searches - are reviewed. The observational evidence for a neutron star origin is discussed. The characteristics of some of the time histories and energy spectra argue strongly in favor of such an origin, but firm evidence, say in the form of an identifiable counterpart, is still lacking.

1. INTRODUCTION

There appear at present to be only a limited number of energy bands in which galactic neutron stars are observable to us: these are the radio, X, and gamma ray regions. In the radio range, over 400 pulsars are now known (Ruderman, 1985; Lyne, 1985). Some 145 galactic supernova remnants have been detected by their radio emission, but not all of these show direct evidence for the existence of a neutron star (Helfand, 1985). In the X-ray range, there are the high mass X-ray binaries in the disk which include about 23 pulsating sources (F. Lamb, 1985; Trümper, 1985) and about an equal number of X ray burst sources (D. Lamb, 1985) all of which are almost certainly associated with neutron stars. In the HEAO A-1 X-ray source catalog (Wood et al., 1984), 93 sources are identified as binaries and 22 as supernova remnants; Helfand (1985) estimates that 36 X-ray supernova remnants have been detected out to about 5 kpc. In the high energy (>100 MeV) gamma ray region, the Crab and Vela pulsars have been detected, and two other sources are suspected to be associated with neutron stars, with around 20 unidentified galactic sources (Bignami, 1985). And then there are the gamma ray burst sources. More than 400 have been observed since 1967, and currently operating instruments detect about 1 every 2-3 days. When their discovery was announced in 1973 (Klebesadel, Strong, and Olson, 1973) it was far from clear that gamma bursters were even objects in our galaxy. Since 1978, however, there has been a growing concensus that this phenomenon is indeed related to

F. Pacini (ed.), High Energy Phenomena Around Collapsed Stars, 317–330.

galactic neutron stars. M. Ruderman (1985) has suggested that this choice is partly by default, and there is some truth to this, since no gamma burst source has yet been identified with any quiescent object in any other energy range (in distinct contrast to the case of many of the radio and X-ray sources cited above). Nevertheless, if, for reasons which will be discussed below, one accepts the gamma burster-neutron star association, then it is clear from the above numbers that observations of gamma bursts provide us with a data base on neutron stars which is comparable in size to those of radio and X ray observations.

Although this paper will deal primarily with the observational aspects of gamma bursts, there are a number of important theoretical topics which are directly related, and will only be touched on here. These include:

a) Stellar evolution. It is not yet known whether gamma bursters are lone objects, or whether they belong to binary systems (there is no observational evidence for the latter). But in either case, the lack of counterparts at other wavelengths imposes unique and stringent constraints which may indicate an unusual evolutionary scenario.

b) The dark matter in our galaxy. There is still little evidence which allows the distance scale to bursters to be estimated. Several recent theories, however, have placed the sources in the distant galactic halo, at distances up to 100 kpc. If these prove to be correct, then burst sources may help us to trace the missing mass of the galaxy which, based on dynamical arguments, should exist. (It is extremely unlikely, however, that these sources constitute more than a very small fraction of it.)

c) The behavior of matter under extreme conditions, and its radiation mechanisms. Observational evidence suggests that the conditions of temperature, gravitational and magnetic fields under which bursts are generated are not only extreme, but perhaps, in their particular combination, unique. The interesting questions here are what the ultimate source of the burst energy is, and how it happens that this energy is released primarily as gamma rays.

d) The neutron star equation of state. E. Liang (1985) has shown that the observations of gamma ray burst energy spectra, discussed below, provide a large data base for estimating the redshifts at or near neutron star surfaces. These, in principle, may be used to choose between hard and soft equations of state.

There are basically four "observables" where gamma bursts are concerned. These are the gamma ray light curves, or time histories, the energy spectra, the statistical properties (spatial distribution, log N-log S relation, and recurrence rate), and the results of deep searches for quiescent counterparts. The four are not equal in quality or quantity, and it is useful to review their general properties before going into greater detail.

a) The time histories are the "purest" information which can be obtained with the detectors now in use (mainly scintillators): there is little that the detector can do to deform them. Moreover, the number of time histories observed with good time resolution is large. But even the simplest ones have no simple, generally agreed upon

interpretation. Some clues exist, but they are not unambiguous.
b) The energy spectra of some bursts contain quite compelling evidence for a compact object origin. The spectra, however, are strongly deformed by the detectors, and the observed spectrum can only be used to define a set of possible true spectra which could have given rise to it. This set may contain both spectra with line features, and spectra without them.
c) Statistical properties, particularly the spatial distribution and the number-intensity (log N-log S) relation, can, in principle, yield the burster distance scale, and considerably limit the choice of theories. However, such studies are still quite limited due to experimental bias and selection effects.
d) It was initially thought that the unique capability to derive accurate locations using the method of arrival time analysis ("triangulation") between widely separated spacecraft would lead to optical or soft X-ray identifications of bursters. It now appears, though, that counterparts to most bursters are extremely faint, and none has been identified with certainty.
In the following sections, each of these observational aspects is treated in detail. For reviews of the more theoretical aspects, the reader is referred to excellent articles by Lamb (1984), Woosley (1984), and Liang (1984), as well as to a conference proceedings edited by Liang and Petrosian (1986).

2. TIME HISTORIES

The time histories of the several hundred gamma bursts which have been observed to date display a wide variety of forms. The shortest lasted around 50 ms.;the longest last well over a minute. The simplest display single peaks, while the most complex may have tens of peaks. Two extremes are shown in Figures 1 and 2. Numerous morphological classification schemes based on these two aspects (duration and complexity) have been suggested (Barat et al., 1981 and 1984a; Desai, 1981; Mazets et al., 1982a; Klebesadel et al., 1982). When the time histories have a relatively simple appearance (i.e., a single prominent peak - Fig. 2) it is possible to fit simple exponential functions to them and characterize them by their e-folding rise and fall times (Barat et al., 1984a). The results of such a characterization show first, that the rise times tend to be equal to or shorter than the decay times. Second, that there is some evidence for a clustering of decay times around 100 ms., and of rise times around 500 ms. And finally, that one and only one very exceptional event had a rise time of <0.2 ms. This was the well known 1979 Mar 5b gamma ray burst. The time history of this event had several other unique features, one of which was that the initial peak was followed by a series of pulsations lasting over 100 s., which display a clear 8 s. period (Cline et al., 1980). Of the many time histories which have been studied, a detection of periodicity has been proposed only for this event and one other (a 4.2 s. period for an event on October 29, 1977: Wood et al., 1981).

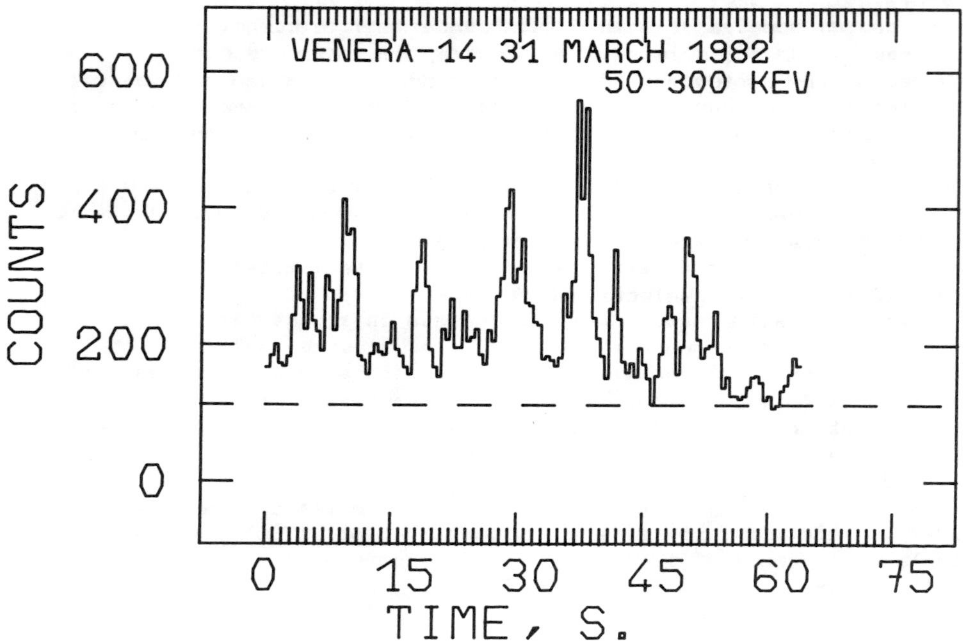

Figure 1. Time history of a long, complex gamma ray burst observed by the SIGNE experiment. Time resolution of the plot is 0.5 s. Dashed line indicates background level.

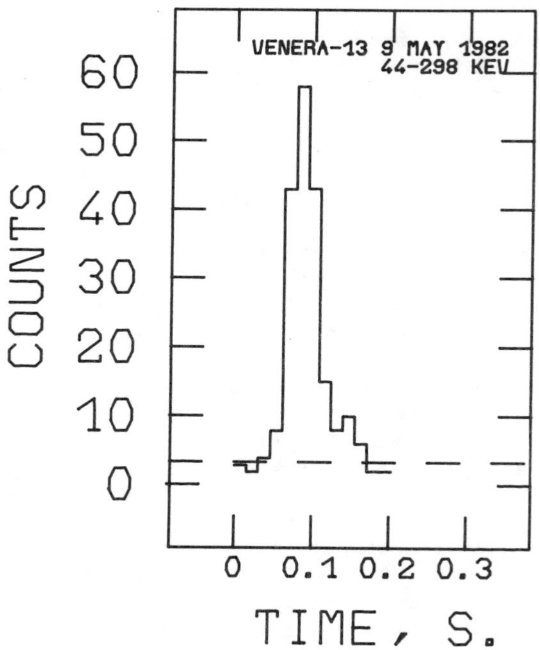

Figure 2. Time history of a short, simple gamma ray burst observed by the SIGNE experiment. Time resolution of the plot is 48 ms. The e-folding rise and fall times are in the 50 ms. range. Dashed line indicates background level.

The various time scales mentioned above have no simple interpretation, although several ideas have been proposed. Tsygan (1975) was the first to point out the possible importance of gravitational radiation for a model in which the burst source was a vibrating neutron star; he predicted burst durations in the 10-30 s range. Ramaty et al. (1980) later constructed a model specifically for the 1979 Mar 5b event in which they showed that gravitational radiation should damp quadrupole and higher mode vibrations of a neutron star responsible for this burst on a time scale of about 0.1 s - the observed decay time for the impulsive phase of this event. This idea may be extended to other bursts having 100 ms decay times as well. Lamb, Lamb and Pines (1973) pointed out that the free fall time scales in neutron star magnetospheres should be in the range 0.1 - 1 s (in agreement with observed rise times), and that dynamical time scales at the surface of a neutron star should be around 0.1 to 1 ms (cf. the <0.2 ms rise time of the 1979 Mar 5b event). Finally, the rare periodicities which have been observed in the 4-8 s range, although longer than radio pulsar periods, are in the same range as that of some pulsating X-ray sources. Thus it can be seen that on the basis of the time histories alone, neutron stars were immediately suspected as a possible source of gamma ray bursts. This idea is reinforced by a consideration of the energy spectra.

3. ENERGY SPECTRA

The energy spectra of gamma ray bursts have been recorded by numerous instruments practically since their discovery; with only a few noteworthy exceptions, however, almost all observations have been made by instruments with intrinsically poor energy and/or time resolution. As will be seen below, this combination of circumstances has probably conspired to give us a view of the energy spectra which is at least rather biased, and possibly even incorrect in some respects. This section will treat three aspects of energy spectra: the continuum, the "line" features, and spectral variability.

Figure 3 shows what might be termed a "generic" gamma ray burst energy spectrum: it displays most of the features of interest found in GRB spectra, although it would be quite rare to find them all in any single event. The medium energy continuum between about 30 keV and 1 MeV (solid line) is often well fit by a simple two-parameter law, e.g. $kE^{-1}\exp(-E/E_0)$, at least when spectra are integrated over 4 s or longer; typical values of E_0 are in the range 150-400 keV (Mazets et al., 1981). When the lower energy data (down to 2-3 keV) are taken into account, good fits over all energies up to about 1-2 MeV appear to be obtained either by a thermal bremsstrahlung spectrum (Gilman et al., 1980), by a double power law spectrum, (Terrell et al., 1982), or by thermal cyclotron or Comptonization models (Laros et al., 1984).

High energy emission (above 1 MeV - dashed line in Figure 3) has recently been shown to be a common feature of burst spectra: Matz et al. (1985), using the data of the SMM Gamma Ray Spectrometer, find that over 60% of all bursts emit above 1.03 MeV, and 24% above 5 MeV.

Nolan et al. (1984a) have demonstrated that it is difficult to fit a simple function to these data above 300 keV, but that thermal synchrotron or power law models fit better than the exponential cited above, which gives too soft a spectrum. Nolan et al. (1983) and Rieger et al. (1982) have shown examples of the count rate (i.e., raw) spectra which appear to be well fit by power laws from 300 keV to energies as high as 10-30 MeV.

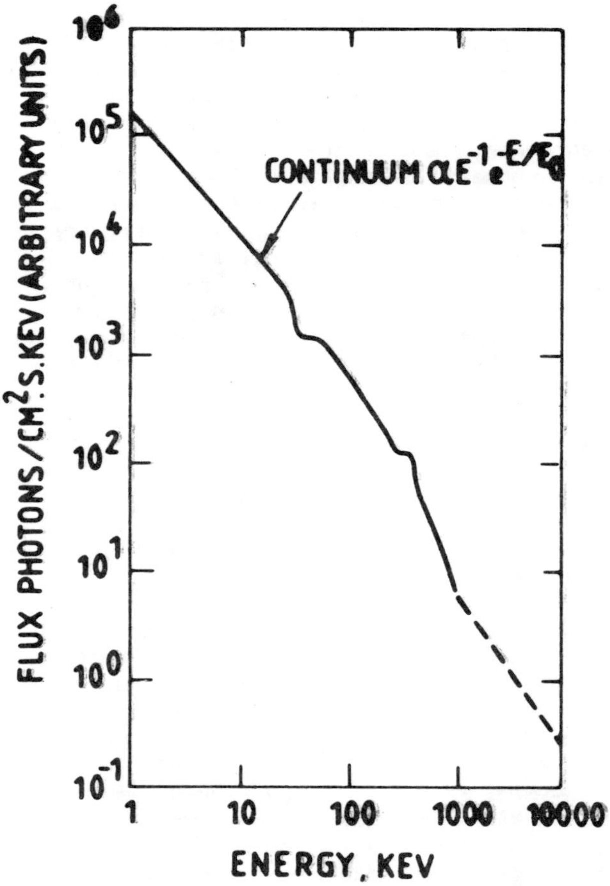

Figure 3. "Generic" gamma ray burst energy spectrum, showing the 1 keV-10 MeV continuum and spectral features around 50 and 400 keV. This spectrum is a composite: a typical single spectrum would not be expected to display all the features shown here.

Evidence for emission features around 400 keV in gamma ray burst energy spectra was first reported by Mazets et al. (1979) and by Teegarden and Cline (1980). Mazets et al. (1981) later presented evidence for 30-70 keV absorption features as well. Both types are indicated schematically in Figure 3, and Table 1 summarizes some of their relevant properties.

Several facts can be added to those in Table 1. The first is that it is quite rare to observe both the low energy absorption and the high energy emission features in the energy spectrum of the same

Table 1. Properties of GRB Line Features

Approximate Energy, keV	Type	Emitted or Absorbed Flux	Width, keV	Frequency of Occurrence	Duration, sec	Confirmation?
50	Absorption	Several % of Continuum	≥ 20	21% of all GRB's	≤ 4	Yes, but not simultaneous (1), and No (2)
400	Emission	Several to 30% of Continuum	≥ 200	10% of all GRB's	$\leq .25$	Yes (3) and No (4)

NOTES: 1) Hueter, 1984
2) 1980 Apr 19 event: Dennis et al., 1982 and Fenimore, Klebesadel and Laros (1983)
3) 1978 Nov 19 event: Teegarden and Cline (1980) and Mazets et al. (1981)
1978 Nov 4 event: Barat et al. (1984b) and Mazets et al. (1981)
4) 1981 Dec 31 event: Nolan et al. (1984b) and Mazets et al. (1983)

burst. The number of such observations is consistent with random coincidences, given the individual frequencies of occurrence. The second is that both features have proven difficult to confirm in any convincing manner. Thus in the few cases where energy spectra of the same burst may be compared in the data of independent experiments, the estimated fluxes of the emission features either differ at the 2-3 sigma level (1978 Nov 4), or by an amount which appears to exceed greatly the quoted standard deviations (1978 Nov 19), or the feature is not confirmed at all (1981 Dec 31). A somewhat similar situation exists for the low energy absorption features (e.g., the 1980 Apr 19 event). Problems such as this might arise for one or more of the following reasons: a) the features are time-variable (see below), b) the spectral continua are incorrectly estimated, and/or c) the line widths are inaccurately estimated. As far as point c is concerned, it must be said that none of the line observations appears to have "over-resolved" or even resolved a feature: the latter sometimes appear as a single deviant point, or at best several points but often with poor statistics.

Mazets et al. (1979) and Teegarden and Cline (1980) suggested that the emission features around 400 keV were due to redshifted electron-positron pair annihilation, and pointed out that the redshift (around 25%) was that which would be expected for a neutron star. This interpretation is generally accepted and indeed has become perhaps the key argument in favor of a neutron star origin for gamma bursts. It has been shown to give rise to an interesting problem, however (Ramaty and Meszaros, 1981), since the observed line widths imply temperatures for the annihilating particles which are considerably less than those of the particles responsible for the continuum emission. This may mean that these two components of the spectrum are generated in different regions.

Mazets et al. (1981) suggested that the low energy features were generated by cyclotron absorption in a magnetic field with a strength of several times 10^{12} gauss. Teegarden (1982) pointed out that the widths of these features were uncorrelated with the temperatures of the underlying continua (a problem analogous to that of the high energy emission features) and that in the presence of the implied field strengths, the continuum was likely to be dominated by synchrotron radiation. The latter point was discussed extensively by Liang (1982), who showed that the optically thin thermal synchrotron spectrum of mildly relativistic electrons provided a good (albeit three parameter) fit to burst spectra. Matz et al. (1985) have pointed out that their observations of bursts above 1 MeV limit the magnetic field strength in the emission producing region to less than 10^{12} gauss, implying either that some sources do not have fields stronger than this, or that this continuum is generated in regions where the field strength is lower than that needed to produce cyclotron absorption.

Just as the time histories of bursts are highly variable, so are their energy spectra. Barat (1983), using data from the SIGNE experiments, showed that spectral variability could be found down to several hundred ms. timescales; Mazets et al. (1983) presented evidence for a correlation between effective spectral temperature and burst intensity at time scales well under 1 s. Laros et al. (1984), while confirming rapid spectral variability in general, found no simply correlation with intensity. Norris et al. (1986) found a basic pattern of hard-to-soft spectral variability in ten bursts studied. Finally, Barat et al. (1984b) showed that the 400 keV emission features were variable on time scales of 250 ms or less.

From this discussion of energy spectra, it is clear why a neutron star origin is now favored for gamma ray bursts. Indeed, if the interpretation of spectral features as lines is accepted, not only do the energy spectra provide us with the largest data base on line emission from neutron stars, but also, the number of high energy line observations from bursters exceeds that from all other astrophysical objects. Many important questions remain unanswered, however. Among them are: What time and energy resolution are needed to provide an accurate picture of burst spectra? Is the relative rarity of spectral features intrinsic to the sources (e.g. can low magnetic field strengths account for cases where absorption features are not seen?) or is it an observational effect (e.g. due to source-observer geometry)? A number of experiments now in the planning or construction phase may eventually provide answers.

4. STATISTICAL PROPERTIES OF GAMMA BURSTERS

Three related issues fall under this heading: the number-intensity relation, the apparent spatial distribution, and the recurrence rate. Figure 4 shows some selected number-intensity (log N-log S) observations.

If all gamma bursters emitted bursts with the same intrinsic luminosity (the "standard candle") assumption, then a uniform unbounded disc distribution would give an S^{-1} law, while a uniform unbounded spherically symmetric distribution would result in a $S^{-3/2}$ law. Thus a careful study of the log N-log S curve might be expected to give an indication of the burster density, and hence the average burster distance; this, in turn, could constrain theories, which variously place bursters at distances from 100 pc to 100 kpc.

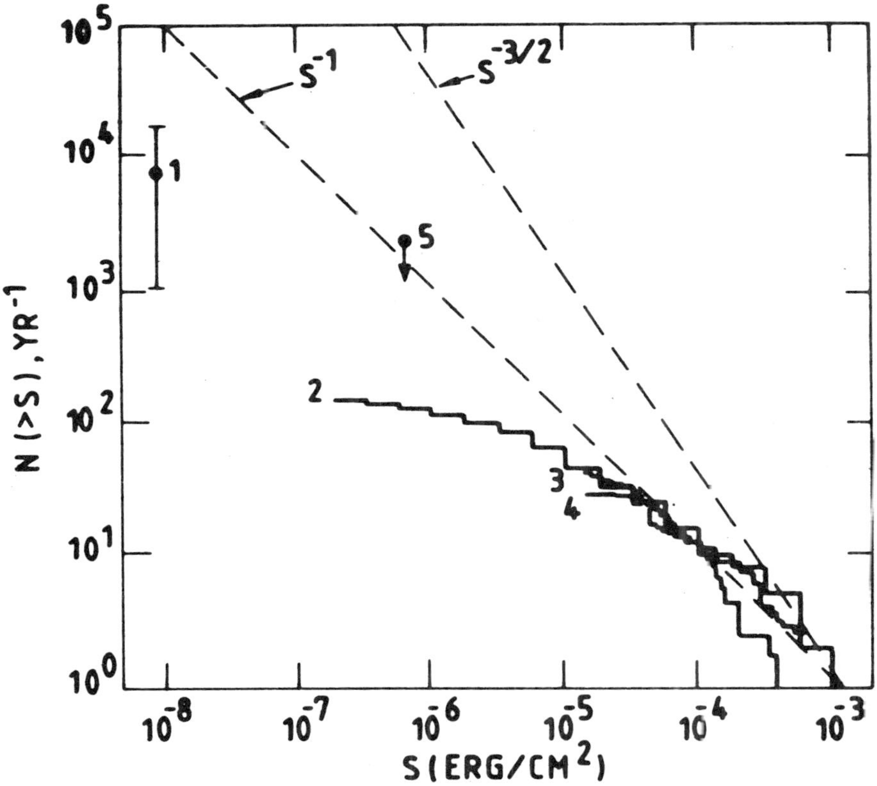

Figure 4. 5 observations of the log N-log S relation. (1) Beurle et al. (1981) (2) Mazets et al. (1982b) (3) Barat et al. (1982) (4) Klebesadel, Fenimore, and Laros (1984) (5) Meegan, Fishman and Wilson (1985). $S^{-3/2}$ and S^{-1} curves are shown for reference only.

At higher fluences, over a limited dynamic range (10^{-3} - 10^{-5} erg/cm^2), the satellite results (solid lines in Figure 4) are generally in agreement with either law. The apparent discrepancy between the curves in the 10^{-3}-10^{-4} range has been shown to be due to statistics and binning procedures (Klebesadel, Fenimore, and Laros 1984). Due to experimental selection effects, these curves bend over at low fluences, but this does not represent the true distribution. At lower fluences, balloon measurements (points 1 and 5 of Figure 4) clearly show a departure from S^{-1} and $S^{-3/2}$ distributions. At the same time, the spatial distribution of bursts in galactic coordinates

(Figure 5) is apparently isotropic, and there is no statistically convincing evidence that weaker bursts are more concentrated towards the equator, as would be expected for a disc distribution. Jennings (1985) has concluded that, if all data are taken at face value, one must conclude that bursters are distributed throughout an extended galactic halo. This conclusion, however, might be expected to change when data from a single experiment covering a wide dynamic range of fluences become available (e.g., the BATSE experiment: Fishman et al., 1984). Thus the burster distance scale remains unconstrained by any hard experimental evidence, and the intrinsic luminosities of bursters may be said to be uncertain by 5-6 orders of magnitude.

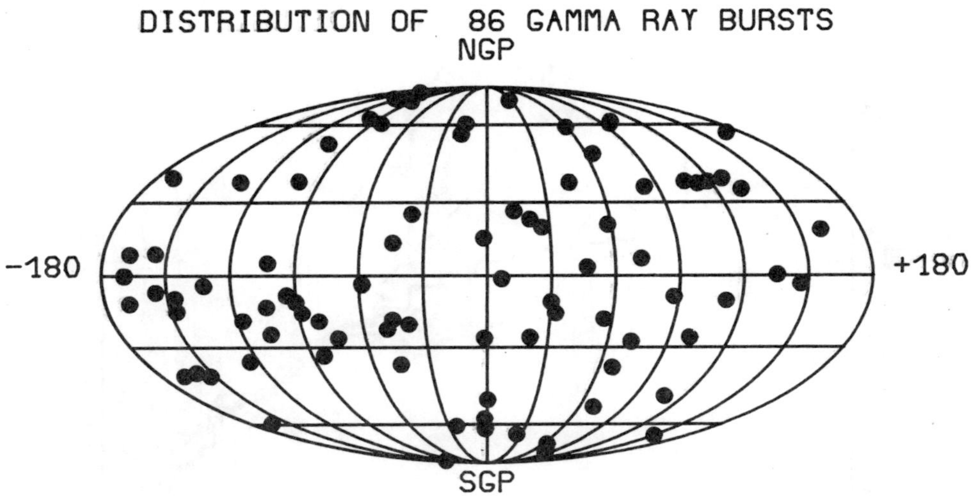

Figure 5. The galactic distribution of gamma burst sources. For details see Atteia et al. (1985a).

Theories might also be constrained by an experimental determination of the burster recurrence rate and luminosity function, i.e. the number of bursts emitted per year and the probability of emitting a burst in a given intrinsic luminosity range. However, only two cases of recurrence have been observed: 3 soft bursts were detected from one source in 3 days (Mazets, Golenetskii, and Guryan, 1979), and 16 bursts (about 0.7/mo.) were detected from the 1979 Mar 5b source (Golenetskii, Ilyinskii, and Mazets, 1984). Schaefer and Cline (1985) and Atteia et al. (1985b), based on a study of the presently available burst localization data, have concluded that these cases of recurrence do not describe the vast majority of bursters. For them, various limits on the recurrence time may be obtained for different assumptions, and the best lower limit to it which can be deduced is around 10 years.

5. BURSTER COUNTERPARTS

Despite numerous deep searches of the smallest gamma ray error boxes, particularly in the optical (e.g., Motch et al., 1985) and soft X-ray (e.g., Pizzichini et al., 1986) ranges, no quiescent burster counterpart has yet been identified. Typical magnitude limits for optical searches are 25. This suggests that bursters, if nearer than 1 kpc, are either lone neutron stars or neutron stars with degenerate companions. The soft X-ray searches, using the Einstein Satellite, reach limits around 10^{-13} erg/cm^2 for the 0.1-4.5 keV range. This leads to model-dependent limits on polar cap accretion rates of 10^{-13} - 10^{-16} $M_\odot$/y.

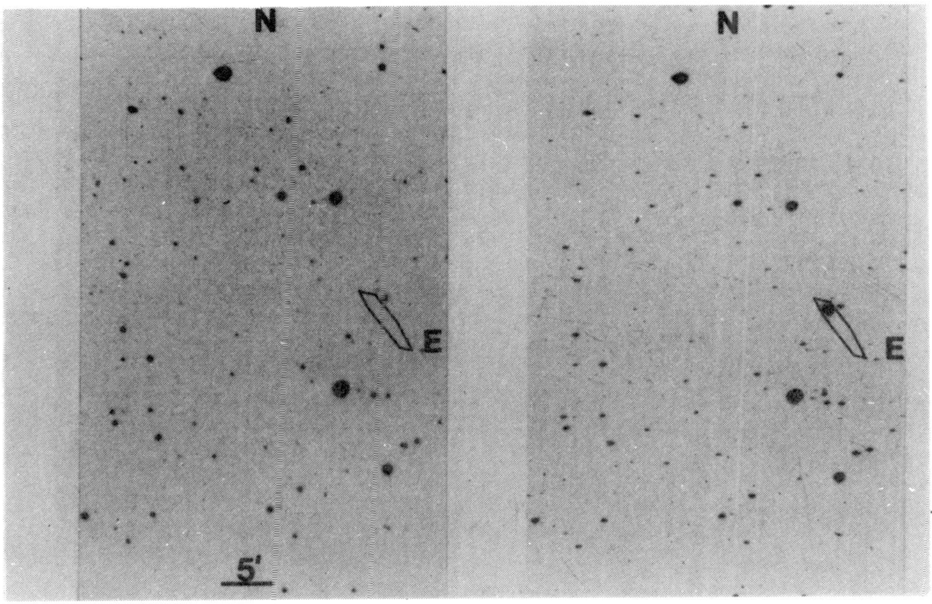

Figure 6. After Schaefer (1981). Two 1928 plates taken at 45 min. intervals, with the 1978 Nov 19 GRB error box superimposed. A transient is visible in the right hand plate. For an assumed duration of 1 s., it reached 3rd magnitude.

Three transient optical counterparts to bursters were found by Schaefer (1981) and Schaefer et al. (1984) on archival photographic plates of the Harvard College Observatory. The brightest of these may have reached 3rd magnitude. As these optical transients were found on plates exposed long before the gamma ray bursts, it is not known whether optical emission generally accompanies a gamma ray burst, or whether the burster emits optical and gamma radiation separately.

Schaefer and Ricker (1983) have shown that if the former is the case, the optical radiation is too energetic to arise simply from the Rayleigh-Jeans portion of a blackbody spectrum peaked in the gamma ray region. The implication in this case is that the gamma radiation is somehow reprocessed and degraded to optical radiation, possibly in the atmosphere of a binary companion (although none has yet been identified in deep searches) or in an accretion disc.

Numerous attempts have been made to detect optical flashes in conjunction with gamma bursts (Pedersen et al., 1984), or on other archival plates (Atteia et al., 1985c), or both (Hudec et al., 1986). None appears yet to have succeeded with any certainty, although continuing searches and new instrumentation hold great promise for a rapid resolution of this problem.

6. CONCLUSION

Although a neutron star origin for bursters seems reasonable, many questions remain unanswered. Perhaps the most fundamental is, What is the ultimate origin of the burst energy? Is it internal, for example due to starquakes, crustquakes, or rotational energy? Or is it external, for example, due to accretion? And if it is external, is the accretion from a disc, a binary companion, or from clouds of solid material (comets, asteroids) surrounding the neutron star? All of these possibilities seem quite capable of furnishing the requisite energies. A second, equally important question is, What causes this energy to be released primarily in the gamma ray region? Another is, What is the nature of the burster? Is the neutron star alone, or in a binary system, and what is its age? No attempt will be made to discuss these questions here; some possible answers may be found in the references cited in Section 1. Thirteen years after their discovery, gamma ray bursts remain in many ways an outstanding astrophysical puzzle.

7. ACKNOWLEDGEMENTS

This work was supported by Centre National d'Etudes Spatiales Contract 86-212.

7. REFERENCES

J.-L. Atteia et al. 1985a, Proc. 19th Int'l. Cosmic Ray Conference, La Jolla, paper OG 1.2-1

, 1985b, ibid, paper OG 1.2-5

J.-L. Atteia et al. 1985c, Astron. Astrophys. **152**, 174

C. Barat et al. 1981, Astrophys. Space Sci. **75**, 83

C. Barat et al. 1982, Astron. Astrophysics **109**, L9

C. Barat 1983, in Positron-Electron Pairs in Astrophysics, Eds. M. Burns, A. Harding, and R. Ramaty, AIP Conf. Proc. No. 101, AIP Press,

American Institute of Physics, New York, p. 54
C. Barat et al. 1984a, Ap. J. 285, 791
C. Barat et al. 1984b, Ap. J. Lett. 286, L11
K. Beurle et al. 1981, Astrophys. Space Science 77, 201
G. Bignami 1985, this meeting.
T. Cline et al. 1980, Ap. J. Lett. 237, L1
B. Dennis et al. 1982, in Gamma Ray Transients and Related Astrophysical Phenomena, Eds. R. Lingenfelter, H. Hudson, and D. Worrall, AIP Press, American Institute of Physics, New York, p. 153
U. Desai 1981, Astrophys. Space Sci. 15, 15
E. Fenimore, R. Klebesadel, and J. Laros 1983, Adv. Space Res. 3(4), 207
G. Fishman et al. 1984, in High Energy Transients in Astrophysics, Ed. S. Woosley, AIP Conf. Proc. 115, AIP Press, American Institute of Physics, New York, p. 651
D. Gilman et al. 1980, Ap. J. 236, 951
S. Golenetskii, V. Ilyinskii, and E. Mazets 1984, Nature 307, 41
D. Helfand 1985, this meeting.
R. Hudec et al. 1986, in preparation, to be submitted to Astron. Astrophys.
G. Hueter 1984, in High Energy Transients in Astrophysics, Ed. S. Woosley, AIP Conf. Proc. 115, AIP Press, American Institute of Physics, New York, p. 373
M. Jennings 1985, Ap. J. 295, 51
R. Klebesadel, I. Strong, and R. Olson 1973, Ap. J. Lett. 182, L85
R. Klebesadel et al. 1982, in Gamma Ray Transients and Related Astrophysical Phenomena, Eds. R. Lingenfelter, H. Hudson, and D. Worrall, AIP Press, American Institute of Physics, New York, p. 1
R. Klebesadel, E. Fenimore, and J. Laros 1984, in High Energy Transients in Astrophysics, Ed. S. Woosley, AIP Conf. Proc. 115, AIP Press, American Institute of Physics, New York, p. 429
D. Lamb, F. Lamb, and D. Pines 1973, Nature Phys. Sci. 246, 52
D. Lamb 1984, in High Energy Transients in Astrophysics, Ed. S. Woosley, AIP Conf. Proc. 115, AIP Press, American Institute of Physics, New York, p. 512
D. Lamb 1985, this meeting.
F. Lamb 1985, this meeting.
J. Laros et al. 1984, Ap. J. 286, 681
E. Liang 1982, Nature 299, 321
E. Liang 1984, in High Energy Transients in Astrophysics, Ed. S. Woosley, AIP Conf. Proc. 115, AIP Press, American Institute of Physics, New York, p. 597
E. Liang 1985, this meeting.
E. Liang and V. Petrosian, eds. 1986, Proceedings of the Stanford Gamma-Ray Workshop, AIP Conference Proceedings Series (in press), American Institute of Physics Press, New York
A. Lyne 1985, this meeting.
S. Matz et al. 1985, Ap. J. Lett. 288, L37
E. Mazets, S. Golenetskii, and Yu. Guryan 1979, Sov. Ast. Lett. 5(6), 343
E. Mazets et al. 1979, Nature 282, 587

E. Mazets et al. 1981, Nature **290**, 378
E. Mazets et al. 1982a, Sov. Astron. Lett. 6(5), 318
E. Mazets et al. 1982b, Astrophys. and Space Sci. 82, 261
E. Mazets et al. 1983, in Positron-Electron Pairs in Astrophysics, Eds. M. Burns, A. Harding, and R. Ramaty, AIP Conf. Proc. No. 101, AIP Press, American Institute of Physics, New York, p. 36
C. Meegan, G. Fishman, and R. Wilson 1985, Ap. J. 291, 479
C. Motch et al. 1985, Astron. Astrophys. **145**, 201
P. Nolan et al. 1983, in Positron-Electron Pairs in Astrophysics, Eds. M. Burns, A. Harding, and R. Ramaty, AIP Conf. Proc. No. 101, AIP Press, American Institute of Physics, New York, p. 59
P. Nolan et al. 1984a, in High Energy Transients in Astrophysics, Ed. S. Woosley, AIP Conf. Proc. 115, AIP Press, American Institute of Physics, New York, p. 399
P. Nolan et al. 1984b, Nature 311, 360
J. Norris et al. 1986, Ap. J. (in press)
H. Pedersen et al. 1984, Nature 312, 46
G. Pizzichini et al. 1986, Ap. J. (in press)
R. Ramaty et al. 1980, Nature 287, 817
R. Ramaty and P. Meszaros 1981, Ap. J. 250, 384
E. Rieger et al. 1982, in Accreting Neutron Stars, Eds. W. Brinkmann and J. Trümper, MPE Report 177, Max-Planck Institute, Garching, F.R.G., p.229
M. Ruderman 1985, this meeting.
B. Schaefer 1981, Nature **294**, 724
B. Schaefer et al. 1984, Ap. J. Lett. 286, L1
B. Schaefer and G. Ricker 1983, Nature 302, 43
B. Schaefer and T. Cline 1985, Ap. J. 289, 400
B. Teegarden and T. Cline 1980, Ap. J. Lett. 236, L67
B. Teegarden 1982, in Gamma Ray Transients and Related Astrophysical Phenomena, Eds. R. Lingenfelter, H. Hudson, and D. Worrall, AIP Press, American Institute of Physics, New York, p. 123
J. Terrell et al. 1982, Ap. J. **254**, 279
J. Trumper 1985, this meeting.
A. Tsygan 1975, Astron. Astrophysics 44, 21
K. Wood et al. 1981, Ap. J. 247, 632
K. Wood et al. 1984, Ap. J. Supp. 56, 507
S. Woosley 1984, in High Energy Transients in Astrophysics, Ed. S. Woosley, AIP Conf. Proc. 115, AIP Press, American Institute of Physics, New York, p. 485

GALACTIC COSMIC RAYS : PROPAGATION AND ORIGIN

Catherine J. Cesarsky
Centre d'Etudes Nucléaires de Saclay
DPhG/SAp, 91191 Gif-sur-Yvette Cedex
France

ABSTRACT. I review here some of the current research on the propagation and the origin of galactic cosmic rays. The abundances of secondary nuclei, generated by spallation reactions in the interstellar medium, are well accounted for by simple models where cosmic rays escape from the galaxy and/or from their source in an energy dependent way ; this may be due to the scattering of cosmic rays by resonant irregularities resulting from the general interstellar turbulence. Protons and helium nuclei in cosmic rays may have a different history than the rest of the particles. Supernovae are the most probable accelerators of cosmic rays. The shock acceleration mechanism can account for the energy budget of cosmic rays, and for their spectrum, but only up to a maximum energy $<10^4$-10^5 GeV for protons. Acceleration in an accretion shock surrounding a collapsed star may lead to higher energies. The source composition of cosmic rays is quite similar to that of ordinary matter, except that cosmic rays are comparatively poor in H and He by an order of magnitude. The abundances of the other elements mimick those of solar cosmic rays and of the solar corona, perhaps indicating that cosmic rays are injected by stellar flares. Other interpretations of the cosmic ray source abundance are discussed.

1. Introduction

In these lectures, I will attempt to summarize some of the current research on cosmic rays. The reader is immediately warned that he will find here many questions, many arguments, and comparatively few answers. This may partly reflect the general skepticism of the writer, and partly the difficulties of this field, where each piece of observational evidence is long sought for, and hard to obtain and confirm ,- still, over the past few years very fine data on cosmic ray composition, and on anisotropy have been gathered, and the theory of particle acceleration by collisionless shocks has been developed. Some pieces of the puzzle have been put together, but a complete picture has not emerged yet.

F. Pacini (ed.), High Energy Phenomena Around Collapsed Stars, 331–358.

I first discuss propagation aspects, in the light of recent observation of the cosmic ray composition. In section 3, I present the requirements on cosmic ray sources and on acceleration mechanisms that can be derived from the data, under various interpretations. Section 4 is devoted to a discussion of acceleration mechanisms, and particularly of the acceleration of cosmic rays by collisionless shocks in a scattering medium. Finally, section 5 reviews many of the different proposals that have been put forward to explain the cosmic ray source composition.

Throughout the lectures, we often refer to cosmic ray rigidities, rather than to energy per nucleon E or total energy E_t. The rigidity of a nucleus of charge Z is defined by : R = pc/eZ GV, where p = momentum, c = velocity of light, e = electron charge. The rigidity of a particle is proportional to its gyroradius r_e in the presence of a magnetic field : $R = r_e Bc$, where B is the magnetic field strength , so that the rigidity is the relevant quantity for describing the dynamics of charged particles in the presence of magnetic fields. At relativistic energies ($E \gg mc^2$), for protons R(GV) = E(GeV). For most of the other nuclei, $A/Z \simeq 2$, and $R(GV) \simeq 2E(GeV/n)$.

In the interstellar medium, the magnetic field strength is $\sim 3\mu G$, so that the gyroradius is $r_e \simeq 10^{12} R_{GV}$cm.

2. Galactic propagation of cosmic rays

a) Clues from elemental abundances

As cosmic-ray nuclei travel through interstellar space, they suffer inelastic collisions with interstellar medium nuclei ; in this way "primary" cosmic-ray nuclei emitted by sources break up into lighter "secondary" nuclei. The amount of interstellar matter traversed by cosmic rays can be estimated by measuring the abundances of certain species expected to be absent in the primary spectrum.

At energies greater than a few GeV/nucleon, the effects of solar modulation and of Coulomb interactions in the interstellar medium are negligible and the cross sections of the spallation reactions affecting the cosmic ray composition are energy-independent.

Phenomenologic models of cosmic ray confinement are essentially motivated by the observed isotropy of the radiation, and by composition data. The most widely used model is the "leaky-box" model, which assumes that cosmic rays are trapped between reflecting boundaries surrounding the galaxy, but with a finite probability of escape into extragalactic space. The cosmic ray density is uniform throughout the confinement volume.

In the leaky-box model, assuming that the interstellar gas consists only of hydrogen, and that the energy is high enough (greater than ~ 10GeV/n) that ionization losses can be neglected, the flux f_j of a species i (where i is the atomic number) is simply related to the source term $Q_i(cm^{-3}s^{-1})$ and the mean escape length $\lambda_e(g\ cm^{-2})$ through

$$f_i = \frac{Q_i/n_H m + \sum_{j>i} \sigma_{j,i} f_j/m}{(\lambda_{di})^{-1} + (\lambda_e)^{-1}} \qquad (1)$$

where $\sigma_{j,i}$ is the cross section of the spallation reaction j(p,)i induced by the element j on interstellar hydrogen atoms, and λ_{di} the pathlength for nuclear destruction of nuclei i on hydrogen ; λ_{di} decreases when i increases [e.g. λ_d(He) = 17 g cm^{-2}, λ_d(C)=7 g cm^{-2}, λ_d (Fe) = 2.5 g cm^{-2}] For purely secondary species, such as the light elements lithium, beryllium, and boron, Q_i =0 and the knowledge of the flux f_i and of the nuclear cross sections involved is sufficient to determine the mean escape length λ_e.

Several balloon measurements of cosmic ray composition at energies up to 150 GeV/n have shown that the ratio of secondary to primary abundances decreases as the energy increases (Juliusson et al. 1972, Ormes and Freier, 1978, and references therein) ; also, the observed spectra of heavy primary species are flatter than those of lighter ones. More recently, the french-danish spectrometer C2 on the satellite HEAO3 has provided extremely accurate data on the cosmic ray elemental composition from boron to zinc in the energy range 0.8-25 GeV/n (Koch-Miramond 1981).

This data make it possible to calculate λ_e as a function of R, in the leaky-box model. Koch-Miramond et al. (1983) have corrected the low energy part of their data for the effects of solar modulation, assuming a modulation parameter Φ = 600 MV, which is apropriate for the time in the solar cycle at which the measurements were made. They find that, at R≥5.5 GV, an escape length λ_e=22 $R^{-0.6}$g/cm^2 of pure hydogen accounts for the secondaries of C, O and Fe (fig 1).

Very recently, the american team of the HEAO3-C3 experiment, taking advantage of the relativistic use of signals in the ionization chambers of their instrument, obtained results on the abundances of several elements from 10 GeV/n up to an energy ≥ 100 GeV/n. Their results are consistent with those of the french-danish group in the range (10-25 GeV/n) where both experiments apply ; at higher energies, the HEAO3-C3 data indicates that the power law dependence with energy of the ratios (iron secondaries/iron) derived by the HEAO3-C2 data extends to about 100 GeV/n, or rigidities of about 200 GV (Jones et al. 1985).

What happens at even higher energies, or rigidities, such as R ≥ 1000GV? The ratio of secondary to primary element abundances have not yet been measured at such rigidities. The prediction of the leaky-box

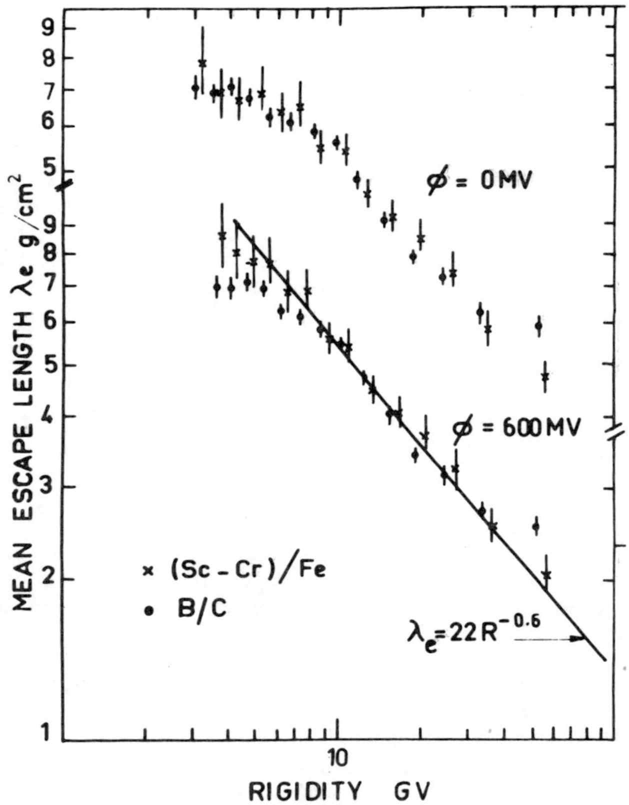

Figure 1. (from Koch-Miramond et al. 1983). Mean escape length as a function of rigidity for a modulation parameter ϕ = 600 MV.

model is rather stringent : even though the all-particle integral spectrum may show some structure at $E_t > 10^{14}$eV , the proton differential spectrum does not appear to suffer any drastic change of slope in the 10-10^6 GeV region. (fig. 2). In the leaky-box model, the mean confinement time of particles, τ_e, is proportional to λ_e.

Neglecting nuclear losses, the cosmic ray density N is related to the source term S through $N = S\tau_e$. Under the plausible assumption that the source spectrum is a power law, $\tau_e(R)$ must also be a power law at least up to $\sim 10^6$GV.

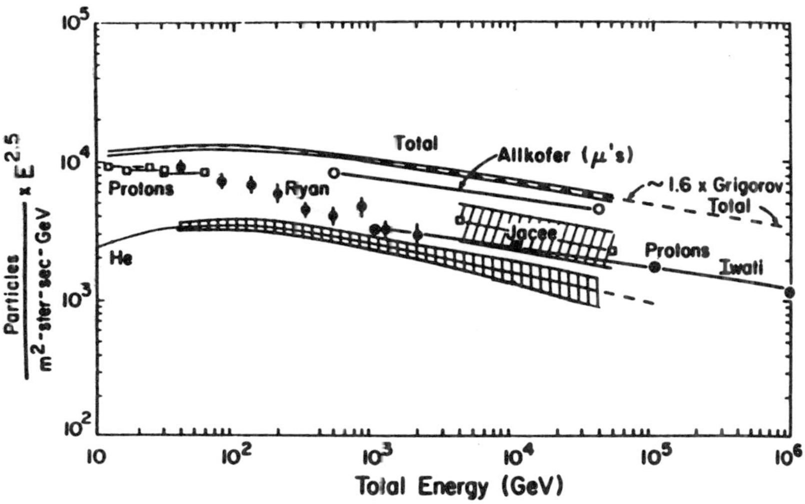

Figure 2. (from Webber, 1983). Direct measurements of total energy spectra between 10-10^5GeV. Measurements of the primary proton only spectrum are shown as well as the He spectrum converted to a total energy spectrum. The total particle spectrum is also shown.

In the nested leaky box model (Cowsik and Wilson; 1973, Meneguzzi 1973), cosmic rays are trapped both near their sources and at the boundaries of the galaxy, with a finite probability of escape from each. The assumption made by these authors is that it is the pathlength traversed in the sources,λ_s, which is rigidity dependent. The composition and the spectra of primaries and secondaries are essentially undistinguishable from those obtained with the energy-dependent leaky-box model, but in this case the galactic proton spectrum is identical to the injection spectrum, independently of the form of $\lambda_s(R)$.

In the leaky-box model, the distribution of pathlengths around the mean is exponential. In contrast, the nested leaky-box model predicts a deficiency of short pathlengths. At R > 5GV, present results of the HEAO3-C2 experiment, together with earlier results, can be accounted for with an exponential distribution of pathlengths.

In most diffusion models, the elemental composition of cosmic rays is also determined almost exclusively by one parameter, λ_e, related to the amount of matter traversed by the particles before escape ; in general, λ_e is inversely proportional to the diffusion coefficient K (in one-dimensional models, or in three-dimensional models with scalar diffusion) or to the component of the diffusion tensor perpendicular to the galactic plane. The constant of proportionality contains all the information on the distribution of the sources and on the boundaries of the containment region. For instance, let us consider one-dimensional models, where the cosmic ray sources are embedded in the gas disk of uniform density n_0 and of height h ; cosmic rays diffuse outwards through a halo of height $H \gg h$ (Ginzburg et al. 1980). The diffusion coefficient K is assumed to be constant in space. Then K is related to the mean escape length λ_e, calculated with the leaky-box formula (1) by :

$$K = (n_{0H} hvm)/\lambda_e. \qquad (2)$$

In terms of diffusion models, then, the elemental composition of cosmic rays can be interpreted as implying that either $K \propto vR^{0.6}$ (my preferred explanation), or that the size of the confinement region decreases with R : $H \propto R^{-0.6}$.

All the models discussed in this section assume that the only energy changes that cosmic rays undergo between production and detection are ionization losses in the interstellar medium, and adiabatic losses during solar modulation. If cosmic rays are accelerated (or decelerated) by some additional mechanism while propagating, secondary particles get transferred to higher (lower) energies, and the (secondary/primary) profile as a function of energy is altered (e.g. Fransson and Epstein, 1980). The fact that the data at $R >$ a few GV is well explained by a simple theory indicates that, at such rigidities, re-acceleration or deceleration is probably only a minor effect.

b) Cosmic ray diffusion and interstellar turbulence spectrum

What may cause cosmic rays to diffuse in interstellar space ? Fermi (1949) has pointed out that large scale ($\gg r_e$), moving inhomogeneities in the magnetic field reflect particles of large pitch angle ; this process can lead to both diffusion and acceleration of cosmic rays. But the Fermi acceleration model has difficulties in satisfying the energy requirements, and in explaining the observed abundances of secondary nuclei. In the last ten to fifteen years, the work on cosmic ray propagation has mostly concentrated on another process : resonant scattering of cosmic rays by hydromagnetic waves whose scales are comparable to their radius of gyration (Wentzel 1974, and references there in). This scattering leads to cosmic ray diffusion along the magnetic field lines ; there is some energy exchange between cosmic rays and h.m. waves, but only to higher order in v_A/c, where

v_A is the Alfven velocity, $(B/\sqrt{4\pi\rho^*})$ ~ tens of km/sec (ρ^* is the density of ionized matter).

Let us define F(k) as the energy density in hydromagnetic waves per logarithmic bandwidth d(log k), relative to the ambient magnetic energy density $(B^2/8\pi)$. Then, in the framework of the quasi-linear theory (applicable if F<<1), the diffusion coefficient along field lines of particles of rigidity R and velocity v is given by :

$$K(R) = \frac{4}{3\pi} \frac{v\ R/Bc}{F(k = 1/r_e = Bc/R)} \quad (3)$$

The spectrum of hydromagnetic turbulence F(k) in the interstellar medium is extremely difficult to determine. Various methods exist that can lead to estimates or upper limits of the density spectrum of irregularities in the distribution of thermal electrons. Presently available results have been compiled by Armstrong et al. (1981). These authors conclude that the data is consistent with a power law spectrum of fluctuations, with an index of -3.6 ± 0.2. If the hydromagnetic spectrum had the same slope, this would be equivalent to :

$$F(k) \propto k^{-0.6 \pm 0.2}. \quad (4)$$

A spectrum of this type may be the result of a cascade of turbulent energy in the interstellar medium from long scales to successively shorter scales ; the turbulence at long scales is fed by cloud motions, which in turn are regenerated by supernova explosions. Kraichnan (1965) has argued that a cascade in an incompressible, weakly turbulent magnetized fluid, leads to a spectrum $F(k) \propto k^{-0.5}$. Such a cascade is energetically feasible in the hot phase ($T \sim 10^6$°K, $n \sim 10^{-3}$ cm^{-3}) of the interstellar medium . If $F \propto k^{-0.5}$ there, then we see from the formula (3) that the cosmic ray diffusion coefficient $K \propto vR^{0.5}$;this is very close to the dependence required to account for the observed variations of the ratio of secondary to primary nuclei with energy. Thus, the present observations of elements heavier than He, at energies lower than ~ 1000 GeV/n, are well accounted for by a model where cosmic rays are scattered by resonant hydromagnetic waves related to the general interstellar turbulence (Cesarsky, 1975, 1980).

c) Radioactive Secondary Nuclei

Measurements of the abundances of unstable secondary nuclei, such as ^{10}Be (with a decay period at rest of τ_d = 2.2 x 10^6 yr), ^{26}Al(τ_d = 0.85 x 10^6 yr), and ^{35}Cl (τ_d = 0.45 x 10^6 yr), can bring some information on the mean age of cosmic rays, and/or help to determine the parameters characterizing the different models.

In the framework of the leaky-box model, such measurements, combined with the determination of λ_e through the elemental composition, permit us in principle to estimate the mean escape time of cosmic rays and the mean gas density in the box. However, because most measurements are done at low energies, solar modulation again complicates the interpretation of the data. Assuming that λ_e is energy independent, and using their own estimates of solar modulation effects, Wiedenbeck and Greiner (1980) deduce from their satellite data on ^{10}Be at 60-185 MeV/n a confinement time of 8.4 (+4.0,-2.4)Myr, and a mean density n_H = 0.33(+0.13,-0.11)cm^{-3}.

The mean age from ^{26}Al is 9(+20,-6.5)Myr (Wiedenbeck 1983), and ^{36}Cl leads to a lower limit to this age of 1 Myr (Wiedenbeck 1985). Since, in the solar neighborhood, the interstellar density (averaged over ~ 1 kpc in the disk) is estimated at 1-2 cm^{-3}, these results are generally interpreted as implying that galactic cosmic rays circulate in a low density halo which is at least 3 times wider than the disk.

In diffusion models with a halo, radioactive isotopes formed in the disk often decay while passing through the halo. In that case, the average confinement time of particles in the galaxy may be much larger than the observed "mean age" (Ginzburg et al., 1980). For instance, in the one-dimensional model described earlier, the abundances of secondary radioactive elements of decay period τ_d are determined by two combinations of parameters : $(n_0 \tau_d)$ and (H/h). In principle, observations of the energy dependence of the abundance of isotopes of mean life at rest of ~ 10^6 years, at energies ≥ 1 GeV/n, should help constraining these parameters (e.g. Cesarsky et al. 1981).

d) Electron Spectrum

According to several recent measurements, the electron spectrum is parallel to the proton spectrum in the energy range 2-10 GeV ; in this range, the electron flux amounts to ~ 1 % of the proton flux. At higher energies ≥ 50 GeV, the spectrum steepens, but electrons are still present at least up to 2000 GeV (Nishimura et al. 1985 and references there-in). A steepening of the high energy spectrum is expected, since the lifetime of a 30 GeV electron against radiation losses in the interstellar medium is ~ 10^7years.

The observed electron spectrum does not impose strong constraints on the models proposed to explain the cosmic ray composition. It is important to remember that the equations describing the behavior and the energy changes of high energy electrons diffusing through the interstellar medium cannot be approximated by results obtained using the leaky-box model. In diffusion models, the distribution of the sources plays an important, or even a predominant role. In addition, the injection spectrum of electrons is not known, and can generally be adjusted to ensure that a given model fits the data.

3. Origin and propagation of different cosmic ray species.

a) Source spectral index : composition or anisotropy ?

After so many years of active research, there is not yet a firm answer to the question : where do cosmic rays come from ? The main problem is, of course, that the arrival direction brings little or no information on the source. Astrophysicists are then left with a less direct set of clues : spectrum, composition, energetics, anisotropy.

Observations must first be corrected for propagation effets ; this is usually done in the framework of the galactic leaky-box model. Once λ_e is derived, at various rigidities, by applying eq. (1) (or its equivalent, including ionization losses, at energies < a few GeV/n) to secondary species, it becomes possible to derive the source abundances Q_i by applying the same equation to primary species. In this way, Engelmann et al. (1985) have derived source spectra of primary species with Z > 5 from the HEAO3-C2 data. If H and He nuclei behave like the other species, to correct for propagation, the observed spectrum must be divided by $\lambda_e(R)$. The source spectra thus obtained are displayed in fig. 3. Data from other experiments are also represented .
In the range R ≃ 2 - 200 GV, Engelmann et al. (1985) find that, while the spectra of primaries with Z > 5 are all similar power laws with index 2.4, the spectrum of protons appears to be flatter, with an index ≃ 2.1. (For He, the situation is somewhat unclear, as can be seen on fig. 3).

The implications of this result, which is awaiting further confirmation, have not yet been studied in full detail. Essentially all of the published work on cosmic ray origin continues to assume that protons and alpha particles originate and propagate as the other species, and that the λ_e derived from studies of heavy nuclei can be used to estimate the energetics. For the local kpsec^2 in the galactic plane, cosmic ray energetics is derived using the fact that, on the average, cosmic rays escape at a rate c λ_{gal}/λ_e,where λ_{gal} is the column density of matter across the galactic disk. The energy requirement to maintain the cosmic ray pool is then ~ 10^{38}erg/Kpc^2 sec. (Alternative derivations, using also the cosmic ray "age" derived from secondary radioactive isotopes, yield similar results). If we retain the same leaky-box model for all species, the results of Engelmann et al. (1985) imply that the local cosmic rays consist of two components : a flat component, with source index ~ 2.1, and a steep component, with source index ≃ 2.7. At rigidities below ~ 100 GV, most of the nuclei heavier than He would belong to the steep component, while at all energies the flat component would be dominant in the proton flux.

The leaky-box formalism, as we have seen, accounts well for the observations relating to the steep component which is rich in heavy

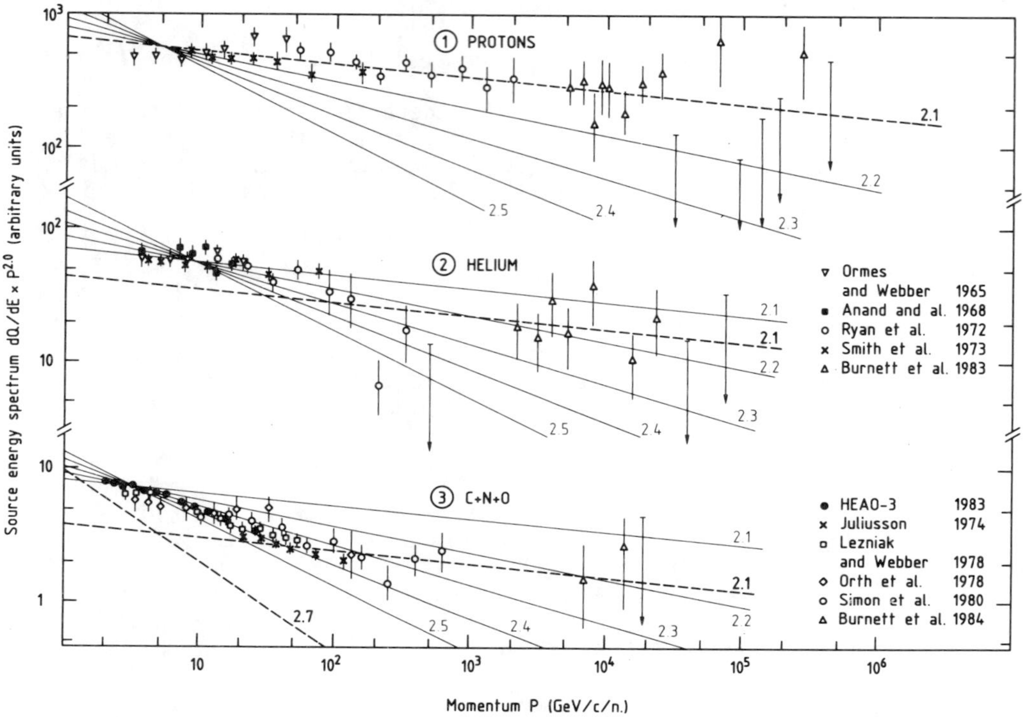

Figure 3. (from Engelmann et al. 1985). Differential source strength dQ/dE vs. momentum for (1) protons, (2) He, (3) CNO nuclei. The spectra have been flattened by multiplication by $P^{2.0}$, where P=p/A. The proton spectrum is well fitted by a power law with index γ = 2.1. The dashed lines for He and CNO are an attempt to fit their high energy data with the same spectral index. In the case of CNO nuclei, a steep component ($\gamma \approx 2.7$) would have to be added to this main component.

nuclei. But there is no compelling reason to believe that the flat component, which is relatively proton-rich, has the same history. The steep component may be just local, and transient ; the determinations of λ_e and of age from radioactive isotopes only relate to this component. But the proton rich component is the only one that counts when discussing energetics, constancy in time of the cosmic ray flux, and isotropy, and the abundances of secondary elements with $Z \geqslant 2$, at energies $\leqslant$ 100 GeV, may simply not be relevant when studying it !

Some light can be thrown on this problem by refining the spectra of hydrogen and helium, and studying carefully their secondaries ^{3}He, D and antiprotons.

b) Antiprotons

Secondary anti-protons are generated in the inelastic collisions between high-energy nuclear cosmic rays and interstellar medium particles. The flux of galactic anti-protons has been measured recently by Golden et al. (1984), by Bogomolov et al. (1979) and by Buffington et al. (1981) at various energies (fig. 4). The p flux observed at 10GeV is lower by a factor ~ 6 than the prediction of the energy dependent leaky-box model that accounts for the abundances of secondary nuclei such as Li,Be,B.

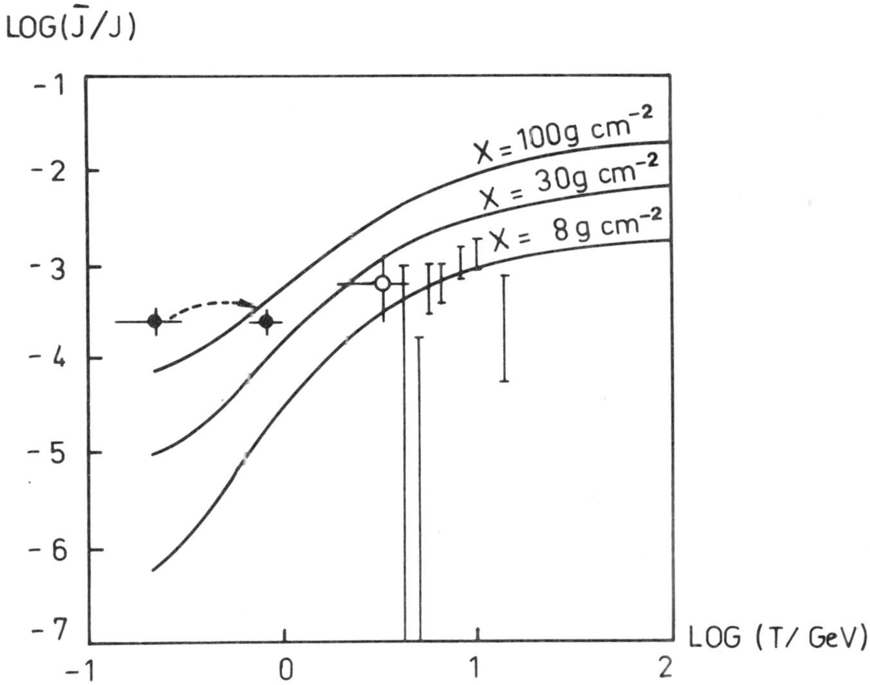

Figure 4. (from Lagage and Cesarsky 1985). Anti-proton flux leaving thick sources with grammages 8,30,100g cm^{-2}, compared with observations at different energies. Black dots, Buffington et al. (before and after modulation); White dots, Bogomolov et al.; Bars, Golden et al. The proton momentum spectrum taken is proportional to $p^{-2.1}$.

Buffington et al. (1981) measured the flux of cosmic ray antiprotons in the range 130-320 MeV, corresponding after demodulation to a interstellar energy of ~ 800 MeV. Their result, which has not yet received an independent confirmation, is more than an order of magnitude higher than the predictions of the standard propagation models.

It may be that this unexpectedly high abundance of antiprotons is an additional indication that the history of all cosmic rays is not as deduced from the abundances of secondary nuclei alone. For instance, the high energy observations of antiprotons could be accounted for if all cosmic ray protons had a source spectrum of index 2.1 and traversed 7.5 g/cm^2 in their sources before escaping into the galaxy, or if a fraction x of the cosmic rays traversed a slab of width X at the source, with $xX = 7.5$ g/cm^2 (Lagage and Cesarsky 1985). As noted by these authors, a problem with this "thick-source" model is that, in addition to the antiprotons, neutral pions are produced, which decay into gamma rays. The total galacic gamma ray flux predicted by this model exceeds that observed by COSB by a factor ~ 3. Alternatively, antiprotons may have a more exotic origin. They may be primary particles, e.g. originating in antimatter galaxies, (Stecker et al. 1983) or from evaporating black holes (Kiraly et al. 1981) ; they may be produced by p-p collisions in relativistic plasmas (Dermer and Ramaty 1985), e.g. in the accretion shock surrounding a compact object ; or by decay of photinos making up for the dark matter in the galactic halo (Silk and Srednicki 1984)...

A more complete set of observations of the spectrum of cosmic ray antiprotons than presently available is required to further investigate these hypotheses. For lack of space, I do not discuss here the present results on D and ^3He, but I stress that, there too, additional observations would be very useful.

c) Anisotropy

An alternative point of view has been taken by Hillas (1984), who uses the anisotropy as the main indicator on the propagation of galactic cosmic rays. This can only be done at energies $\geq 10^2$, 10^3GeV, since at lower energies the trajectories of the cosmic rays are perturbed by the solar wind. Hillas notes that, at energies $\geq 10^3$GeV, the amplitude of the first harmonic of the cosmic ray anisotropy is, very roughly, proportional to the product (cosmic ray differential flux . $E^{2.47}$) (fig. 5). Now, if τ_e is the confinement time, the anisotropy is expected to be ~ t/τ_e, where t is the time for escape in a straight line. Hillas proposes a simple interpretation of fig. 5 : that the source spectrum is a power law of index 2.47 over the whole energy range, and that all the features in the spectrum are due to propagation effects. At 10^3GeV, the amplitude of intensity variation is of ~ 0.06%. If the boundary of the cosmic ray confinement region is at x kpcs, $\tau_e(10^3 \text{ GeV}) \simeq 5x$ Myr. Since the spectrum of protons does not appear to change significantly between 5 and 10^3 GeV (see fig. 3), the mean age at 5 GeV would be ~ $(1000/5)^{2.7-2.47}\ \tau_e(10^3 \text{ GeV}) \simeq 17x$ Myr (where 2.7 is the observed index of the proton spectrum at these energies). This is comparable to the age derived from radioactive secondary isotopes, so the global energetics of galactic cosmic rays is not very much changed.

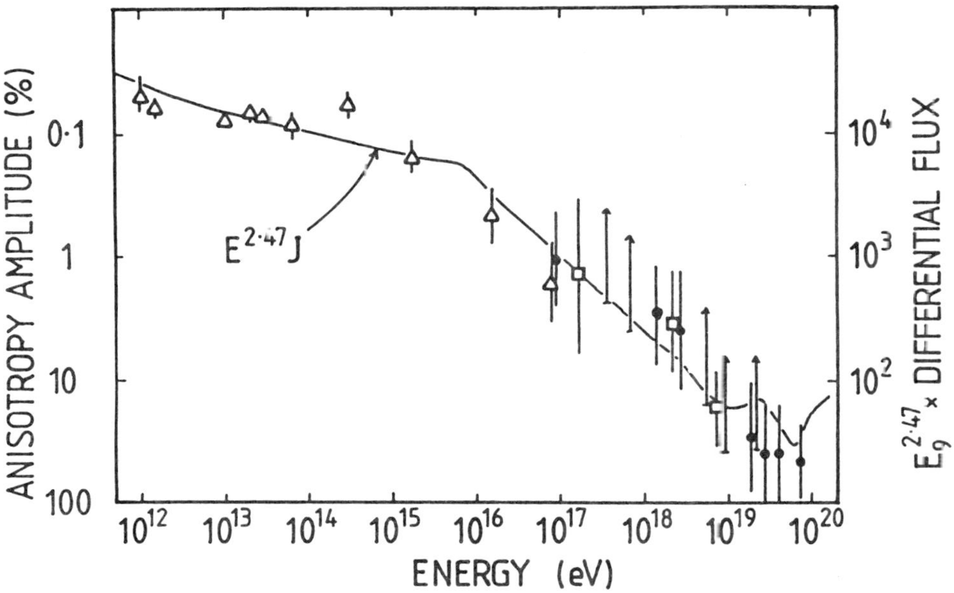

Figure 5. (from Hillas, 1984). Amplitude of first harmonic as a measure of residence time : variation in anisotropy compared with variation in flux.

d) Cosmic ray sources

In summary : what are the requirements on cosmic ray sources ?

i) energetics : the order of magnitude of the power required to replenish cosmic rays "within" a cylinder of base 1 Kpc^2 within the galactic disk, of height 1 to ~ several Kpc, is ~ 10^{38} ergs/sec.

ii) source spectrum : most probably a power law, at least in the range from a few GeV/n to ~ 10^6GeV/n, perhaps up to 10^8 or even 10^9 GeV ! Spectral index : 2.1 ? 2.4 ? or 2.7 ? Or somewhere in this range.

iii) source composition : well determined now, for most elements, in the GeV range. (See later). May give clues to the origin of the cosmic radiation or at least, as we have seen, to a component of it.

Within a radius of 3 Kpc from the sun, the average energy input from supernovae is estimated to be $\sim 10^{39}$ erg $s^{-1}kpc^{-2}$; supernovae are widely believed to be the main accelerators of cosmic rays. Stellar winds expend $\sim 10^{38}$ erg $s^{-1}kpc^{-2}$ in the interstellar medium, and they may also contribute (Cesarsky and Montmerle 1983). Composition arguments have often been invoked to eliminate another likely candidate, from the energy budget standpoint : pulsars. But the debate on the role of pulsars in cosmic ray acceleration is not closed. (see lectures by M. Ruderman).

4. Acceleration mechanisms

Let us now turn to acceleration mechanisms capable of producing a power law spectrum.

a) Fermi acceleration

The basic concept of acceleration of particles via encounters with "moving magnetic walls" was introduced by Fermi as early as 1949. Consider a fast particle of velocity v, that encounters a magnetic wall moving at a velocity V. The wall reflects the particle. In the frame of the wall, there is only a static magnetic field, and thus no energy exchange : just an elastic collision. Transforming back to laboratory space, however, one finds that the particle energy has changed by an amount :

$$\Delta E \simeq E\ (1 + 2Vv \cos \Theta/c^2 + 2\ V^2/c2) \qquad (5)$$

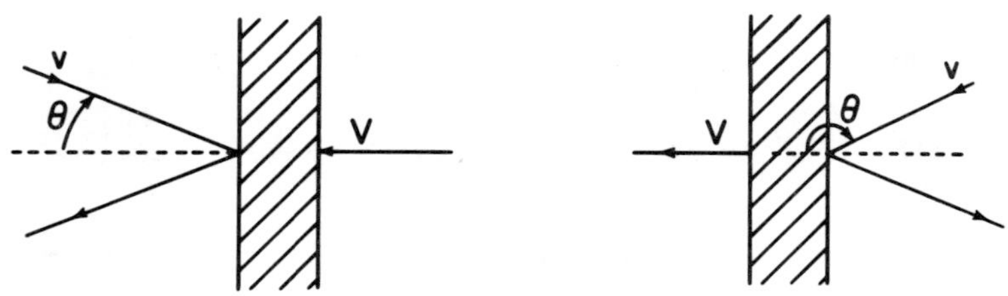

Figure 6 . Head-on and overtaking encounters of a particle of velocity v with a magnetic wall of velocity V.

To first order in V/c, particles gain energy if $\cos \Theta > 0$ (head-on encounters) and loose energy if $\cos \Theta < 0$ (overtaking encounters).

Let λ be the mean distance between moving walls. If somehow the particle only makes head-on collisions --e.g., if it is trapped between two magnetic walls approaching each other, the rate of energy gain, for relativistic particles, is $dE/dt = a\, VE/\lambda$, where a, of order 1, depends on the distribution of incidence angles θ. This type of process is called "first order Fermi mechanism".

If the walls move at random, the first order terms in equation (5) cancel each other, when summing over all encounters. The second order term in eq. (5) guarantees that the particles still gain some energy ; in addition, for a given value of $|\cos\Theta|$, head-on encounters are more probable than overtaking encounters by a factor $(v|\cos\Theta|+V)/(v|\cos\Theta|-V)$. Consequently, the rate of energy gain of relativistic particles is only :

$$\frac{dE}{dt} = \frac{2V^2}{c\lambda} E = \alpha E \tag{6}$$

(where we have assumed that particles and clouds have isotropic distributions of velocities). This is called second-order or statistical Fermi mechanism.

One of the main reasons why this process has enjoyed such an endurable popularity among astrophysicists is that, under very simple conditions, it predicts that the energy spectrum of the colliding particles should be a power law, and power law spectra are extremely frequent in non-thermal sources of radiation, all over the universe. Assume that a group of particles is undergoing Fermi 1st or 2nd order acceleration in a box, from which they have a constant probability to escape, while particles of energy E_0 are being supplied, at a constant rate. For particles in the box,

$$\frac{dE}{dT} = \alpha E, \tag{7}$$

so that at age t, their energy is :

$$E(t) = E_0 \exp(\alpha t) \tag{8}$$

If T is the mean escape time, the probability of having an age t is :

$$P(t)\, dt = (dt/T) \exp(-t/T) \tag{9}$$

Therefore, the particle spectrum is :

$$N(E)dE = P(t)dt = \frac{1}{E} \exp\left(-\frac{1}{\alpha t} \ln \frac{E}{E_0}\right) dE \tag{10}$$

or :

$$N(E) \propto E^{-\gamma}$$
$$\gamma = (1 + 1/\alpha\ T)$$

The great drawback of this theory is that the observed exponents γ in cosmic rays and cosmic sources are within a narrow range (most frequently $2 < \gamma < 3$), while here γ and T appear to be completely independent of each other (but see Syrovatskii, 1961).

b) Particle acceleration by parallel shocks in a scattering medium.

This attractive mechanism must have been in the air several years ago, as it has been discovered simultaneously by astrophysicists all over the world (Krimsky, 1977, Axford et al. 1977, Blandford and Ostriker 1978, Bell 1978). This is somewhat surprising, as the tools used in the various derivations, and the motivation, have been around for a much longer time.

Let us consider a strong shock, propagating at a velocity V in the direction of the magnetic field lines. We assume that V v_A, where v_A is the Alfven velocity . In the shock frame, the gas is flowing in at a velocity $u_1 = V$. At the shock, the gas is compressed by a factor r, so that the velocity downstream, relative to the shock, is $u_2 = V/r$.

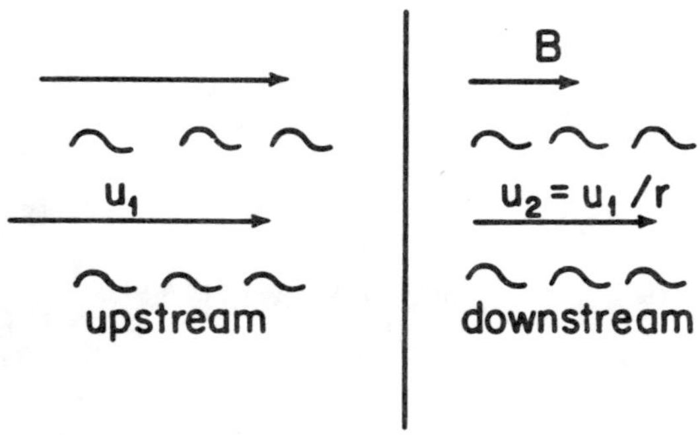

Figure 7 . Parallel shock in a scattering medium

The presence of scattering centers of cosmic rays is postulated, so that cosmic rays diffuse on both sides of the shock ; the diffusion coefficient is, in general, a function of space, of particle momentum, and of time. In any case, the scattering centers act as cosmic ray traps, ensuring that the particles will be reflected back and forth across the shock a large number of times. Every passage through the shock is equivalent to running head on into a "magnetic wall" of velocity $V = u_1 - u_2 = V(1 - 1/r)$; averaging over incidence angles, we obtain the mean gain per passage :

$$\Delta E = (4/3)\ (V/c)\ (1-1/r)E$$

Assume that particles of energy E_0 are continuously fed to the shock. To complete a picture "a la Fermi", we need to know the mean probability of escaping the system. Bell (1978) has put forward the simplest derivation of this probability. His argument goes as follows : the diffusion mean free path of cosmic rays, λ, is much larger than the width of the shock. Consequently, the cosmic ray density is continuous across the shock. The scattering ensures that, far from the shock, the distribution of the momentum vectors of cosmic ray is isotropic in the frame of the gas. The shock generates anisotropies of order V/c ; but, to zero order in V/c, the distribution at the shock is isotropic. Thus, the flux of cosmic rays crossing the shock, at any given time, is simply (1/4) n(o) v, where n(o) is the cosmic ray density at the shock, and v the c.r. velocity. The number of particles swept back downstream, and thus escaping the system every second, is $n(o)u_2$.Thus, the escape probability per cycle is :

$$\eta = \frac{n(o)u_2}{n(o)v/4} = \frac{4u_2}{v} \qquad (11)$$

After ℓ passages through the shock, the cosmic ray energy is :

$$E\ell = E_0 \exp\left\{\frac{4}{3}\ \ell\ \frac{V}{c}(1-1/r)\right\} \qquad (12)$$

$$P_\ell = (1 - \eta)^\ell \qquad (13)$$

Folding E_ℓ and P_ℓ, we obtain the time independent spectrum :

$$N(E) \propto E^{-\mu},\ \mu = (2+r)/(r-1) \qquad (14)$$

For strong adiabatic shocks, r=4 and μ=2. Weaker shocks generate steeper spectra.

The remarkable property of this mechanism is that, in the time-independent limit, the slope of the power law it generates depends only on the shock, and not at all on the diffusion coefficient (assumed "small enough") or the dimensions of the scattering region (assumed "large enough"). This is a consequence of eq. (11).

The study of shock acceleration of cosmic rays is now an active area of research. A fundamental review of the subject has been written by Drury (1984). A detailed application of the linear, time independent mechanism I have just described to the acceleration of galactic cosmic rays is given in Blandford and Ostriker (1980) ; see also Axford (1981).

Many aspects of this mechanism have been studied since, and it is impossible to review this rich field here. Let us just emphasize some of the main problem areas :

i) This problem has always been treated in the framework of the quasi-linear theory, which assumes that the turbulent energy in the hydrodynamic waves acting as particle scatterers is much less than the energy density of the magnetic field. However, the anisotropies induced by supernova shocks in the pre-existing population of galactic cosmic rays are sufficient to render these waves extremely unstable ; the wave amplitudes predicted by the quasi-linear theory are too high to be fully consistent with this theory.

ii) If cosmic rays extract so much energy from the shock, their pressure can become the dominant one. For instance, this will inevitably occur if cosmic rays are getting accelerated by a strong shock, to a spectrum E^{-2},for a sufficiently long time. Even if the shock is not so strong (r 4), the cosmic ray pressure can become dominant if the rate of injection of particles in the system is sufficiently rapid. The expectation is that, eventually, the cosmic rays broaden the shock, making it a less efficient particle accelerator. If the shock becomes wider than the particle mean free path λ, all particles of a given energy obtain the same amount of adiabatic acceleration as they cross the shock region.

A complete, but approximate solution to the problem of cosmic ray-dominated shocks has been developed, over the years, by Ellison and Eichler (1985 and references therein). The main assumptions are :

- free injection of particles into the mechanism, from the thermal tail
- a diffusion coefficient which increases with the energy. As the shock is broadened by the effect of the cosmic rays, particles of different energy "see" a different shock strength ; only the very high energy particles see the full shock.
- fast dissipation of the Alfven waves by an undefined damping mechanism

\- particles of energy greater than some E_{max} are assumed to escape from the system. This makes it possible to study the process in a time-independent fashion. But then the shock is no longer adiabatic ; in principle the compression ratio can exceed its maximum adiabatic value (7 in this case, since for the cosmic ray gas the adiabatic index is 4/3), and become arbitrarily large. Nevertheless, Ellison and Eichler (1985) find that this mechanism still produce a universal spectrum which is very similar to a power law of index ~ 2. This is because the compression ratio is tempered by the fact that, upstream, the scattering centers of the cosmic rays are not attached to the fluid, but propagate with respect to it at the Alfven velocity. The efficiency of cosmic ray acceleration by this mechanism is of 25 %.

iii) An important problem of the theories of shock wave acceleration is that the maximum energy that can be attained is limited, either by the lifetime of the shock itself or by its curvature radius. This problem was treated in detail by Lagage and Cesarsky (1981,1983). In the case of supernova shocks, the limiting factor is the shock lifetime ; under most optimistic assumptions, the maximum energy E_{max}, for particles of charge Z, is only ~ 10^5 Z(B/10^{-6}G) GeV, where B is the strength of the magnetic field in the most diffuse phase of the interstellar medium.

This result holds whether the shock is linear or cosmic-ray dominated. Taking into account the non-linearity introduced by the fact that, upstream, the Alfven waves are generated by the cosmic rays, so that the diffusion coefficient is space and time dependent, limits E_{max} to values which may be as low as 2000 Z (B/10^{-6})GeV. Invoking supernova shocks propagating in the galactic halo does not alleviate the problem (Lagage and Cesarsky 1986).

The possible acceleration of high energy cosmic rays by stellar wind terminal shocks is still controversial. If shock acceleration is operating there over long times, stellar winds have the advantage that the shock is a standing shock, which remains strong for longer times than supernova shocks. The maximum energy is then determined by the shock curvature, and the strength of the magnetic field : E_{max} ~5.10^5Z(B/10^{-5}G) (D/5 pc) GeV, where D is the shock radius. In a recent paper Kazanas and Ellison (1985) attempt to model the binary X ray source Cygnus X-3, which may be emitting ultra high energy gamma rays (Sammorsky and Stamm 1983, Watson 1985 and references there in), and thus be a source of cosmic rays of energy up to 10^7-10^8 GeV. Assuming the presence of a collisionless, spherical accretion shock around the compact object in Cygnus X-3, and assuming that the magnetic field strength is in equipartition with the accretion flow, Kazanas and Ellison argue that protons of energy as high as 7.10^6 GeV may be accelerated by the shock in this system.

5. Interpretations of the elemental cosmic ray source composition

a) Source abundances

The study of systematic trends in elemental abundances is important for unfolding the nuclear and/or atomic effects that should govern the shaping of source abundances and in constraining the parameters of cosmic ray acceleration models (for reviews see Cassé, 1984, Simpson 1983). Recent data on the composition of the Galactic Cosmic Ray Sources (GCRS), derived as explained in section 2a, mainly from HEAO3-C2 data, are summarized in table 1 (from Meyer 1985b) together with careful estimates of elemental abundances in our Local Galactic (LG) environment, in Solar Energetic Particles (SEP), Solar Corona and in the Solar Wind (SW).

ELEM.	LOCAL GALACTIC (LG)	SOLAR CORONA	SOLAR WIND (SW)	SEP (m.u. baseline)	GCRS	ELEM.
H	$2.71.10^6$ (1.10)	$2.55.10^6$ (1.40)	$1.43\ .10^6$ (1.30)	-	$0.077\ .10^6$ (1.12) [a]	H
He	$0.26.10^6$ (1.25)	$0.25.10^6$ (3.00)	$0.056.10^6$ (1.30)	[$0.038.10^6$ (1.27)] [b]	$0.0118.10^6$ (1.07)	He
C	1260. (1.26)	600. (3.00)	-	290. (1.30)	420. (1.08)	C
N	225. (1.41)	100. (1.70)	-	81. (1.33)	34. (1.34)	N
O	2250. (1.25)	630. (1.60)	650. (1.50)	650. (1.13)	505. (1.04)	O
F	.093 (1.60)	-	-	-	< 2.5	F
Ne	325. (1.50)	90. (1.60)	100. (1.40)	84. (1.32)	62. (1.14) [c]	Ne
Na	5.5 (1.18)	7. (1.70)	-	8.5 (1.47)	4.6 (1.90)	Na
Mg	105. (1.03)	95. (1.30)	-	123. (1.28)	105. (1.06) [c]	Mg
Al	8.4 (1.05)	7. (1.70)	-	8.9 (1.55)	10.2 (1.45)	Al
Si	≡ 100. (1.03)	≡ 100. (1.30)	≡ 100. (3.00)	≡ 100. (1.37)	≡ 100. (1.06) [c]	Si
P	.94 (1.24)	-	-	-	< 2.5	P
S	43. (1.35)	22. (1.70)	-	20. (1.80)	14.1 (1.20)	S
Cl	.47 (1.60)	-	-	-	< 1.6	Cl
Ar	10.7 (1.50)	5.4 (1.80)	2.6 (1.50)	3.8 (1.70)	3.2 (1.30)	Ar
K	.34 (1.41)	-	-	-	< 1.9	K
Ca	6.2 (1.14)	7.5 (1.60)	-	7.6 (1.55)	6.8 (1.34)	Ca
Sc	.0035 (1.15)	-	-	-	< 0.8	Sc
Ti	.27 (1.16)	-	-	-	< 2.4	Ti
V	.026 (1.21)	-	-	-	< 1.1	V
Cr	1.29 (1.10)	-	-	2.25 (1.90)	< 2.9	Cr
Mn	.77 (1.24)	-	-	-	< 3.7	Mn
Fe	88. (1.07)	100. (1.50)	57. (2.00)	99. (1.47)	91. (1.06)	Fe
Co	.21 (1.15)	-	-	-	0.28 (1.55)	Co
Ni	4.8 (1.13)	5.5 (1.70)	-	4.5 (1.75)	4.7 (1.17)	Ni
Cu	.052 (1.60)	-	-	-	0.060 (1.20)	Cu
Zn	.098 (1.22)	-	-	-	0.058 (1.20)	Zn

TABLE 1. Elemental abundances in local samples (local galaxy, solar corona, solar wind, solar energetic particles) and in galactic CR sources (Meyer 1984 a,b). In parenthesis : error factors ("within a factor of ..."). The entries are normalized to Si.

a) Based on H/He ratio at given magnetic rigidity. (A value close to 1 is obtained if H and He are taken at the same momentum per nucleon, after correction for propagation effects as in Engelmann et al., 1985).

b) Not really significant, since the "mass unbiased baseline" SEP abundance HE cannot be properly defined (Meyer 1985a).

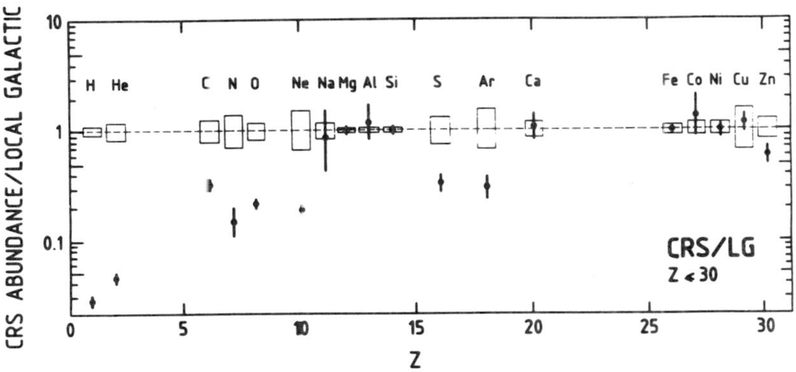

Figure 8. Ratios of elemental cosmic ray source (CRS) to local galactic (LG) reference abundances, versus Z, for Z<30, normalized to Si. The boxes represent the uncertainties in the local galactic abundances, and the bars those in the cosmic ray source composition as derived mainly from the french-darish instrument aboard HEAO-3.

The abundance of elements of higher Z, from zinc to uranium have been determined by the companion experiment on HEAO3 : HEAO3-C3. The results of this experiment have shown that the abundance ratio of the actinides (uranium et al.) to the lead-platinum group of elements are consistent with the solar system abundances (Binns et al. 1982). The ratio Pb/Pt observed is equal to 0.25 ± 0.09 (Binns et al. 1985). This value is lower than would be expected if the source composition had been that of the solar system as given by Cameron (1981) (leading, after propagation, to Pb/Pt ~ 0.4). Another interesting result of Binns et al. (1985) is that the cosmic ray abundance of germanium relative to iron is down by a factor of about 2 compared to the solar system.

The most obvious idfference in composition between cosmic rays and the ordinary matter to be found in our galactic heighborhood lies in the ratio of elements with Z > 5 to hydrogen and helium ; cosmic rays are comparatively H and He poor by about one order of magnitude - and we have seen that the local cosmic rays may contain at least two components, one of which would be even poorer in H and He !

The pattern of the heavy element abundances, other than H and He, is reminiscent of that of galactic matter, although some elements differ by factors of up to 6 (see fig. 8).

b) r process ?

The anomalies in cosmic ray source composition may reflect an acceleration of particles out of sites of current nucleosynthesis. The resulting pattern of abundances should then be directly related to the detailed nuclear properties of the various species. Alternatively , they may reflect any kind of selective process governing the composition of the particular media from which the particles have been extracted, or operating during the acceleration process itself. Most conceivable selection effects are related to atomic properties.

Until recently, the most widespread interpretation of the cosmic ray elemental source composition was that the bulk of the cosmic ray nuclei were accelerated out of supernova matter as a result of supernovae explosion. As such, the major elemental composition anomalies should correlate with the expected yields of specific nucleosynthetic processes in massive stars, or resemble the expected global yield of a supernova. However, inspection of the cosmic ray source elemental composition, and in particular of the recent results which I just quoted on the actinides, does not support this hypothesis at all. Some specific anomalies, such as the low abundances of sulphur and argon relative to neighbouring elements, are particularly difficult to explain in such a framework.

c) First ionization potential, solar particles and stellar coronae

The existence of a simple correlation between the elemental abundance anomalies of cosmic ray sources and the first ionization potential (FIP) was noted several years ago, and has consistently been confirmed by new data (see fig. 9). Heavy elements with a first ionization potential higher than ~ 9eV, difficult to ionize, have a tendency to be comparatively underabundant in cosmic ray sources (fig.9 for $Z<30$; a similar trend is found for $Z>30$) which would be the result of a selection process between neutral and ionized atoms in a dilute warm gas at around 8000K (Cassé and Goret, 1973, Arnaud and Cassé 1985). The elemental composition of the energetic particles accelerated by the Sun, although highly variable with time and energy, shows a basic pattern which is remarkably similar to that of cosmic ray sources.

According to Meyer (1985a,b) this common feature suggests that galactic cosmic rays are extracted from the coronae of solar-like stars (F and G stars only) and that stellar flares, similar to solar flares, inject subrelativistic particles into the surrounding medium, presenting on the average the elemental composition of the coronal medium. Occasionaly some of the injected particles are boosted to relativistic energy by an external agent of acceleration, namely a passing supernova shock wave. This is the simplest way to explain why the composition of

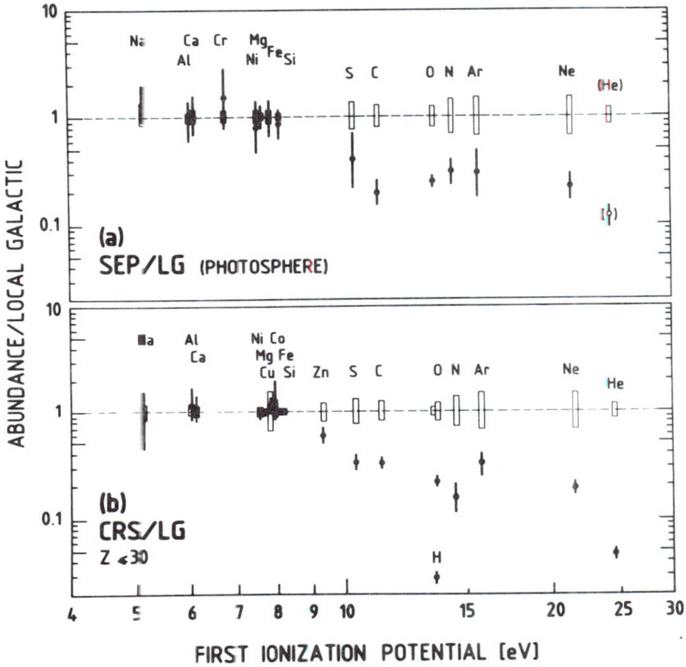

Figure 9. (from Meyer 1985b). Ratios of elemental abundances in (a) solar energetic particles (SEP), and (b) cosmic ray sources (CRS) to local galactic reference abundances, versus first ionization potential, for Z < 30, normalized to Si ; Boxes and error bars as in Fig. 8.

the solar corona and that of the galactic cosmic ray sources are so similar. Meyer (1985b) further argues that the source composition is generated by a single process (leaving aside a ^{22}Ne-rich component) : a selective filtering of particles rising from stellar chromospheres to stellar coronae, sensitive to whether a given element is ionized or not. This scenario has to be confronted with an important additional observation : in the solar corona, H has a normal abundance relative to the heavy low-FIP elements, and thus behaves at variance to the heavy elements with similar FIP. Qualitatively this feature would be expected if thermal diffusion was responsible for the selective filtering between solar chromosphere and corona.

The energy of cosmic rays that is observed requires a main acceleration stage that is much more powerful than can be provided by the flaring of ordinary stars. But, if many ordinary stars produce suprathermal stellar energetic particles with the right composition, re-acceleration of some of these particles by interstellar shocks can lead to the observed high energy galactic cosmic rays.

d) Grain destruction products

Most refractory elements have a low first ionization potential, and tend, as we have seen, to be overabundant in cosmic ray sources. There are, nevertheless, a few elements which have a low first ionization potential and are volatile. Ge and Pb are among those, and we have seen that they are underabundant in cosmic rays. This suggests that volatility, rather than first ionization potential, is the important parameter for the selection of cosmic rays. If this is the case, there is probably a link between cosmic rays and cosmic dust particles.

Epstein (1980) has proposed that interstellar dust grains are picked up by the supernova shock acceleration mechanism, just like ordinary charged particles are. He argued that grains are accelerated to relativistic velocities, and later on break up. This scenario would yield a cosmic ray component almost entirely devoid of hydrogen, helium and other noble gases, so it could only account for part of the cosmic radiation. Cesarsky and Bibring (1981) suggested an alternative scenario. They consider a shock wave sweeping matter within an interstellar cloud. Behind the shock the gas cools rapidly and is compressed ; the grains, whose interaction length with the gas is much longer than the mean free path of the particles, do not "see" the shock, so that their velocity with respect to the shocked gas is of the order of the shock velocity. Interstellar particles interacting with the grains acquire the grain velocity ; they then leave the grain togther with sputtered particles. In this way, a distinct kinematical population is produced, which is a mixture of gas and grain particles, and whose composition matches quite well that of cosmic rays (although the abundances of rare gases heavier than He remain a problem). These particles are then preferentially picked-up by a second-order Fermi acceleration mechanism operating in the shocked gas.

A drawback common to both these models is that, if the grain scenario does apply, the striking similarity between the composition of Solar Energetic Particles and cosmic ray sources is purely accidental.

e) Injection out of the thermal tail in cosmic ray dominated shocks

Let us consider again the cosmic-ray dominated shock model proposed by Eichler (1976) and Ellison and Eichler (1985). These authors assume that the particles involved in the acceleration mechanism have been directly picked-out of the thermal tail.

	Scenarios for origin of galactic CR's	Acceleration mechanism	H, He	Relative proportions of heavy elements	Spectrum	High energies?	Specific advantages	Specific problems (other than composition)
STELLAR MODELS	Selective injection in warm stellar plasmas	unspecified	extra-assumption needed. (rigidity dependent injection needed).	Problems for S (Ar) N ? (1) Ge/Fe (2) Pt/Pb (2)	not specified	not specified	Simplicity	Phenomenological model only, physics lacking
STELLAR MODELS	Non-selective injection in stellar coronae	Selective filtering of ions from stellar chromosphere to coronae; 1st stage acceleration by stellar flares, 2nd stage by interstellar shocks	Not applicable (H and He behave at variance to heavy elements in SEP)	accurate fitting except N ? (1) Ge/Fe (2) Pt/Pb (2)	given by shock wave acceleration	limited by shock lifetime	accounts for striking similarity between abundances in solar cosmic-rays and galactic CR sources	2nd acceleration stage before slowing down to thermal energies. Scenario restricted to F and G stars
INTERSTELLAR MODELS	Shock wave acceleration with injection from thermal plasma	shock wave acceleration	OK	OK except for S, Ar (if grains absent in the hot ISM Ne-Mg, N(?) (1) Ge/Fe (2), Pt/Pb (2)	possibly OK Effect of escape; of non-linearities under study	Supernova shocks can only accelerate protons up to less than 10^5 GeV (stellar wind shocks ?)	fully consistent model	-First stage of injection not well understood. -Grains in the ISM ? -Non-linearities ? -Similarity with SEP
INTERSTELLAR MODELS	Injection of grain debris	2 stages: injection by 2nd order Fermi on grain products, 2nd stage unspecified	OK	OK except Ne, Ar N ?	unspecified	-	-Ge/Fe (2), Pt/Pb (2) -Injection stage fully specified	Similarity between composition of solar cosmic-rays and of galactic CR sources

Table 2. Relative merits of the main cosmic ray source models intending to account for the source composition.
(1) The source abundance of nitrogen is still controversial (see e.g. Meyer 1985b)
(2) The solar system abundances of Ge, Pt and Pb (Cameron, 1981) need to be confirmed.

An important point in this mechanism is that low energy particles in the early stages of acceleration "mediate" the shock, making it less sharp, and thus a less efficient accelerator. Particles of higher rigidity have a longer mean free path ; they "see" a stronger shock and therefore profit more from the interaction with the shock. The rigidity of particles at the onset of acceleration is inversely proportional to their initial effective charge Z*, which is determined by the physical conditions prevailing in the unshocked gas. It is clear that particles with high A/Z* ratios are favoured by the acceleration process (Ellison et al., 1981). Cesarsky et al. (1986) have studied the variation of the abundance ratio (CR sources/Local Galactic) as a function of the mean (A/Z*) ratio at different temperature assuming ionization equilibrium in a thin plasma. They find that, while this model can account for the general trend of the cosmic ray composition if particles are extracted from a plasma at a temperature of $\sim$ several 10^5 °K, some abundance ratios, such as H/He, or S/Si or Ar/Si, cannot be matched.

Moreover, this theory may encounter another obstacle : it is known that in the denser and cooler regions of the interstellar medium, refractory elements are almost entirely condensed into interstellar dust grains ; they are therefore highly deficient in the gas phase. It seems quite possible, although not certain, that the refractory cores of these grains do not evaporate even in the hottest regions of the interstellar medium. In this case, cosmic rays extracted directly out of this hot interstellar gas phase should be depleted in refractory elements, contrary to observation.

The arguments on most of the models discussed in this section are summarized in Table 2 (from Cassé and Cesarsky, 1984).

References

Armstrong,J.W., Cordes,J.M., and Rickett,B.J.:1981, Nature 291, 561
Arnaud,M., and Cassé,M.: 1985, Astron.Astrophys. 144, 64
Axford,W.I.: 1981, 17th Int.Cosmic-Ray Conf.,12, 155
Axford,W.I., Leer,E., Skadron,G.: 1977, 15th Int.Cosmic-Ray Conf., 11, 132
Bell,A.R.: 1978, M.N.R.A.S. 182, 147
Binn,W.R., Fickle,R.K., Garrard,T.L., Israel,M.H., Klarmann,J., Stone,E.C., and Waddington,C.J.: 1982, Ap.J.Lett. 261, L117
Binns,W.R., Brewster,N.R., Fixsen,D.J., Garrard,T.L., Israel,M.H., Klarmann,J., Newport,B.J., Stone,E.C., Waddington,C.J.: 1985, Ap.J. 297, 111
Blandford,R.D. and Ostriker,J.P.: 1978, Ap.J.Lett. 221, L129
Blandford,R.D. and Ostriker,J.P.: 1980, Ap.J. 237, 793

Bogomolov,E.A., Lubyanaya,N.D., Romanov,V.A., Stepanov,S.V., Shulakova,M.S., Ioffe,A.F.: 1979, 16th Int.Cosmic-Ray Conf., Kyoto, 1, 330
Buffington,A., Schindler,S.M., Pennypacker,C.R.: 1981, Astrophys.J. 248, 1179
Cameron,A.G.W.: 1981, in "Essays in Nuclear Astrophysics", ed. C.A.Barnes,D.C.Clayton,D.N.Schramm, Cambridge Univ.Press,p.23
Cassé,M.: 1984, in "Stellar Nucleosynthesis", ed.C.Chiosi, and A.Renzini, Reidel, p.55
Cassé,M., Goret,P.: 1978, Ap.J., 221, 703
Cassé,M., and Cesarsky,C.J.: 1984, in "High Energy Astrophysics", ed.J.Audouze and J.Tran Thanh Van, Frontières,p.363
Cesarsky,C.J.: 1975, 14th Int.Cosmic-Ray Conf.,12, 4166
Cesarsky,C.J.: 1980, Ann.Rev.Astron.Astrophys.,18, 289
Cesarsky,C.J., Koch-Miramond,L., Perron,C.: 1981, 17th Int. Cosmic-Ray Conf.,2,22
Cesarsky,C.J., Bibring,J.P.: 1981, in "Origin of Cosmic-Rays", ed.G.Setti,G.Spada,A.W.Wolfendale, Reidel,361
Cesarsky,C.J., Lagage,P.O.: 1981, 17th Int.Cosmic-Ray Conf.,9, 250
Cesarsky,C.J., Montmerle,T.: 1983, Sp.Sci.Rev. 36, 173
Cesarsky,C.J., Rothenflug,R., Cassé,M.: 1986, in preparation
Cowsik,R., Wilson,L.W.: 1973, 13th Int.Cosmic-Ray Conf.,1, 500
Dermer,C.D., Ramaty,R.: 1985, 19th Int.Cosmic-Ray Conf.,2, 338
Drury,L.O.C.: 1983, Rep.Prog.Phys.46, 973
Eichler,D.: 1979, Ap.J. 229, 419
Ellison,D.C., Jones,F.C., Eichler,D.: 1981, J.Geophys.50, 113
Ellison,D., Eichler,D., 1985: Phys.Rev.Lett., in press
Epstein,R.I.: 1980: M.N.R.A.S. 193, 723
Fermi,E.: 1949, Phys.Rev. 75, 1169
Fransson,C., Epstein,R.I.: 1980, Ap.J. 242, 411
Ginzburg,V.L., Khazan,Y.M., Ptuskin,V.S.: 1980, Ap.Sp.Sci.68, 295
Golden,R.L., Mauger,B.G., Nunn,S., Horan,S.: 1984, Astrophys.Lett. 24, 75
Hillas,M.: 1984, Ann.Rev.Astron.Astrophys.22, 425
Jones,M.D., Klarmann,J., Stone,E.C., Waddington,C.J., Binns,N.R., Garrard,T.L., Israel,M.H.: 1985, 19th Int.Cosmic-Ray Conf.,2, 28
Juliusson,E., Meyer,P., Muller,D.: 1972, Phys.Rev.Lett.29, 445
Kazanas,D., Ellison,D.: 1985, Nature, in press
Kiraly,P, Szabelski,J.,Wdowczyk,J., Wolfendale,A.W.: 1981, Nature 293, 120
Koch-Miramond,L.: 1981, 17th Int.Cosmic-Ray Conf.,12, 21
Koch-Miramond,L., Engelmann,J.J., Goret,P., Juliusson,E., Masse, P., Soutoul,A.: 1983, 18th Int.Cosmic-Ray Conf.,9, 277
Kraichnan,R.H.: 1965, Phys.Fluids 8, 1385
Krymsky,G.F.: 1977, Dok.Akac.Nauk.SSSR 234, 1306
Lagage,P.O., Cesarsky,C.J.: 1983, Astron.Astrophys.125, 249
Lagage,P.O., Cesarsky,C.J.: 1985, Astron.Astrophys.147, 127
Lagage,P.O., Cesarsky,C.J.: 1986, in preparation
Meneguzzi,M.: 1973, 13th Int.Cosmic-Ray Conf.,1, 378

Meyer,J.P.: 1985a, Ap.J.Suppl.S.57, 151
Meyer,J.P.: 1985b, Ap.J.Suppl.S.57, 173
Nishimura,J., et al.: 1985, 19th Int.Cosmic-Ray Conf., in press
Ormes,J., Freier,P.: 1978, Ap.J.222, 471
Samorski,M., Stamm,W.: 1983, Ap.J.Lett.268, L17
Silk,J., Srednicki,M.: 1984, Nature 53, 624
Simpson,J.A.: 1983, Ann.Rev.Nuc.Part.Sci.33, 323
Stecker,F.W., Protheroe,R.J., Kazanas,D.: 1983, Astropnhys.Space Sci.96, 1971
Syrovatskii,S.I.: 1961, Soviet Physics, JETP 13, 1257
Watson,A.A.: 1985, 19th Int.Cosmic-Ray Conf., Rapporteur Paper,in press
Webber,W.: 1983, in "Composition and Origin of Cosmic Rays", ed.M.M.Shapiro, Reidel, p.25
Wentzel,D.G.: 1974, Ann.Rev.Astron.Astrophys.12, 71
Wiedenbeck,M.E., Greiner,E.D.: 1980, Ap.J.Lett.239, L139
Wiedenbeck,M.E.: 1983, in "Composition and Origin of Cosmic Rays" ed.M.M.Shapiro, Reidel, p.65
Wiedenbeck,M.E.: 1985, 19th Int.Cosmic-Ray Conf.,2, 84

QUASIPERIODIC OSCILLATIONS IN GALACTIC BULGE SOURCES: INFORMATION ON THE BOUNDARY LAYER FROM POWER SPECTRA

M.Ali Alpar
Scientific and Technical Research Council of Turkey
Research Institute for Basic Sciences
TÜGAM-TBAE, P.K.74 Gebze
KOCAELİ, TURKEY

ABSTRACT. The angular location of clumping, as well as the width of the boundary layer and the radial velocity in the disk can be extracted from QPO power spectra under certain conditions.

INTRODUCTION. Quasiperiodic oscillations (QPO's) have been interpreted as a beat frequency phenomenon[1,2] between a rotating neutron star and clumps[3] in the inner boundary of an accretion disk. The aim of this paper is to delineate the signature of various physical properties as reflected in simple model power spectra, rather than to fit the observed power spectra of the QPO sources known at present. Two effects will be discussed. 1) If the clumps are formed at one specific azimuthal angle (or narrow range of angles) from the neutron star magnetic axis, the value of that angle will effect the qualitative form of the power spectra. 2) Velocity shear among the clump will cause a distinct set of features in the power spectrum. The observation of such features would lead to a determination of the radial velocity of the clumps.

MODEL POWER SPECTRA. Consider an accretion luminosity that consists of the contributions of a large number of clumps i

$$L(t) = \sum_i F(t-t_i) \tag{1}$$

where t_i is the time at which the ith clump started contributing to L(t). It is natural to assume that the first appearance of a clump is equally likely to occur anytime during the observations (each t_i is uniformly distributed) and that there is a characterictic rate R of appearance of clumps, so that the number of arrivals in a time interval T is Poisson distributed with mean RT. The power spectrum of the luminosity time series L(t) is then given by the simple expression[4]:

$$P(f) = R|S(\omega)|^2 \tag{2}$$

where P(f) is the power per unit interval in frequency (Hz) and $S(\omega)$ is the Fourier transform of the individual "event" F(t):

F. Pacini (ed.), High Energy Phenomena Around Collapsed Stars, 359–366.

$$S(\omega) = \int_{-\infty}^{\infty} e^{-i\omega t} F(t)\, dt. \tag{3}$$

A simple model for the contribution of each clump is:

$$F(t-t_i\,;\,\phi_i) = \{1+\beta\cos(\omega_o(t-t_i)\}\,\theta(t-t_i)e^{-(t-t_i)/\tau} \tag{4}$$

This describes the periodic modulation at the beat frequency ω_o and the finite lifetime of a clump. The lifetime envelope $\theta(t)e^{-t/\tau}$ (sudden turn-on and exponential decay) is chosen for simplicity. The effects we seek to explore are not dependent on the form of this envelope. Only the fundamental of the beat frequency has been included since we shall concern ourselves with structure around a single peak in the power spectrum. The results can be generalized to higher harmonics. The full generalization of Eq.(4) can be found in Ref.3. As we shall be interested in the <u>shape</u> of the power spectrum, the amplitude of F(t) is suppressed, and we shall set R=1 in the following. The DC term is essential since F(t) must describe a contribution to the accretion rate, a positive quantity. Thus, $|\beta|\leq 1$ is required. The DC term then leads to a peak at zero frequency, while the oscillatory term gives a peak at $\omega=\omega_o$. Due to the finite lifetime τ, both peaks will have a HWHM $\Delta\omega=\tau^{-1}$. At the low frequency end, for $\omega>\tau^{-1}$ the power spectrum will be one of red noise. Such low frequency noise must accompany QPO's as a necessary consequence of the positiveness of the accretion rate in these models.

The Fourier transform of Eq.(4) is:

$$S(\omega;\phi) = \frac{\tau}{1+i\omega\tau} + \frac{\beta\tau}{2}\left[\frac{e^{i\phi}}{1+i(\omega-\omega_o)\tau} + \frac{e^{-i\phi}}{1+i(\omega+\omega_o)\tau}\right] \tag{5}$$

The phase ϕ_i of the ith clump is simply its azimuthal position (at the time t_i when that clump started contributing to the accretion luminosity) with respect to a "special direction"[2,3] of the magnetic field where the accretion is most efficient. To take account of the distribution of ϕ_i, we must replace P(f) with:

$$\overline{P(f)} = \int_0^{2\pi} P(f;\phi)\rho(\phi)\, d\phi \tag{6}$$

where $\rho(\phi)$ is the probability distribution for ϕ. If the phases are random, $\rho(\phi) = 1/2\pi$, and

$$\overline{P(f)} = \frac{1}{2\pi}\int_0^{2\pi} P(f;\phi)d\phi. \tag{7}$$

Physically, the phases will be random if the magnetic field of the neutron star does not influence the formation of the clumps at all. In this case, all terms $e^{\pm i\phi}$, $e^{\pm 2i\phi}$ will have zero averages - the interference terms do not contribute to a random phase averaged, which is simply:

$$\overline{P(f)} = \tau^2 \left[\frac{1}{1+\omega^2\tau^2} + \frac{\beta^2}{4}\frac{1}{1+(\omega-\omega_o)^2\tau^2} + \frac{\beta^2}{4}\frac{1}{1+(\omega+\omega_o)^2\tau^2}\right] \tag{8}$$

THE EFFECT OF PHASE ANGLES. At the other extreme from completely random phases, we can envisage all clumps being formed at random times t_i but within a narrow range of angles with respect to the magnetic axes. Physically, this situation would arise if the magnetic field of the neutron star determines the location where the clumps first start to contribute to accretion. If we take $\rho(\phi) = \delta(\phi-\phi_o)$, then

$$P(f;\phi_o) = |S(\omega;\phi_o)|^2 = \overline{P(f)} + P_{int.}(f;\phi_o) \tag{9}$$

where the interference contribution $P_{int}(f;\phi_o)$ consists of terms in $\beta \cos\phi_o$, $\beta \sin\phi_o$, $\beta^2 \cos 2\phi_o$ and $\beta^2 \sin 2\phi_o$ which can be obtained straightforwardly from Eqs.(2) and (5). A series of power spectra having the same ω_o, τ, β but each characterized by a different phase ϕ_o are plotted in Fig.1. The value of ϕ_o clearly effects the qualitative shape of the spectrum. This dependence can be understood without much algebra if one neglects the third term in Eq.(5) (the peak at $\omega=-\omega_o$). This is a good approximation for $\omega>o$ given that $\omega_o\tau >>1$. The power spectrum in Eq.(10) can then be written as

$$\overline{P(f)} \cong \tau^2 \left[\frac{1}{1+\omega^2\tau^2} + \frac{\beta^2}{4}\frac{1}{1+(\omega-\omega_o)^2\tau^2}\right] \tag{10a}$$

$$P_{int}(f;\phi_o) \cong \beta\tau^2[N_1(\omega;\omega_o,\tau)\cos\phi_o + N_2(\omega;\omega_o,\tau)\sin\phi_o] \tag{10b}$$

where

$$N_1(\omega;\omega_o,\tau) = \frac{\{1+\omega(\omega-\omega_o)\tau^2\}}{(1+\omega^2\tau^2)(1+(\omega-\omega_o)^2\tau^2)} \tag{11a}$$

and

$$N_2(\omega;\omega_o,\tau) = \frac{-\omega_o\tau}{(1+\omega^2\tau^2)(1+(\omega-\omega_o)^2\tau^2)} \tag{11b}$$

By examining N_1 and N_2 one can see easily that $P(f;\phi_o)$ will have the sharpest minimum between $\omega=o$ and $\omega\cong\omega_o$ for $\phi_o\cong 35^o$. For $\phi_o \sim 180^o$, the reversal of sign for $N_1(\omega;\omega_o,\tau)$ leads to a second minimum in the power spectrum at $\omega>\omega_o$. As the plots in Fig.1 show, the shape of the power spectrum is a sensitive indicator of ϕ_o.

THE EFFECT OF VELOCITY SHEAR. Suppose a clump starts contributing to the accretion when it is at a Keplerean rotation rate $\Omega_{K,o}$ and a corresponding beat frequency ω_o. As it subsequently moves radially inward through the boundary layer the Keplerean frequency increases. Using $r \cong r_o - V_r t$, where V_r is the radial velocity of the clumps, we find

$$\omega[r(t)] \cong \Omega_K(r(t)) - \Omega_* = \omega_o + \frac{3}{2}\omega_o \frac{V_r}{r_o} t \ . \tag{12}$$

The phase of the oscillation at the beat frequency is given by:

$$\phi(t) = {}_o\int^t \omega(t')dt' = \omega_o t + \alpha t^2 + \phi_o \tag{13}$$

$$\alpha \equiv \frac{3}{4}\omega_o \frac{V_r}{r_o} \equiv \frac{3}{4} \frac{\omega_o}{\tau} \frac{\delta}{r_o} \tag{14}$$

Here δ is the width of the boundary layer within which a clump contributes to the accretion luminosity and τ the lifetime of the clump through that layer. The time dependence of the contribution of each clump now becomes:

$$F(t;\phi_o) = (1+\beta \cos(\omega_o t + \alpha t^2 + \phi_o))\ \theta(t)e^{-t/\tau}. \tag{15}$$

In this section, we shall use random phase averaged spectra, $\overline{P(f)}$, to isolate the effects of velocity shear.

The shape of the power spectrum will have structure on scales $\Delta\omega = (4\pi\alpha)^{\frac{1}{2}}$. Writing

$$F_{\omega_o}(t) = \frac{\beta}{2} e^{i\omega_o t} e^{i\phi_o}\ \theta(t)e^{-t/\tau} .e^{i\alpha t^2} \equiv F_o(t)e^{i\alpha t^2} \tag{16}$$

the corresponding term in $S(\omega)$ for the peak at $\omega=\omega_o$ becomes

$$S_{\omega_o}(\omega;\phi_o) = F\tau(F_{\omega_o}(t)) = \frac{\beta}{2} \frac{\tau e^{i\phi}}{1+i(\omega-\omega_o)\tau} * F\tau(e^{i\alpha t^2}) \tag{17}$$

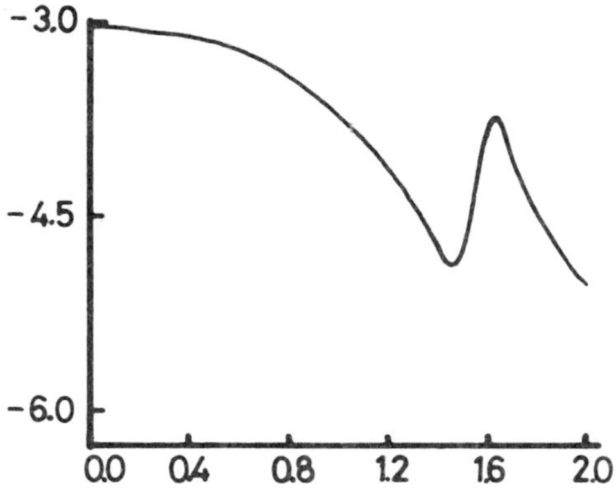

Fig. 1a

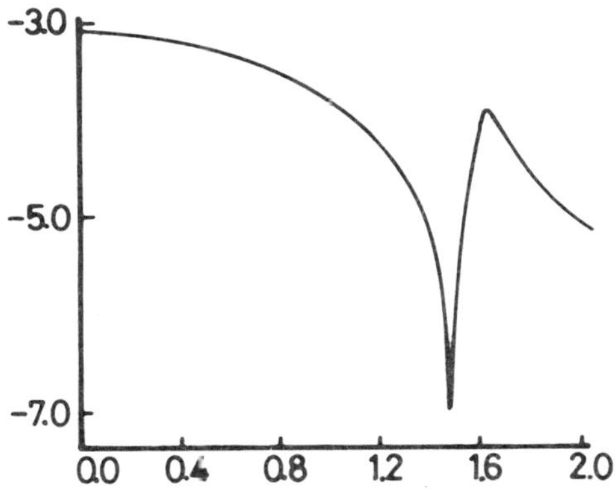

Fig. 1b

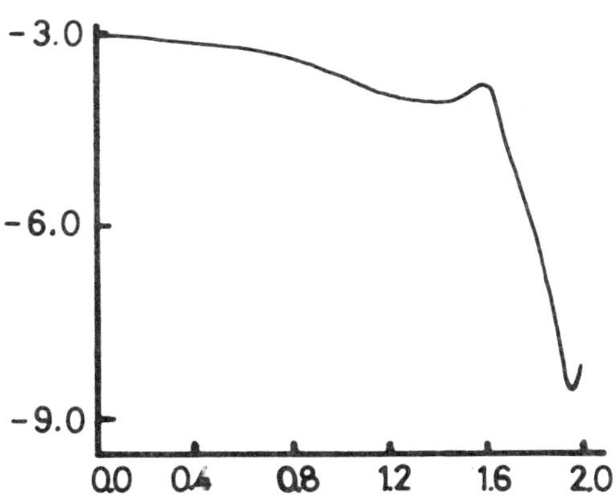

Fig. 1c

Fig. 1. Power spectra for various values of the phase angle ϕ_0. Log P(f) in arbitrary units is plotted against log f, for $\tau=0.03$ s, $\nu_0= 40$ Hz. ($\omega_0=2\pi\nu_0$), $\beta=0.8$ and $\alpha=0$. a) $\phi=0^\circ$, b) $\phi=35^\circ$, c) $\phi=180^\circ$.

where $F\tau$ denotes the Fourier transform defined in Eq.(3) and * denotes a convolution. Now

$$F\tau(e^{i\alpha t^2}) = \int_{-\infty}^{\infty} e^{i\alpha t^2} e^{-i\omega t} dt = \exp(-\frac{i\omega^2}{4\alpha}) \int_{-\infty}^{\infty} \exp[i\alpha(t-\frac{\omega}{2\alpha})^2] dt \quad (18)$$

This shows that $\overline{P_{\omega_0}(f)} = \overline{|S_{\omega_0}(\omega;\phi_0)|^2}$ has an oscillatory dependence on $\omega^2/2\alpha$, so that near the peak maximum (which will be at $\omega \cong \omega_0$, as can be found explicitly from the convolution in Eq.(17)), there will be further maxima at $\omega_n \cong \omega_0 + \Delta\omega_n$, such that

$$\frac{\Delta\omega_n^{\ 2}}{2\alpha} = 2\pi n \ , \ \Delta\omega_n = (4\pi\alpha)^{\frac{1}{2}} n^{\frac{1}{2}} \ . \quad (19)$$

To understand this periodicity, consider two frequencies, ω_1 and ω_2, which differ by $\Delta\omega$. These will "fit" the phase $2\alpha t^2$ at different times t_1 and t_2 (the factor 2 enters because the power spectrum relates to the <u>square</u> of the time series) such that $\omega_i t_i = 2\alpha t_i^2$. Thus, $t_2 - t_1 \equiv \Delta t = \Delta\omega/2\alpha$. The Fourier components at ω_1 and ω_2 will be in phase in representing $\cos(2\alpha t^2)$, say, if $\Delta\omega\Delta t = (\Delta\omega)^2/2\alpha = 2\pi n$. Since the width of the convolution will be larger that the widths of its factors, the periodicity in ω due to velocity shear will effect the shape of the peak if

$$(4\pi\alpha)^{\frac{1}{2}} >> \tau^{-1} \ . \quad (20)$$

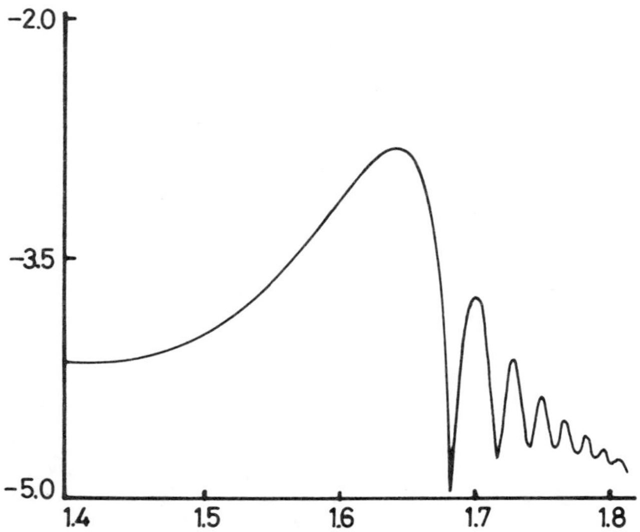

Fig.2. Power spectrum illustrating the oscillatory behaviour due to velocity shear when $(4\pi\alpha)^{\frac{1}{2}} >> \tau^{-1}$. The peaks near ν_0= 40 Hz. in random phase averaged spectra with β=1, $\tilde{\tau}$=0.1 s, and α=125 are shown.

This is indeed observed in the power spectra. Fig.2 shows a typical power spectrum in the regime $(4\pi\alpha)^{\frac{1}{2}} >> \tau^{-1}$. The exact expression for $\overline{P(f)}$ is

$$\overline{P(f)} = \frac{\tau^2}{1+\omega^2\tau^2} + \frac{\beta^2}{4} (|I_1|^2 + |I_2|^2) \tag{21}$$

where

$$I_1 = \frac{\tau}{1+i(\omega-\omega_o)\tau} \pi^{\frac{1}{2}} Z_1 \exp(Z_1^2) (1-\mathrm{erf}(Z_1)) \tag{22}$$

$$Z_1 = \frac{1+(\omega_o-\omega)\tau + i(1+(\omega-\omega_o)\tau)}{2\sqrt{2}\alpha^{\frac{1}{2}}\tau} \tag{23}$$

and

$$I_2 = I_1(\omega;-\omega_o, -\alpha) \tag{24}$$

corresponds to the peak at $\omega= -\omega_o$.

The range of α for which the oscillatory signature of the velocity shear will be observed in the power spectrum is:

$$\frac{1}{4\pi\tau^2} << \alpha << \frac{\omega_o}{\tau} \tag{25}$$

where the first inequality is the requirement given in (20), and the second is the condition for expanding the time-dependent frequency to first order. Using Eq.(14), we obtain

$$\frac{1}{4\pi\omega_o\tau} << \frac{3}{4} \frac{\delta}{r_o} << 1 . \tag{26}$$

For $\tau \gtrsim 1$ s., $\omega_o = 2\pi\nu_o$, $\nu_o \sim 40$ Hz. , one sees that values of δ/r_o in this range are quite plausible. If power spectra like those in Fig.2 are indeed observed, one can in principle deduce α from the separation of peaks, and combine this with a knowledge of r_o for the observed ω_o (obtained from fitting the QPO frequency - intensity relation to the data) and τ (obtained from the behaviour of $\overline{P(f)}$ at low frequencies) to deduce V_r and δ. The prospect of deducing V_r observationally is particularly exciting since, within a factor describing the density enhancement on clumping, V_r in the clumpy boundary region will give V_r in the continuous accretion disk. Note that although the periodicity in frequency is independent of the choice of envelope function, in our particular model the oscillatory behaviour occurs at frequencies higher than ω_0. This is because the envelope $\theta(t)\, e^{-t/\tau}$ makes the oscillations start sharply at the initial frequency ω_0. Only higher frequencies are realized at subsequent times. If for example, a gaussian in the time domain is chosen as the envelope function, subpeaks occur symmetrically to both sides of the primary maximum at $\omega=\omega_0$. Thus, the distribution of subsidiary maxima around $\omega=\omega_0$ carries information about the shape of the envelope, that is, about the symmetry between the turn-on and decay parts of the contribution of an individual clump to accretion.

CONCLUSIONS. Important physical parameters of a QPO source, in particular the azimuthal location of any preferred clump formation sites with respect to the magnetic axis, the width of the boundary layer and the radial velocity of the clumps in the boundary layer will lead to simple signatures in the QPO power spectra. The practical applications of these ideas will be complicated by the necessity to fit the data with a combination of physical parameters like ω_0, β, τ, ϕ_0 and α.

ACKNOWLEDGEMENTS. I thank F.Lamb, C.Pethick and N.Shibazaki for useful discussions, and S.Erbay and F.Yarman-Vural for introduction me to the computer facilities at TÜGAM.

REFERENCES.

1. Alpar,M.A., and Shaham,J., IAU Circular No.4046 (1985).
2. Alpar,M.A., and Shaham,J., Nature 316, 239 (1985).
3. Lamb,F.K., Shibazaki,N., Alpar,M.A. and Shaham,J., Nature 317, 681 (1985).
4. Rice,S.O., Bell System Technical Journal 23 and 24 (reprinted in Selected Papers on Stochastic Processes, ed.N.Wax,Dover,1954).

THE 35-DAY CYCLE OF HER X-1

J. Truemper
Max-Planck-Institut fuer Extraterrestrische Physik
Giessenbachstrasse 1
D-8046 Garching

The 35-day cycle of the binary X-ray source Hercules X-1/HZ Herculis which was discovered by Uhuru (Tananbaum et al. 1972) has not yet been satisfactorily explained. Among the various mechanisms proposed, that of a precessing accretion disk obscuring the primary X-rays has been most successful (Gerend and Boynton 1976, Katz 1973, Roberts 1974, Petterson 1975, 1977). However, the physical feasibility of disk precession has been doubted (Papaloizou and Pringle 1982, Kondo et al. 1983).

We have observed Her X-1 with EXOSAT throughout a 35-day cycle at about four day sampling intervals. These measurements include for the first time, an accurate determination of the 1.24 sec pulse profiles during the short-on state at the middle of the 35-day cycle (Fig. 1). The data show a large change in the pulse profiles between the main and short-on states. In particular the relative importance of the pulse and interpulse of the 1.24 sec variation changes with 35-day phase. It seems extremely unlikely that obscuration by an accretion disk having an extension of $\gtrsim 10^5$ neutron star radii affects the two polar beams (corresponding to pulse and interpulse) in a different way. We conclude that if the modulation is caused by obscuration it must take place much closer to the neutron star. The most straightforward interpretation of our result, however, is in terms of free precession of the neutron star. We examine the consequences of this precession with respect to the structure of the inner accretion disk.

A more detailed discussion of the observations and their implications is given in Truemper et al. 1985, 1987.

REFERENCES:

Gerend, D., and Boynton, P. E. 1976, Ap. J. **209**, 562
Katz, J. I. 1973, Nature Phys. Sci. **246**, 87
Kondo, Y., Van Flandern, T. C., and Wolff, C. L. 1983, Ap. J. **273**, 716
Papaloizou, J., and Pringle, J. E. 1982, M.N.R.A.S., **200**, 49
Petterson, J. A. 1975, Ap. J. (Letters) **201**, L61
Petterson, J. A. 1977, Ap. J. **218**, 783

F. Pacini (ed.), High Energy Phenomena Around Collapsed Stars, 367–368.

Roberts, J. W. 1974, Ap. J. **187**, 575
Tananbaum, H., Gursky, H., Kellog, E. M., Levinson, R., Schreier, E., and Giacconi, R. 1972, Ap. J. (Letters) **174**, L143
Truemper, J., Kahabka, P., Oegelman, H., Pietsch, W., Voges, W. 1985, Ap. J. (Letters), in press
Truemper, J., Kahabka, P., Oegelman, H., Pietsch, W., Voges, W. 1987, Proceedings of the NATO-ARW on Evolution of Galactic X-Ray Binaries, Reidel (Dordrecht), p. 239.

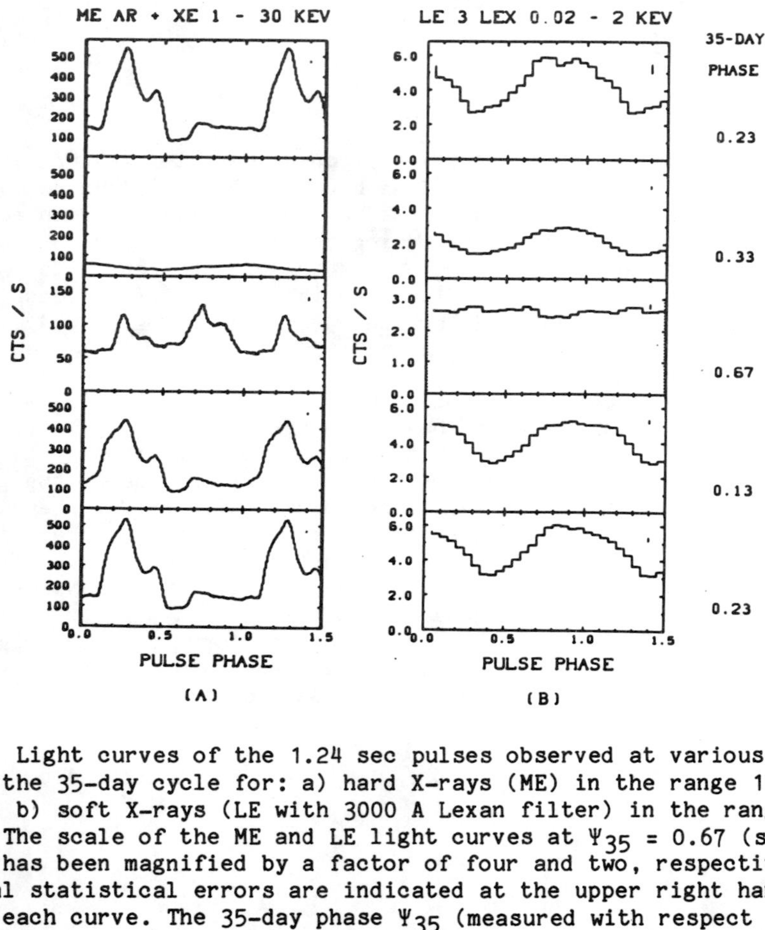

Figure 1. Light curves of the 1.24 sec pulses observed at various phases of the 35-day cycle for: a) hard X-rays (ME) in the range 1 to 30 keV; b) soft X-rays (LE with 3000 A Lexan filter) in the range .02 to 2 keV. The scale of the ME and LE light curves at Ψ_{35} = 0.67 (short-on state) has been magnified by a factor of four and two, respectively. The typical statistical errors are indicated at the upper right hand corner of each curve. The 35-day phase Ψ_{35} (measured with respect to turn-on), which progresses from top to bottom, is also indicated for each pair of ME, Le pulse shapes.